Fundamentals of Internet of Things for Non-Engineers

Fundamentals of Internet of Things for Non-Engineers

Edited by
Rebecca Lee Hammons and Ronald J. Kovac

CRC Press is an imprint of the
Taylor & Francis Group, an Informa business
AN AUERBACH BOOK

CRC Press
Taylor & Francis Group
6000 Broken Sound Parkway NW, Suite 300
Boca Raton, FL 33487-2742

Printed on acid-free paper

International Standard Book Number-13: 978-1-138-61085-9 (Hardback)

Visit the Taylor & Francis Web site at
http://www.taylorandfrancis.com

and the CRC Press Web site at
http://www.crcpress.com

Contents

Preface

It all started, in our minds, with a discussion some years ago. The discussion was energetic and dealt with the then-known Internet. Why didn't the Internet connect "everything," not just people and computers? Little did we know, this was the start of our involvement with IoT, the Internet of Things, and not what we called it years ago, the "Internet with Everything."

The IoT is the next manifestation of the Internet. The trend started by connecting computers to computers, progressed to connect people to people, and is now moving to connect everything to everything. The movement started like a race—with a lot of fanfare, excitement, and cheering. We're now into the work phase, and we have to figure out how to make the dream come true. What is the dream? It manifests in different ways for each of the participating vertical industries, but its essence is the same—to enhance the effectiveness and efficiency of processes and procedures. IoT can help with the safety of an older person in their home or can help produce better and cheaper products. IoT can reap significant amounts of valuable, and not so valuable, data to enhance our strategies and tactics. The value is in how we embrace the basic concepts of the Internet and the IoT.

The IoT will have many faces and involve many fields as it progresses. It will involve technology, design, security, legal policy, business, artificial intelligence, design, Big Data, and forensics; about any field that exists now. This is why we decided to write the book. We saw books in each one of these fields, but the focus was always "an inch wide and a mile deep." We wanted a book that would introduce the IoT to the non-engineer and allow them to dream of the possibilities and explore the work venues in this area. We wanted a book "A mile wide and a few inches deep." We think we have met this goal by engaging experts from a number of fields and asking them to come together to create an introductory IoT book. We hope you agree.

Co-editors:
Drs. Rebecca Lee Hammons and Ronald J. Kovac

Acknowledgments

An adventure such as this is not possible without the sincere help, assistance, and hard work of many people. The list of people who we would love to thank can go on forever, but we will limit it to two groups who helped immeasurably to get this book started and launched.

Research Assistants: We are lucky enough to have many research assistants at Ball State University in the Center for Information and Communication Sciences who help us with our academic and project workloads. For this book, our assistants dug out research facts, performed editing, created figures and diagrams, and performed a lot of the yeoman's work that is necessary for a venture of this size. We sincerely thank the following CICS research assistants: Zoey Spangler, Ethan Atkins, Rashida Peete, Alex Peczynski, Nick Benedict, Eric Satterthwaite, Konnor Miller, and Tiago DeMesquita.

Outside Reviewers: After writing each chapter, it was reviewed by each co-editor, the research assistants, other authors, and finally the whole book was reviewed by "Outside Reviewers." These people had a *fresh* look at the whole book and used their extensive experience in the ICT world to read hundreds of pages, look at innumerable diagrams, and critique our work from the lowest level (spelling and grammar) to the highest (communication and accuracy). We would like to thank the following people who served in this role and to whom we owe a big debt of gratitude: Judy Off, Jaki Lederman, Rayford Steele, Ana and John Combs, and William A. Howard.

Acknowledgments

Editors

Dr. Rebecca Lee Hammons has extensive technology industry experience in establishing and leading software quality assurance, development lifecycle services, and project management teams. She has worked for Ontario Systems, Apple, Raytheon, Tivoli Systems and Wang, in addition to several niche software firms. She is a certified quality manager and certified software quality engineer with the American Society for Quality (ASQ) and a Certified Scrum Master and Certified Scrum Product Owner with Scrum Alliance. Dr. Hammons is an associate professor at Ball State University. Her technology research interests include Slow Tech, Usability, Burnout Theory, and Gold-Collar Workers. Dr. Hammons thrives on leading organizational change initiatives and coaching individuals and teams to reach their full potential.

Dr. Ronald J. Kovac, is a full professor in the Center for Information and Communication Sciences at Ball State University in Indiana. The center prepares graduate students in the field of telecommunications. Previous to this position, Dr. Kovac was the telecommunication manager for the state of New York and a CIO for a large computing center located on the east coast. Dr. Kovac's previous studies included electrical engineering, photography, and education. Dr. Kovac has published two books and over 50 articles and has completed numerous international consulting projects in both the education and telecommunications field. Additionally, he speaks worldwide on issues related to telecommunications and holds numerous certifications, including the CCNA, the CCAI, and the almost complete CCNP. Dr. Kovac is also a Fulbright Scholar and loves life, education, and technology.

Contributors

Dr. Biju Bajracharya is an Assistant Professor in the ISOM department at Miller College of Business, Ball State University. Dr. Bajracharya's areas of interest include daily systems operation, networking/system administration, application development, cybersecurity, advanced computing, numerical modeling, optimization, simulation, and data mining.

Dr. Christopher Davison is an Assistant Professor in the ISOM department at Miller College of Business, Ball State University. Dr. Davison has published over 20 research papers and presented in more than 50 research conferences. His research interests include organizational resiliency, autonomous moral agents, disaster response, and privacy enabling technologies. He recently received DARPA funding for privacy preservation in Internet of Things spaces.

Dr. Gerald DeHondt joined the Ball State University faculty in December of 2017 from Oakland University in Michigan. He previously worked in the ICT industry as a project manager and IT professional.

Dr. Vamsi Gondi is an Assistant Professor in the ISOM department at Miller College of Business, Ball State University. Dr. Gondi's research interests include cyber-physical systems, wireless communications, software-defined networking, virtualization, wireless sensor networks, Internet of Things, and cloud computing.

Dr. Frank Groom spent many years in the telephony side of life and came to academia to help the field move ahead. Dr. Groom is a prolific author, a brilliant researcher, and a noteworthy speaker at international events.

Dr. David Hua is an Associate Professor in the ISOM department at Miller College of Business, Ball State University. Dr. Hua founded the undergraduate Computer Technology program at Ball State University in 2000. He is a Co-Founder of the Intelligent Networked Devices Institute (INDi). The mission of the institute is to promote the responsible development of smart objects and their interactions with mankind in the Internet of Things.

Dr. Rebecca Lee Hammons joined Ball State University in May 2017 after 40 years of corporate leadership in the software development industry. She focuses on Smart Cities/IoT, Slow Tech, and Software Quality research initiatives.

Dr. Thomas Harris has a 30+ year career that spans the ICT and business side of life. He is currently concentrating on the IoT world and how technology and business can be more closely aligned.

Dr. Timon Heinis, Dr. Johannes Heck, Dr. Filippo Fontana and Prof. Dr. Mirko Meboldt are Researchers at ETH Zurich, Switzerland, at the chair of product development and engineering design held by Prof. Dr. Mirko Meboldt. In their research they focus on design for new technologies—such as Internet of Things—and human-centered design to improve the process of product and service creation. The learnings from their research find application in industry and academic education. In addition, Timon Heinis and Filippo Fontana run the development and innovation agency EMBRIO. The agency supports business organizations in identifying and realizing value-adding digital innovations and Internet of Things applications.

Dr. David Hua is an associate professor in the ICT field specializing in networking, hardware, and software. He is a co-founder of the INDI, the regional Internet of Things Interdisciplinary Institute.

Dr. Stephan Jones began his career in telephony but expanded his interests to the whole ICT field. He has published many books in the field and has served on international committees.

Mr. Sagar Kaja was a graduate assistant for the Computer Technology program. He completed his Master of Science in Computer Science in 2018 and is now a Software Engineer at eMPiGO Technologies.

Dean Kevin Keathley is a Dean, School of Information Technology at Ivy Tech Community College in Indianapolis, Indiana. He has several decades of Cloud and IT industry experience in addition to some great skills as a musician and performer.

Mr. Aaron Khoury is a recent graduate of the Center for Information and Communication Sciences. He currently works in the technology industry as a consultant for DeLoitte.

Dr. Ronald J. Kovac has worked in the ICT field (for corporate and academia) for 30 years and has explored most areas of this field. He is a prolific publisher and has put on many international presentations.

Mr. Jared Linder is the Chief Information Officer for the State of Indiana's Family and Social Services Administration. He currently leads enterprise efforts for health and human services technology solutions, focusing on IT modernization, data and analytics, and systems interoperability. Jared also represents the state by serving on committees and boards both locally and nationally, focusing on furthering the delivery of health and social programs through advancements in technology and data. Jared holds a bachelor's degree in English and a master's degree in Information and Communication Sciences, both from Ball State University, and an MBA from Butler University.

Lauren McNally is a previous Graduate Assistant for the Center for Information and Communication Sciences at Ball State University and currently a Channel Manager at one of the world's largest Telecommunication provider and Fortune 500 companies.

Alex Peczynski is a 2018 graduate of the Center for Information and Communication Sciences, Ball State University, and a Software Engineer in DevOps at the Ford Motor Company in Detroit, Michigan.

Timothy Roe is an information technology (IT) project manager and a Certified Scrum Professional who has 20 years of career experience in higher education, as well as seven years of public broadcasting experience. When he is not leading project teams to new solutions, Tim plays drums semi-professionally for various groups in and around the Indianapolis area.

Ruth Schwer, PMP, BRMP, has worked as an information technology (IT) professional for more than 20 years. Ruth holds a master's in library science from Kent State University, and has served in Butler University as an IT project manager, analyst, researcher, trainer, technical communicator, and as an IT business relationship manager. She works to bring clarity and efficacy into the implementation of technology for learning organizations.

Dr. Dennis Trinkle has worked in industry and academia for many years and specializes in innovation, entrepreneurship, and the neurosciences. Dr. Trinkle joined Ball State University in July 2017 and is the Director of the Center for Information and Communication Sciences.

Jerry Walker is the Director of IT for the City of St. Louis Treasurer's Office. Mr. Walker has over 16 years of IT experience in both non-profit and profit organizations that includes LAN/WAN engineering and architecture, DRP policy and planning, cloud conversion, strategic information technology (IT) business planning, network security, PCI compliance, IT policy and procedure governance and planning, project management, IT health-care business start-up consulting, server systems management, and overall leadership strategy planning. Mr. Walker is the President/CEO and Owner of Nucomm Technology Consulting Firm. Mr. Walker is a proud graduate of Ball State University where he earned his bachelor's and master's degrees in Information and Communication Science. Mr. Walker is an active and proud member of Omega Psi Phi Fraternity.

Dr. Angelia Yount is a recent graduate from a doctoral program and a member of Ball State University's iLearn team, Dr. Yount provided us with a literature review segment for our Data Analytics chapter.

THE BASICS OF IoT

I

Chapter 1

The Yin and Yang of IoT

Rebecca Lee Hammons
Ball State University

Contents

1.1 Introduction

The Internet of Things (IoT) industry shows promise for providing internet-based devices that enable an improved quality of life for individuals, families, and communities. While in recent years, IoT innovations have been embraced by early adopters who relish the opportunity to become familiar with new technology, we are shifting into an era in which IoT devices will be mainstream along with the data analytics that are applied by the vendors, administrators, and users. This chapter provides a look at how IoT innovations are used to create smart homes and cities and to improve healthcare services.

The applications of IoT are limited only by vision, imagination, and wireless capabilities. However, the growth in this technology market also brings interdependent issues and challenges that must be considered by consumers, designers, manufacturers, and vendors. The chapter also examines the implications of integrating a preponderance of such devices into our businesses, schools, homes, and cities through a discussion of the concept of Slow Tech and the importance of being socially responsible in our use of information and communication technologies (ICTs).

1.2 Smart Homes

According to the Central Intelligence Agency (CIA) Factbook, 76% of the U.S. population had internet access in 2016. That percentage has undoubtedly grown in the past 2 years. About 77% of the U.S. population have smartphones (Pew Research Center, 2018). When you put together a smartphone and home internet access, along with a wireless router, you get the opportunity to create a smart home environment by connecting various sensors to commonly used household items for the purpose of controlling them over the wireless network. A smart phone, tablet, or laptop can be used as a controller for the smart home network and devices; this chapter will center on smartphone-centered control of home IoT.

For many years, home security systems have been available that utilize sensors and an alerting system (internet or cell phone based) for home protection. Vendors like Koorsen and Comcast have built strong business models in this space. Recently, it has become possible for an individual to create their own home-centered security system without the use of a third-party service provider. For example, one can implement sensors on doors and windows along with interior and exterior cameras with motion detection to monitor their own home while away from a smartphone. This is one of the most common ways a homeowner might enter the smart home world.

Another common way to begin a smart home project is to upgrade existing home lighting with smart lighting products (smart bulbs) and use a monitoring solution like an Amazon Echo to use voice commands to turn lights on, off, or dim. Smart lighting can also be controlled remotely via a smartphone application.

Smart doorbells with cameras also allow homeowners to monitor traffic to their doors while away, and smart door locks make it possible for the homeowner to allow someone access to the interior if needed. Smart garage door openers likewise enable a homeowner to monitor approaches to the garage and open or close the garage door while away from home (or without leaving the couch!).

Smart thermostats like the Nest™ make it possible to remotely control the temperature setting of one's home via a smartphone app; enhancements now provide for flooding alerts. Tech savvy consumers can instrument their second homes with such tools to provide for warm homes upon arrival and peace of mind in a weather crisis. More recent entrants to the IoT for smart homes marketplace include smart refrigerators and smart window air conditioners. The refrigerators are remotely viewable to assess what items one might need to pick up at the store on the way home; smart window air conditioners enable a user to begin cooling down their home while commuting. Some of the IoT-based smart home tools are clearly "nice to have" items while others may provide cost and energy savings over time that will someday be considered essential to our lifestyles.

Implementations are moving toward home automation that is self-learning to optimize efficiency over time through learning one's habits of use. Compatibility of smart home devices is a consideration when designing a smart home strategy given that there are no standards requiring such compatibility and manufacturers do not have aligned standards. Other considerations include effective home network security including strong passwords and regular system updates to prevent outside hacking of the smart home devices for malicious purposes. Figure 1.1 provides an overview of the various Smart Home Possibilities. A homeowner must weigh several factors when contemplating a shift to a smart home infrastructure:

- One's technical skills to configure and maintain the infrastructure
- The cost to implement versus the time and cost savings
- The value proposition for the implementation
- Performing a smart home upgrade on one's own versus using a smart home service provider
- Privacy considerations, given the volume of personal usage data that will be shifted to the cloud
- Health or disability issues that could benefit from home automation assistance.

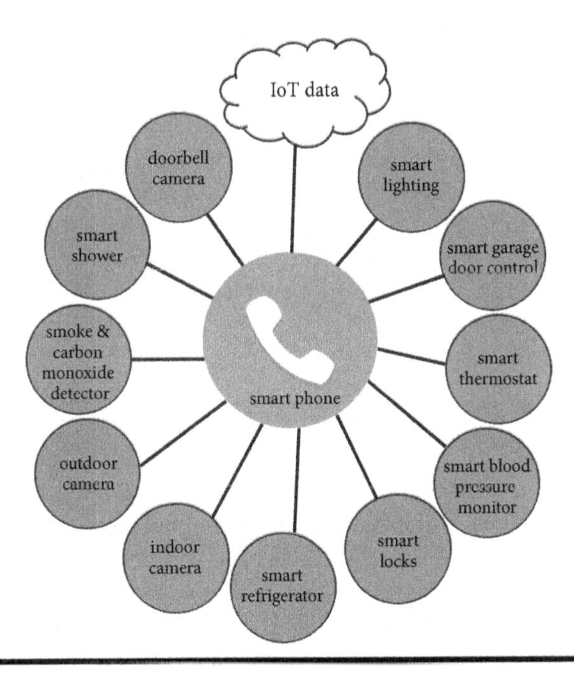

Figure 1.1 IoT smart home possibilities.

Both Amazon and Google are frontrunners in the marketing of smart home assistants—the Amazon Echo and the Google Home. Although not essential to a smart home deployment, these IoT devices make it possible to use voice commands to control many other IoT features within a home, such as lighting, temperature, and door controls. As a freestanding device, the smart home assistant centralizes and supports a number of helpful functions like checking the weather, listening to music, placing calls via one's contact list, reminders for activities or medicines, and such. A new "remember" feature makes it possible for the device to remind you of an important task or note as well. As with other cloud-based smart home gear, these smart home assistants gather considerable data on the user. They are always in "listening" mode unless muted, so there are a few downsides such as intrusiveness from machine learning algorithm-produced messaging. The application programming interfaces are fairly easy to use libraries of preconfigured commands that do not require coding skills. This puts home automation of a variety of different IoT devices within the grasp of many.

For those who desire a smart home environment but lack the skill or patience to design and implement, there are numerous companies that provide such services. For example, alarm.com provides customized and comprehensive home security systems that are monitored 24/7. Users receive alerts to activities occurring within the system, such as when children return home, water or smoke sensors are triggered, windows or doors are opened. QuietCare® Health Harmony home sensing provides caregiver alerts when motion activity within a home, such as for an elderly or disabled person, triggers an anomaly based on machine learning of daily routines and patterns. Vivint is the number one smart home provider, according to Sensors Online (2018). Vivint offerings focus on security with smart locks, cameras, and doorbells.

In-home providers of services for lifestyle enhancement and pleasure are not yet commonplace, yet this implies a potential business opportunity for entrepreneurs. Early adopters of tools like smart blinds and lighting tend to learn from trial and error at this point in the IoT lifecycle. Expect to see some of these devices become mainstream in the years to come, especially elements that improve energy efficiency. Some utility companies are now offering rebates for homeowners who implement smart thermostats.

1.3 Healthcare

IoT devices, and the associated data analytics and monitoring services, promise to meet the needs of a variety of users who suffer from health and disability concerns. For example, an Apple watch, an IoT fitness and communication wearable device, was recently cited as the reason a teenager's surprising kidney disease was identified, through monitoring and alerts of her rising heart rate (Friedman, 2018). A variety of devices are now available or under development to benefit the user and also address the problem of rising healthcare costs. IoT devices allow for independent living for those who are disabled—smart pill bottles, assistive robots, smart blood pressure monitors, even smart shoes are available. According to the United Nations, 15% of the world's population lives with some type of disability, with 18.65% of the U.S. population having a disability (R-Style Lab Blog, 2016). Figure 1.2 illustrates the prevalence of common disabilities.

The field of IoT wearable devices (wearables) is growing, well beyond the fitness and health trackers made popular by health insurance plans in recent years. Here are a few examples:

- Nokia wireless blood pressure monitor (https://health.nokia.com/us/en/blood-pressure-monitor)
- Accu-Chek Aviva Connect blood sugar meter (www.accu-chek.com/meters/aviva-connect-meter)
- CardioNet Mobile Cardiac Outpatient Telemetry to monitor for heart arrhythmia (www.cardionet.com/patients_01.htm)

According to Wikipedia (2018), "Wireless Health is the integration of wireless technology into traditional medicine, such as diagnosis, monitoring and treatment of illness, as well as other tools that can help individuals improve their personal health and wellbeing." Through wireless synchronization of monitoring output to a smartphone application and subsequent sharing of the data to the cloud, it is possible for companies and medical practitioners to track activity and provide alerts to users or healthcare providers if data points are outside of established control limits. For example, "smart pillboxes" are now available that monitor patient adherence to treatment schedules and provide reminders to patients.

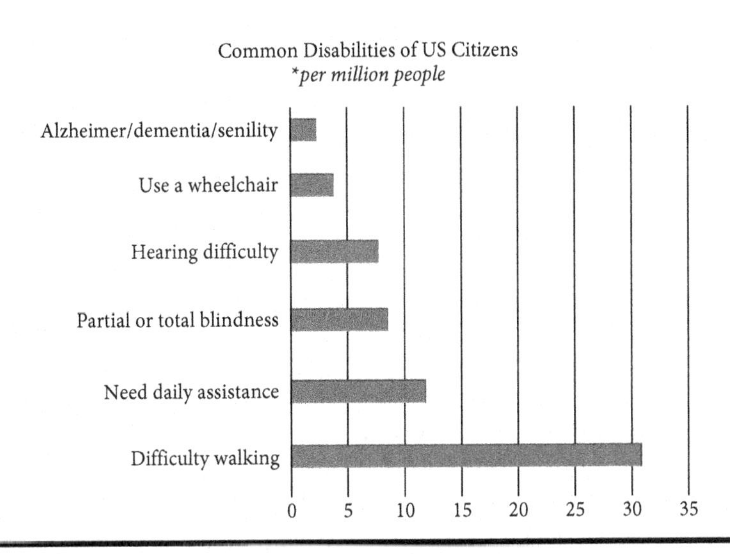

Figure 1.2 Common disabilities of U.S. citizens (Pew Research Center, 2018).

In 2015, Fortune reported that there are 57 million Americans with disabilities, referencing the spokesman for the American Association of People with Disabilities (Fortune Magazine, 2015). That said, only 5.6 million with disabilities have smart home platforms due to a lack of insurance coverage for smart home technology and the high prices for smart devices (e.g., $250 for a Nest thermostat or an August lock). A nonprofit built a home for six people with disabilities, which cost $600,000—the technology itself cost $100,000. Another company built a smart home for war veterans with disabilities at a cost of $400,000–$500,000. As the IoT industry matures, there is the possibility that smart home "packages" or "footprints" of technology solutions can be optimized and produced in a lower cost format for implementation in smart homes to support the vast number of disabled Americans. As noted in Chapter 8, *Empowering Older Adults with IoT*, a large percentage of elders are disabled and would benefit from such solutions.

Some more basic IoT solutions that target everyday needs, like the smart pill system, are under development using Cloud computing. E-commerce IoT solutions provide much promise to disabled and elderly consumers according to Das et al. (2017); the authors described many opportunities for innovation in independent living such as automated doors, key chains, pillboxes, movement trackers, and monitoring/alarm systems. Das et al. (2017) also described the possibilities for improving hospital stays from the patient standpoint through use of eye-tracking equipment to control the hospital room functions lost to those who cannot move.

Research conducted by Ukil et al. (2016) on healthcare analytics stated that, "The ubiquity of smartphones, cheaper storage, highly powerful edge-processors enable in-house, remote as well as preventive healthcare a near-future reality" (p. 994). The authors' research in the area of anomaly detection in cloud-based systems points to the critical importance of predictive analytics capabilities including accuracy for the use of IoT devices for epidemic early warning, preventive healthcare and monitoring, predictive analysis of disease, and other benefits to the world. Nguyen et al. (2017) described three groups of IoT healthcare applications including monitoring, self-care, and clinical support. The accuracy of monitoring improves over time with the increased amount of monitoring data and thus the ability to identify anomalies more readily. The authors of this research propose that, "The results of the review show self-care, data mining and machine learning as areas where next generation of IoT applications for healthcare should focus on" (Nguyen et al., 2017, p. 260–261). It is not the single connected product or device, such as a fitness tracker, that makes the significant user impact but instead the data derived over time from a large volume of such products or devices in use by many users. The data provides the means for building models that can be used for predicting issues or reacting to anomalies in usage trends.

IoT healthcare solutions have the potential to be disruptive to many existing healthcare practices and services. As an example, Boric-Lubecke et al. (2014) described physiological radar monitoring systems that can produce diagnostic data from noncontact sensors for remote sleep monitoring. This IoT solution may displace brick-and-mortar healthcare facilities that conduct obstructive sleep apnea testing today. Cardiac care monitoring once only available through an in-patient hospital stay can now be achieved remotely with IoT connected sensors and monitoring equipment. These types of healthcare solutions are beneficial to patients by eliminating the stress of an in-patient experience and reducing healthcare costs; insurance providers likewise benefit from lower healthcare monitoring costs. The middlemen, the existing brick-and-mortar medical facilities who have historically provided such services, must innovate and reimagine their healthcare services and business models to remain in business. Such is the nature of change in healthcare as with many types of industries and services due to the innovations coming with IoT.

At the heart of many of the healthcare service improvements that will result from the IoT, the key to a successful solution or service involves a user-centric focus. Entrepreneurial (or intrapreneurial) approaches to innovation with the IoT will succeed if the user experience (UX) influences those seeking healthcare services to accept or even embrace the changed experience. People are often very resistant to change, especially in established habits or practices. However, if the new UX through connected devices is based on a proper UX-driven design and evaluation, user adoption of IoT-based practices is more likely.

1.4 Smart Cities

According to the United Nations Economic Commission for Europe (UNECE, 2018), smart cities should be "inclusive, resilient, safe, sustainable and 'more connected'." Characteristics of a smart city include:

- "Ensure access to adequate and affordable housing
- Provide access to safe, affordable, and sustainable transport systems
- Enhance inclusive and sustainable urbanization
- Safeguard the world's cultural and natural heritage
- Reduce the number of deaths, displacements, and losses caused by disasters
- Reduce its environmental impact
- Provide universal access to safe and accessible green and public spaces
- Support positive economic, social, and environmental links between urban and rural areas
- Integrate innovative technologies and ICT within its different sectors" (UNECE, 2018).

Smart cities are needed due to the shift of the world's population to urban areas and the changes required to support densely populated urban areas in a sustainable manner.

The IEEE's Smart Cities Consortium (2018) states, "As world urbanization continues to grow with the total population living in cities forecast to increase by 75%

by 2050, there is an increased demand for intelligent, sustainable environments that offer citizens a high quality of life. This is typically characterized as the evolution to Smart Cities. We believe a Smart City brings together technology, government and society and includes but is not limited to the following elements:

- A smart economy
- Smart energy
- Smart mobility
- A smart environment
- Smart living
- Smart governance"

Each of these elements is comprised of ICT solutions that are tailored to meet the needs of an urban population. For example, smart energy can include sensors for moderating the use of electricity and fuel within an urban environment. Data analysis of usage patterns can be developed to predict and manage future usage trends. Smart governance can include sensors and monitors to assure the cleanliness and safety of a community, as is done in Singapore, the current leader in the use of smart cities IoT technology. Smart traffic sensors and monitoring support the flow of traffic within urban environments; this use of ICT is already commonplace in the United States. Figure 1.3 identifies many of the elements of a Smart City.

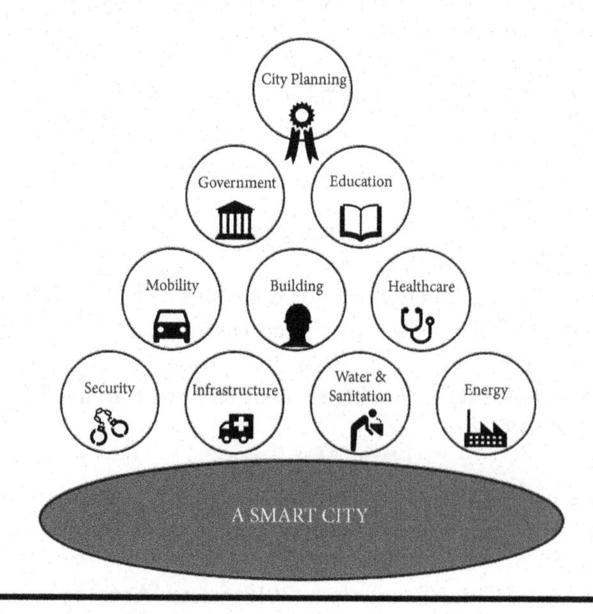

Figure 1.3 Elements of a smart city.

Singapore has revolutionized its urban environment over the past 15 years through a strategic focus led by government. A totalitarian society, Singapore is in a position to identify policies and processes, supported by laws that target public infrastructure management. Democratic communities face more difficult journeys to align on smart city strategies, policies, and funding due to the give and take nature of democratic governance. In Singapore, for example, it is unlawful to urinate in an elevator. This is likely true for many communities. However, in Singapore, there are sensors in elevators that can detect urine and, if present, the elevators will lock and retain the offender until the police arrive to arrest them (Business Insider, 2015). This is one example of how IoT is used to enforce smart cities policies. Singapore has extensive networks of cameras and monitoring in place to detect and address anomalies in behavior, such as vandalism, chewing gum, and public smoking (all illegal activities subject to penalties). With tight control over the laws to govern a population and the ICT infrastructure to monitor and control compliance to the laws, an urban environment can become a smart city "from the top down."

Democratic urban environments face a more uphill battle to become smart cities and, most likely, embrace ICT and IoT innovations with grassroots initiatives. Some communities in India, for example, are setting an intention to compete with Singapore to become renowned smart cities. Additional benefits of being a smart city, aside from the marketing advantages, are using real-time data to improve the way heating and cooling systems perform, manage the flow of people through cities, and improve beauty through landscaping. On a minor level, many U.S. urban communities already use IoT instrumentation to manage some services such as traffic flow, traffic tickets, toll passes, and the electric grid. Some communities are beginning to use smart trash cans, which have sensors which alert sanitation crews when it's time to empty them. This saves transportation and labor costs in comparison to making routine "runs" to empty all trash cans whether they need to be serviced or not.

Some of the considerations for building a smart city strategy include network infrastructure, having a strategy for working with the resulting data sets, sharing information across services, and understanding the needs of the community. Targeted areas for instrumentation and data collection include communications, energy, transportation, and infrastructure. Smart cities require an investment in ICT talent to create and implement a strategy and work with the resulting data to optimize outcomes. Strategies should reflect the inevitable staging of new IoT-based infrastructure to replace aging infrastructure over time; it is rare that a city can fund an "all new" approach to implementing IoT. Especially for emergency services, solutions should include the ability to share information across various services to benefit the community.

While governments can undertake the journey to become a smart city, the voice of the customer is an essential element to consider. Voice of the customer is a quality management concept that reflects the integration of the needs and views of the user of a product or service. Smart city services should reflect the concerns and needs expressed by those who reside within the community or urban area, whether through research and advocacy by elected officials or through the use of a mobile application for civic engagement that empowers the residents to weigh in on service priorities and concerns. While change is inevitable in this world, and a community is either evolving to be better or not, people tend to be more supportive of change if their voice is heard throughout the process. Modest research has been done in the area of applications for civic engagement, and these tend to focus on very specific issues such as reporting wildfires or potholes. Significant opportunities exist to create robust civic engagement applications to gather community input and provide direct communication to stakeholders, especially with the waning of the printed newspaper as a means of reaching community constituents. A truly smart city will emphasize human connectedness in its design and implementation.

Other smart city considerations include security, recycling and reuse, and service quality. There are currently no standards for assuring data and communication security within a smart city context; security is presumed to be the responsibility of the purchaser, such as creating strong passwords, routinely applying patches and updates, and assuring the integrity of "insiders" who have access to the network and devices. Recycling and reuse of ICT equipment is an essential consideration for sustainability; only about 27% of U.S. electronics are recycled each year within the United States. The explosive use of sensors, cameras, monitors, routers, servers/storage, and cabling to build smart cities requires meticulous policies and plans for recycling and reusing equipment. Likewise, with the implementation of a robust smart cities infrastructure to provide services, an emphasis on service quality including uptime monitoring and reporting is critical. Configuration or service errors within the infrastructure could inadvertently result in significant downstream rework and costs, including the potential for legal action by constituents. Businesses that currently host applications in the cloud through services such as Amazon Web Services or Azure already face these types of issues with customers when there are unplanned service outages.

Business considerations for smart cities include significant revenue opportunities, UX design, integration methods, data analytics support, as well as the role ethics should play within the product or service design.

It's undeniable that the application of IoT to develop smart cities has created a significant business innovation opportunity that could result in revenue growth for companies that participate in this segment. We are likely at the very early stages of this industry focus and the future may be unimaginable. As companies move into this market segment, it is important that they invest in the design of a good UX for those who conceive of, implement and maintain IoT infrastructure. Without a good UX, work practices will be riddled with user-induced defects and training costs will be unsupportable. Imagine the effect that employee turnover might have on the ability of an urban government to maintain infrastructure, if the UX is poor. Integration methods including secure data exchange must also be considered by designers and manufacturers as we will see many companies enter this space.

Standards such as OneM2M are emerging through collaboration by industry, and participation in such standards bodies by industry is essential to the efficient and effective use of ICT by smart cities.

Data analytics, sometimes called data science, is an emerging field that will produce professionals essential to the management of the tremendous amount of data that results from the use of IoT in smart cities. Strategies for fulfilling this critical need must be established by smart cities, and this need creates significant career opportunities for those who are able to invest in this education. This may be one of the more challenging roles to fill in the management of smart cities, mainly due to a shortage of skilled workers. Policies will need to address the definition of data types to be collected, limitations on the collection, specification of the purpose, limitations on the use of the data, and openness or transparency with the public.

As smart cities use ICT to improve quality of life in urban communities, it is important to consider ethical issues as well. For example, what are the data privacy rights of the individuals who use the services? Are advertising and commercials allowable within the ICT service delivery process as a means to recoup expenses? If increased taxation is the means to implement the ICT infrastructure and provide smart services, what voice should the constituents have in the priorities and costs? How will smart cities benefit marginalized populations? These, and many more, ethical questions will be important to discuss in building products, services, and smart cities infrastructures. From the individual's point of view, the perspective may be that "it's all about me!" Unless people see a strong value proposition for smart cities enhancements, pushback is probable by some segments of the population. Individuals will be concerned about privacy, security/vulnerability, cost, and usability.

1.5 Slow Tech

Whatever the intended benefits of an IoT device or infrastructure, it's appropriate to emphasize the need for Good, Clean, and Fair ICT in the product lifecycle and supply chain. Slow Tech (Hammons, Patrignani & Whitehouse, 2017) is a concept modeled after the Italian Slow Food movement and incorporates computer ethics, business ethics, and corporate social responsibility into the business and usage models for ICT solutions. Little attention has been paid to the hidden costs of supply chain management practices in the ICT space such as abusive labor practices, electronic waste, illegal mining of minerals, child labor, or other social issues related to the ICT industry. We are just now waking up to the challenges of electronic device/internet addictions, especially to smartphones, and researching the effects and implications for users of all ages. ICT products lack sustainable designs that enable a longer, useful lifecycle; many products are made to be used over a couple of years and thrown away, to be replaced by a "better, newer model." The many billions of new connected devices that promise to bring many benefits to users through smart nations, smart cities, smart homes, and smart healthcare will also become tomorrow's electronic waste unless corporations are influenced to embrace Slow Tech design, development, and manufacturing (lifecycle) practices. As well, communities will benefit from adopting a Slow Tech approach to the integration of connected devices in community infrastructure, homes, schools, and other environments.

The IoT trend will be wide and deep and affect individuals, teams, communities, nations, and the global community. Consideration in the design and planning stages for the tens of billions of new devices to be manufactured should focus on attributes such as meeting diverse user needs, sustainable sourcing and manufacturing practices, reuse/recycling/repurposing, and other Slow Tech considerations.

1.6 Case Study

A Midwest producer of canned vegetable products has embraced the IoT as a means to remain competitive in a global market. Figure 1.4 depicts at a very high level the key steps to manage the lifecycle.

The unique and competitive advantage of this business is its use of sensors and data for timely decision-making throughout the entire lifecycle, from determining when to plant the crops, to the management of crops in the field, to the oversight of the pace of picking the crops leading to the management of transportation of crops from field to processing center. The processing facility is aware of the progress of inbound trucks to better manage the production lines and handling of incoming produce. The materials used for the production process are likewise monitored with smart sensors to determine replenishment needs to ensure the lines run smoothly and thus utilize labor efficiently. Customers of this business are made aware of the status of their orders through the instrumentation and communication vehicles. The shipping process is handled effectively with smart logistics related to transport needs and status.

From beginning of the planting cycle to the movement of product to customers, this business has made a conscious decision to implement IoT infrastructure as it becomes available (sensors, wireless communications, monitoring, and data analytics). This strategy has supported them in optimizing their operating cost model in addition to providing timely and efficient service to the stakeholders in their business lifecycle. Many other types of businesses are integrating IoT solutions as a means of reducing costs, improving cycle time, and remaining competitive. Examples include the trucking industry, the management of overseas shipping containers, food delivery services, and Uber and Lyft transportation services. Entrepreneurs will continue to find ways to improve products and services in creative and innovative ways through the implementation of IoT solutions in their ICT strategies.

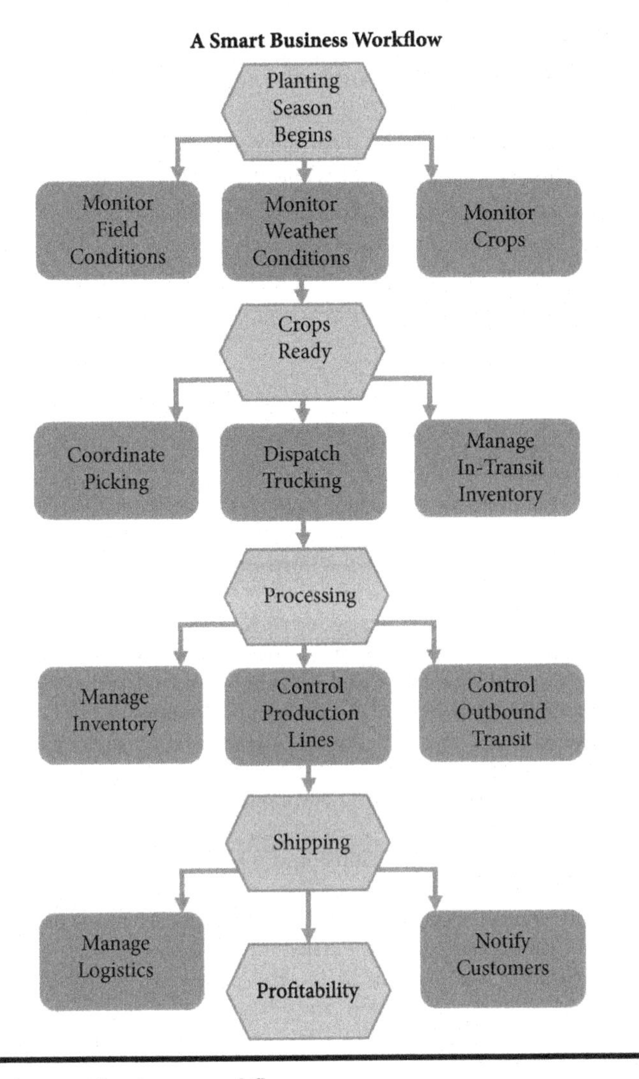

Figure 1.4 A smart business workflow.

1.7 Discussion Questions

1. Why would a community want to embrace smart technology? What are some of the smart technology solutions that might be beneficial in your community?
2. What are the available electronic waste programs in your community? If there are none, brainstorm several methods for addressing the problems of e-waste that result from the proliferation of connected devices and their associated ICT infrastructure.

3. Describe the types of IoT devices that currently exist within your business or school environment. What purpose do they serve and what practices, if any, did they displace?
4. If you were to embrace smart technologies in your home environment, what devices would you acquire and why? What type of wireless infrastructure would you need? Draw a household network topology.
5. Consider the healthcare needs of someone close to you. How are their needs currently being met, and how could IoT solutions address their needs in a way that provides them with more independence and an improved quality of life?

References

Boric-Lubecke, O., Gao, X., Yavari, E., Baboli, M., Singh, A., & Lubecke, V. (2014). E-Healthcare: Remote Monitoring, Privacy, and Security. *IEEE MTT-S International Microwave Symposium (IMS2014).*

BusinessInsider. (2015). Retrieved from www.businessinsider.com/things-that-are-illegal-in-singapore-2015-7

Das, R., Tuna, A., Demirel, S., & Yurdakul, M.K. (2017). A Survey on the Internet of Things Solutions for the Elderly and Disabled: Applications, prospects, and challenges. *International Journal of Computer Networks and Applications (IJCNA).* Vol. 4, Issue 3, pp. 84–92, May-June 2017. EverScience Publications.

Fortune Magazine. (2015). Retrieved from http://fortune.com/2015/02/01/disabled-smart-homes/

Friedman, A. (2018). Retrieved from www.phonearena.com/news/Apple-Watch-saves-another-life_id104551

Hammons, R., Patrignani, N., & Whitehouse, D. (2017). *The Slow Tech Journey: An Approach to Teaching Corporate Social Responsibility.* Retrieved from http://sites.ieee.org/futuredirections/tech-policy-ethics/november-2017/the-slow-tech-journey-pt1/

IEEE Smart Cities Consortium. (2018). Retrieved from smartcities.ieee.org

Nguyen, H.H., Farhaan, M., Naeem, M.A., & Nguyen, M. (2017). A Review on IoT Healthcare Monitoring Applications and a Vision for Transforming Sensor Data into Real-time Clinical Feedback, p. 260. *Proceedings of the 2017 IEEE 21st International Conference on Computer Supported Cooperative Work in Design.* 978-1-5090-6199-0/17

Pew Research Center. (2018). Retrieved from www.pewinternet.org/fact-sheet/mobile/

R-Style Lab Blog. (2016). Retrieved from http://r-stylelab.weebly.com/blog/internet-of-things-for-people-with-disabilities

Sensors Online. (2018). Retrieved from www.sensorsmag.com/components/top-10-smart-home-service-providers-us

UNECE. (2018). Retrieved from www.unece.org/housing-and-land-management/projects/housingsmartcities/smart-cities-characteristics.html

Ukil, A., Bandyoapdhyay, S. Puri, C., & Pal, A. (2016). IoT Healthcare Analytics: The Importance of Anomaly Detection, p. 994. *2016 IEEE 30th International Conference on Advanced Information Networking and Applications.* DOI: 10.1109/AINA.2016.158

Wikipedia: Wireless Health. (2018). Retrieved from https://en.wikipedia.org/wiki/Wireless_health

Chapter 2

IoT System Integration

David Hua, Biju Bajracharya, Sagar Kaja, Christopher Davison, and Vamsi Gondi

Ball State University

Contents

2.1 Introduction

A common misconception of the Internet of Things (IoT) is that it is primarily comprised of smart devices or "things." The reality is that there is a great deal more than just a thing for an IoT-enabled solution to be successful. The process of creating these solutions includes the creation of the device or application, communications, data storage, and analytics. Lack of foresight on any of these factors can doom an IoT project to failure. This chapter will explore the requirements of creating an IoT-enabled solution. It will also discuss issues associated with integration among the many moving parts of these endeavors.

2.2 IoT Solution

You have been introduced to the world of the IoT. An environment that is ripe with opportunities to embed sensing, computational, and communication capabilities into just about any "thing." At this point, you have been inspired to develop an IoT-enabled solution that will transform operations in your organization, will offer your customers with newfound capabilities and efficiencies, or provide consumers with new experiences. The world is yours if you could just make your idea a reality.

That is when the questions start running through your mind. Where do I start? What is going to be required? What if I don't have the technical expertise to get this done? You realize that there is a single answer to each of these questions, "I don't know." With your hopes dashed, another brilliant idea is lost to the ages. But wait. You think back to what your parents said when you were a child facing a seemingly insurmountable school project, "How do you learn to walk: One step at a time."

With newfound hope, you decide to break down your idea into its component parts and deal with them one step at a time. To put the complexity of developing a complete IoT-enabled solution into perspective, consider the following scenario. You want to develop a consumer service to help people prepare for an evening meal. In the morning, you announce to a voice-enabled home assistant (e.g., Amazon Echo, Google Home, or Apple Homekit w/Siri) that tonight's menu for two people will be beef stroganoff over noodles, steamed asparagus, home-baked bread, and a nice Merlot. You then run off to get to work on time, secure in the knowledge that your home assistant has things well in hand.

Your voice command triggers a series of events. The home assistant looks up the recipes for each of the items on the menu. It then determines if you have the ingredients for the dishes. Any ingredients you are out of are placed on a shopping list. The home assistant then polls local stores and sends you a message notifying you of the ingredients you need and the alternatives for getting them for dinner. The alternatives include delivery services (e.g., Peapod), pickup services (e.g., Walmart Grocery Pickup, Kroger Clicklist), or stores with the best price that will be on your driving route home where you can stop in to purchase yourself.

You respond to the home assistant with a choice to pick up the ingredients on your way home. When you get into your car to head home, markers have been added to the navigation map in your car indicating where you can stop to purchase the ingredients and their prices. As you drive home, the intelligent system in your car notifies you that you are approaching a store until you have purchased your ingredients. Once you purchase the remaining items needed to prepare your dinner, the home assistant and intelligent system in your car consider their job complete and let you enjoy the rest of your drive home without any further reminders.

The provided scenario presents an IoT-based solution that contains many moving parts. The remainder of the chapter will explore each of these parts.

2.3 Devices aka "Things"

The devices in the IoT are comprised of three major components: sensing, computing, and communications (Bojanova 2015; Dixon 2018; Ray 2016). The sensors attached to intelligent devices interact with the physical environment, other IoT devices, and applications through a digital communication mechanism. Sensors enable these intelligent devices to collect data from the physical environment and transfer that data to other devices. In the consumer market, there are the myriad assortments of "smart" devices that include everything from smart watches that you wear, voice-assisted systems that can serve as an audible interface to IoT, to self-driving cars. Applications are a software/computing component that can reliably transfer or exchange sensor data into useful information for people, businesses, or other IoT devices. Digital communication mechanisms are the networking systems that interconnect IoT devices and software components.

The concept of the IoT is represented in an architectural framework (Figure 2.1) which is an integration of IoT components that allow data exchange between the physical environment and computer systems over digital communication systems. Based on this architectural framework, IoT components are grouped into:

a. Transducers (sensors and actuators)
b. IoT devices/smart things/embedded devices
c. Networking infrastructure/communication medium
d. Application processing/IoT platform
e. End user interface/M2M

2.3.1 Transducers (Sensors and Actuators)

A transducer is a device that responds to the change in physical environment which is then converted to an analog or digital electronic signal and vice versa. These analog or digital electronic signals are recorded by IoT devices and transferred to the application processing or application which can send these signals to the transducer. Sensors and actuators are forms of transducers.

Sensors are devices that capture or detect physical events that transduce or generate signals which are then used for actions, analysis, monitoring, or measurement.

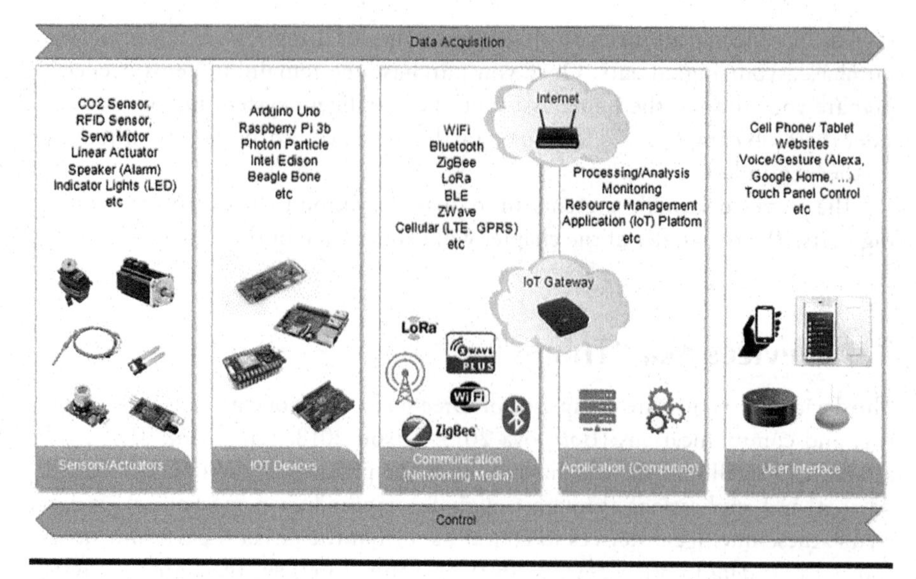

Figure 2.1 Data acquisition.

To accurately measure the data, the right sensors need to be selected. IoT devices can be equipped with sensors on device or can be attached separately.

Actuators are components that receive signals from applications to respond with a physical action or motion. Light-emitting diodes (LED), LCD displays, and sound output from speakers are indicator devices which are used to alert someone about the status of the device. There are many sensors and actuators available. Some of them are listed below:

List of Sensors

a. Position/presence/proximity sensors
b. Motion/velocity/displacement sensors
c. Accelerometer
d. Magnetometer
e. Temperature sensors
f. Humidity/moisture sensors
g. Acoustic/vibration sensors
h. Light/color sensors
i. Chemical/gas sensors
j. Flow sensors
k. Force/load/torque/strain/pressure sensors
l. Leaks/levels sensors
m. Electric/magnetic sensors
n. Light sensors

List of Actuators

a. DC motor
b. Stepper motor
c. Servo motor
d. Pneumatic actuator
e. Piezoelectric actuator

See the other chapter in this book that deals specifically with Sensors and Actuators.

2.3.2 IoT Devices

The IoT device is the intelligent component that is embedded in to another device to convert it into a "smart" device. It is a hardware component consisting of a computing unit, storage unit, general purpose input/output (GPIO) unit, and communication unit. Each of these units can be expanded by adding external modules or drivers. These devices are available in different forms:

a. Microcontroller units (MCUs)
b. Microcontroller development boards
c. Single board computer (SBC)

Microcontroller unit is a compressed microcomputer manufactured to control the functions of embedded systems in office machines, robots, home appliances, motor vehicles, and several other gadgets. Examples of microcontroller units are:

a. PIC16F886-I/SO from Microchip
b. ATMEGA328P-MU from Atmel
c. STM32F103C8T6 from ST Microelectronics
d. MSP430G2402IN20 from Texas Instruments

Microcontroller development board consists of all hardware units along with IO pins and sockets necessary to connect sensors, actuators, and communication media. Development boards are also called prototyping boards. These boards require a programming interface, which consists of software running on a PC, to access the development board through a connection cable, such as a USB cable. Examples of development boards include:

a. Arduino Uno
b. Particle Photon
c. Adrafruit Flora
d. LightBlue Bean
e. pcDuino
f. Beagle Bone
g. Intel Galileo

SBCs have all the units along with interface for monitor, mouse, and keyboard which make it a full computer. Usually they run on Linux distributions or a dedicated IoT operating system. SBCs do not require other PCs to access it. As with most computers, an SBC only needs a monitor, mouse, and keyboard. Example of SBCs are:

a. Raspberry Pi 3b
b. Asus tinker board
c. Qualcomm Dragonboard
d. Raspberry Pi Zero W
e. Rock64 Media board
f. ODROID-XU4
g. Orange Pi PC 2

These boards can be expanded or extended with additional functions and features with add-on devices like modules or drivers. These are stackable or modular circuit boards that piggyback onto the main microcontroller board to instill it with additional functionality. Usually modules are extended components connected via jumper wires. Drivers are extended components that generally handle high power needs like stepper motor and linear actuators.

2.3.3 Networking Infrastructure/Communication Medium

The communication medium consists of protocols and technologies that enable two physical objects (IoT devices) to exchange data. IoT devices equipped with arrays of sensors and actuators require a communication medium to interact with humans or other machines in real time. IoT devices are generally resource constrained devices. Constraints that limit communications are low power consumption, low bandwidth, low computation power, and seamless connectivity between different communication technologies. There are various communication standards, protocols, and data formats that address these constraints. Because of these various standards, the IoT environment is heterogeneous, decentralized, and complex. These communication technologies are used before their connectivity into mainstream networks (Internet). For example, how does a smart watch know that you have an incoming phone call or an appointment? Depending on the device, it may have to receive communications using Bluetooth from a cell phone to provide you with the desired information. Other IoT devices may directly connect to the Internet using Wi-Fi, wired Ethernet, or they need to connect to the Internet via IoT gateway devices.

In general, the IoT devices are resource constrained and, therefore, are traditionally designed to support specific applications. The data collected by the devices are limited by storage constraints. Analyzing power is limited by computing resources. Cloud computing helps to avoid these constraints by providing storage and compute services. This allows IoT devices to offload their task and storage to the cloud for infinite computational, analyzing, and decision-making power.

Depending on the application, factors such as range, bandwidth requirements, security, and battery life dictate the choice of communication technologies. This communication technology choice will have direct impact on the IoT device requirement and cost. Some of the major communication technologies are listed below:

a. Bluetooth
b. Zigbee
c. Z-Wave
d. 6LoPAN
e. NFC
f. Sigfox
g. Neul
h. LoRaWAN
i. BLE
j. Wi-Fi
k. Cellular

2.3.4 Application Processing

You were introduced to the need for processing earlier. Now it is time to dive a little deeper so that you will be able to decide the type of processing your IoT-enabled solution will require. An MCU differs from a microprocessor unit (MPU) (see Figure 2.2). An MPU contains only a CPU and therefore requires added peripherals to perform tasks. MCUs, on the other hand, contain random access memory (RAM), read-only memory (ROM), and similar peripherals, which allow them to perform (simple) tasks independently.

Despite having similar names and appearances, microcontrollers and microprocessors differ widely in their applications. Microprocessors are more powerful but must be employed as single components in larger systems to function. Microcontrollers, meanwhile, are limited in power and functionality but can

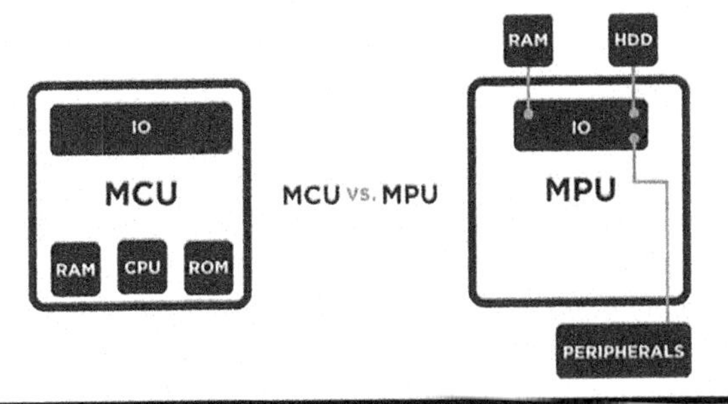

Figure 2.2 MCU versus MPU.

perform simple functions independently. An Apple TV, for example, requires a microprocessor to handle all the varied and demanding tasks it performs. A connected coffeemaker, on the other hand, only needs to perform simple routines and tasks and therefore employs a microcontroller.

In our current scenario, the voice assistant must utilize a microprocessor to handle all the sensors and process the data whereas the RFID scanner in the fridge can get away with a controller because it has only one job to scan an RFID tag and send the data to the voice assistant. RFID stands for radio-frequency identification. It is a relatively inexpensive wireless strategy for communicating small amounts of information over short distances. It is important to consider RFID's cost-effectiveness. A processor, although not overtly expensive, is still pricier than a controller.

There are functions the voice assistant can implement locally like activating itself on voice command, checking time, and sending data to the cloud and acting on the results, like understanding your full sentence and replying with the result. With enough memory, you can implement the computational part locally on the assistant. But implementing the computing process locally makes it process intrinsic because it must keep track of sensor data processing and keep track of your position from the smartphone app to check grocery stores for the needed groceries and compare prices then map the shortest route on your way back home. The ideal implementation of this would be to send the RFID data to a desktop computer and let it handle the continuous stream of data from your phone and the voice assistant process the information and send the result back to the assistant or your phone which can then act upon the result like show the route on the phone. This is viable if we are talking about implementing a single system. When multiple instances of this system are being implemented as a business product then we must consider the scalability of processing power, storage, and network capacity. The next option is to use a server. But a server is expensive to maintain, so the viable solution is to use a cloud-computing service.

This brings us to the topic of cloud computing. In the simplest terms, cloud computing means storing and accessing data and programs over the Internet instead of your computer's hard drive. It means that instead of all the computer hardware and software on your PC or somewhere inside your network, it is provided for you as a service by another company and accessed over the Internet. Exactly where the hardware and software are located and how it all works does not matter. To the user it is just somewhere up in the nebulous "cloud" that the Internet represents. Advantages of this approach are that you only pay for the services you use and don't have to worry about maintenance costs.

Cloud computing is better and fast, but there is always network delays that we must factor into consideration. The need for enormous amounts of data to be accessed quickly and locally is ever growing as more people are moving towards IoT. That is where "fog computing" comes to play. Fog computing, or "fogging," is a distributed infrastructure in which certain application processes or services are managed at the edge of the network by a smart device, but others are still managed in the cloud. It is, essentially, a middle layer between the cloud and the hardware to enable more efficient data processing, analysis, and storage, which is achieved by reducing the amount of data which needs to be transported to the cloud.

In our current, scenario because we have a processor in the voice assistant, we can use that as a middle layer to process simple and immediate responses like responding back with your name, checking if an item taken out of the fridge has been replaced like a milk jug. The complex processing like looking up the RFID for a matching item, checking the best prices for the item, and mapping a route can be done by the cloud. The same applies to the GPS tracking your route to home. Your phone can calculate your current speed and show your position synchronized with the GPS signal while cloud handles finding the nearest store with best prices.

2.3.4.1 End-User Interface/M2M Interface

In the context of IoT/cyber-physical systems two entities, humans and machines, coordinate and compute in multiple ways to envision smart applications from multiple environments. In this section, we will be discussing how humans and machines interact and how "data," which is extracted from the physical environment is processed into machine and human readable formats.

Users and applications typically integrated into the IoT world with sensors and actuators. As discussed in the previous sections, the perception of the monitoring environment is extracted in machine-readable format and application's processes execute the necessary algorithms to extract the knowledge at various locations of the systems. The extracted knowledge is used in various applications, and the end results are communicated to human entities, machine entities, or executed in applications. The extracted knowledge is processed and integrated into the system in four different ways.

1. **Device to Device:** In this scenario, two or more user devices directly connect, communicate with each other rather than communicating through intermediary application servers in cloud or in fog computing scenarios. The knowledge extraction, processing, and the application hosting is accomplished locally, a typical example scenario includes home automation systems - garage openers, lighting systems, and heating/cooling systems.
2. **Device to Cloud-based applications:** In this scenario, the devices (sensors and actuators) are connected directly connected to Internet-based cloud application services. Typically, data is collected in the physical environment and sent to back-end servers, data is then processed or aggregated through application servers, if there is any feedback for the actuators; the application servers push the necessary actions to the application queues, where these application actuators are registered to receive the instruction sets. A typical application scenario is shown in Figure 2.3. Typical application includes; nest systems, health monitors, and smart speakers (e.g., Amazon Echo, Google Home, and Apple HomePod).

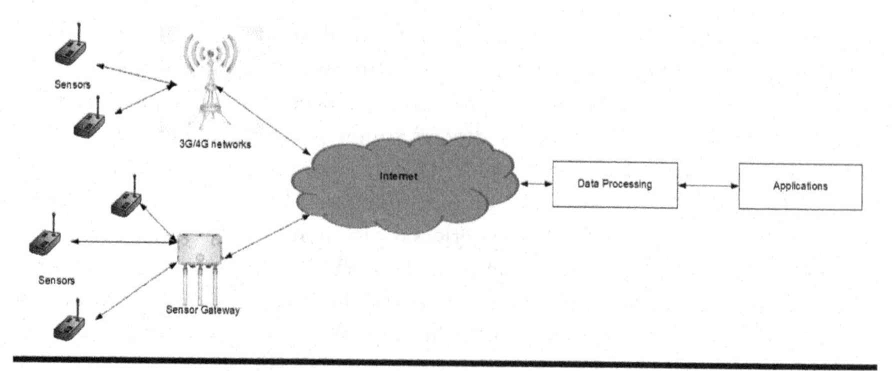

Figure 2.3 Cloud-based applications.

3. **Device to Smart gateways / cloud-based applications:** In this scenario, the application intelligence is hosted not only at the cloud but also at the smart gateways. The device data and actuation are controlled locally with the help of the cloud application entity. The device collects the data and send it to the smart gateway, the gateway processes the application data and send the necessary feedback directly to the device, thus reducing the network overhead (latency). This kind of application processing in the critical event applications, where the processing time is critical in an event scenario. Some of the applications include smart transportation systems, where event process timing is critical to assess an event such as an accident and cars that need to take necessary evasive actions to avoid the collisions.
4. **Machine to Machine:** In this application scenario, data is collected and processed by not only by one application but also by multiple applications. The data is shared by multiple devices to the cloud, and the intelligence is shared and processed by different applications to extract the knowledge by multiple applications to make a necessary decision. A typical example application scenario can be a smart sprinkler system, where the soil moisture sensor can detect what is the current moisture content in the soil that is needed by the grass and how much is needed for watering the grass in a lawn. The sensor collects the data and sends the data to the back-end cloud. An application programming interface (API) is a set of tools that allow developers to create software to interact with and pull data from devices. A weather station or other data sources such as IBM's weather API can provide the current weather condition and future weather patterns. A smart sprinkler system can access both these data sets at the cloud level and asses if lawn needs to be watered, based on the application outcome the necessary instructions such as watering the lawn and how much quantity of the water is needed for lawn is relayed to sprinkler. Based on the instruction from the cloud, the sprinkler does provide the water to lawn. Typical application scenario is shown in Figure 2.4.

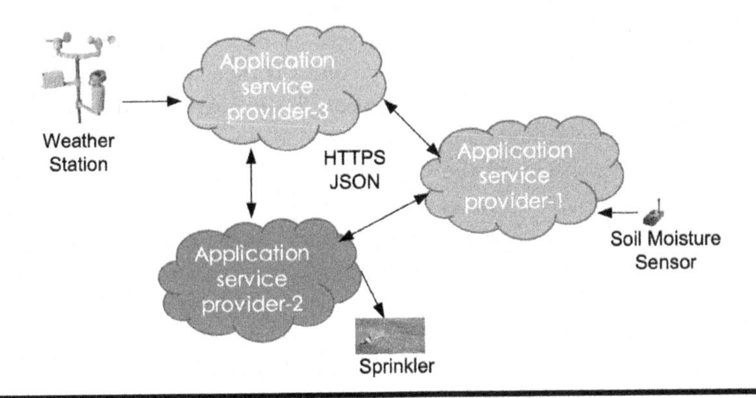

Figure 2.4 Smart sprinkler system.

2.4 Programming

So far, you have been presented with the hardware components of an IoT solution. To implement any IoT solution, there is programming involved in every part of the system, from making the refrigerator read RFID tags and sending the information to the voice assistant to your voice assistant acting on the results from the cloud. Programming is a necessary skill for any IoT implementation. Below are a few skills that are necessary.

- Business Intelligence: Implementation of IoT is all about collection, storage, and analysis of streams of data from smart devices. Needed skill sets include sensor data analysis, data center management, predictive analytics, and programming in Hadoop and NoSQL.
- Information Security: Security can't be an afterthought for IoT devices. The desire to keep information safe means that developers who are familiar with vulnerability assessment, public key infrastructure (PKI) security, ethical hacking, and wireless network security will be key players.
- UI/UX Design: Nothing sells if consumers cannot use it. The interfaces between the device and the consumer must be effective and user-friendly. An IoT solution that may be capable of great things will likely fail if the user interface is too complicated or not intuitive. Responsive web design and service design are the valued skills in this aspect.
- Mobile Development: Most IoT devices will be managed through smart phones. The ability to develop apps that communicate with external hardware and sensors is helpful.
- Hardware Interfacing: Hardware programming is essential if you want to develop yourself into an embedded engineer. Knowing at least one operating system, such as Linux, and one embedded system such as Contiki would be a nice asset to have.

- IP Networking: In IoT, embedded sensors interact with the environment around them. The information they collect is then sent to be analyzed. The network through which the information flows must be designed to be flawless, set up for enormous traffic, and must be secure and reliable. Developers must know the basics of the Open Systems Interface (OSI) stack, how the connectivity protocols work, and the latest standards in IoT communications.
- Automation: Given the large amount of data and interfaces, developers who know how to connect automatic API testing with manual testing will be the ones who get their products to market.
- Design for Data: Big Data drives IoT, and the job of software engineers, network engineers, and UX engineers is to make the data work seamlessly for users. Ability to read and interpret data in a meaningful way will be a most helpful skill.

All the above skills can be implemented in numerous ways using different programming languages and databases. Also, it's not necessary for one person to have all the skills but when assembling a group these are the skills one must look to fulfill.

2.5 Data Storage

There are billions of "things" in the world engaged in sensing or monitoring. You have already seen how this activity can impact the communications or network connectivity of an IoT-based solution. The next dilemma is where to put all of the data that these devices are collecting and transmitting. You will find that in some instances, the best choice will be to simply dump the data after it has been received and responded to. After all, is it necessary to store your recipes, the ingredients that were used, how you choose to purchase those ingredients, or even the day of the week in which you prepare these meals?

What if there are additional insights to be gained beyond the face-value information provided by the device output? Could that information be used to create a better product or service? With that possibility, it may be prudent to store the data for ongoing analysis. The question becomes how and where.

Srinivasan (2018) presented a series of questions that will help you determine the storage requirements of an IoT solution. The first question asks the type of database you want. A database is a software application designed to input and organize data into a framework that allows for retrieval of that data in whole or in part. There are many different database applications to choose from. You may choose to pay for a developed and supported database such as Oracle or Microsoft SQL. Instead, you may choose to use an open source database application that is available for free, such as MongoDB and Crate DB. Keep in mind that open source software may have limited to no support available by the creators of the software.

If the software has an active community, there may be a wealth of independently developed tutorials and guides available online. In addition to cost and support, the choice of database software will be impacted by the need for a SQL, NoSQL, or time series database.

The second question to ask concerns the type of workloads that will need to be processed. There are several workload types, including batch, transactional, analytic, high-performance, and database workloads. Within IoT, the most common of these workload types are transactional or analytic. Transactional workloads focus on the automation of processes. This is frequently seen in purchasing and billing. For example, transactional workloads would be used when you tell your voice assistant to order more coffee from your favorite e-commerce vendor. Analytic workloads are better suited for making sense out of vast amounts of data.

The next question has to do with scalability. During the development phase of your IoT solution, you anticipate thousands of devices will be collecting and then transmitting data for storage. A dedicated server hosting the database is used to receive that data. Your market forecast failed to predict the overwhelming demand for your IoT solution. Instead of selling thousands of your devices, the demand is for hundreds of thousands and then millions of them. All of a sudden, that on-premise database is not an appropriate data storage mechanism. How are you going to adjust as your data needs expand?

2.6 Data Analytics within the Context of the IoT

The IoT promises to greatly impact the science of Data Analytics. Few things can deliver the volume, velocity, and variety (the three Vs of Data Analytics) of data as the IoT. Small, connected devices, churning out vast amounts of data promise to push the limits of the bandwidth and capacity of database systems. With data analytics, the challenges are to make sense of the vast array of data, in a timely fashion, which provides useful knowledge to humans.

Traditional database system models use tuples with attributes to store and manipulate related data. Information systems for student enrollment and healthcare patient information systems are groupings of related data that make sense when compiled into a system. These systems provide information about people (objects) that have characteristics (attributes) and perform operations (methods) such as enroll in school and pay invoices. However, in the not always connected and often abstract scenarios of the IoT, it is challenging to model events such as the scenario presented: Some entities desire a specific meal. Disambiguation of such a request is a daunting task.

Some obvious and some unobvious questions arise from the meal request. The obvious questions such as: "How many entities are dining? When does the meal need to be ready? When will the meal take place?" are some examples. However, there are some less obvious questions that arise: "Who are these entities? Are they humans, pets, or perhaps non-biological?" The answer to the questions impacts the data being utilized for the request.

Food for humans is quite different than food for canine companions. Beef stroganoff-flavored dog food would most likely not be the preferred choice of humans. The preferred *food* for an artificial intelligence (AI) companion might be a high-resolution image of the meal with plenty of bytes.

Relatedly, such questions as "how much food?" are asked. Perhaps some *a priori* knowledge of the requestor's eating capacity and the capacity of the fellow diners is required. If this data is unavailable it would be difficult to arrive at an optimum food quantity and stochastic modeling algorithms would only provide a best guess.

From a conceptual viewpoint, how is "hungry for beef stroganoff" stored in a database? While the capacity to tell a home assistant that users are "hungry for beef stroganoff" exists, the capacity to properly model and store that concept requires further research. At best, the current level of technology will return beef stroganoff recipes when it is told the user is "hungry for beef stroganoff."

Another important aspect to the IoT is data communications. Consider the connectivity of the IoT regarding the meal request. As networks and systems go online and off-line, this impacts all three Vs of data analytics. Furthermore, using the data to make decisions such as price, availability, and delivery timings becomes challenging in unstable networking conditions.

There is the additional question of utilizing data analytics to model user preference and taste. Using analytics to make decisions on taste outcomes presents a problem. Selecting the correct data and models to choose the right ingredients becomes problematic. Which variables are more important and how does the analytics system weight the variables? Several weighted variables include online reviews, past ordered ingredients, new producers of ingredients, and prince among many others. Are those enough variables and the correct ones to produce an optimum outcome? How can the data model account for user preferences and taste changes over time? Modeling and storing subjective concepts such as taste preferences is a difficult challenge in data science.

2.7 IoT Integration Issues

As you have seen in this chapter, there are many moving parts. It starts with all of component decisions for the device or "thing" that will enable it to capture, react to, and transmit what you desire. It also includes many parts that extend beyond the individual devices: communications, cloud providers, data storage, and analysis. Each point in the solution feeds into the other.

This brings us back to where we started. How do you begin to develop an IoT solution? One approach is to piecemeal together the components from various providers to build an integrated solution. One of the first issues is whether the disparate components will be compatible with each other. Does the communications component in a device utilize a protocol that is compatible with a receiving component on the network? Is the data format compatible with the choice of database? The success of an IoT solution depends on all of the pieces working together in harmony. You will need a team of engineers, developers, database administrators, and more to develop an IoT solution.

Putting together this research and development team can be a costly endeavor. So again, we are faced with a dilemma. Do you give up on your idea due to the cost and complexity of entry or do you find a way to forge ahead? Maybe there is an alternative in the research and development of your IoT solution. Maybe there is somebody out there who can help you get your IoT solution off the ground.

This is where IoT integration providers come into play. IoT integration providers offer a platform of products and services to aid in the development of your IoT solution. These providers have gone through the trial and error of determining what and how the various pieces in an IoT solution work together effectively. There are providers prepared to offer whatever level of assistance you require to make your idea a reality.

For example, consider the initial prototyping of the device. The development of an IoT-enabled device contains both hardware and a software component. Let us say that your in-house team includes a programmer who has tinkered around with the Raspberry Pi and Arduino. That is a start because you have someone who realizes that these devices are building blocks and not the final solution. It is not unlike working with Lego blocks as a child (or child at heart). These SBCs are like that large square plate with the rows of bumps. This, like the Raspberry Pi and Arduino, was not very much fun until you started to add the blocks. In this case, the blocks are the multitude of components that you can connect to the SBC. So now you are faced with the dilemma of deciding what blocks to use. The programmer is able to identify and assemble a rough model from the sensors, actuators, and communication modules. But what has been created looks like a mash-up of wires, breadboards, and components. It has some of the basic functionality of what you would like to do, but it needs more. It certainly does not look like a full-featured, marketable product.

Realizing that you need help, you contact an integration provider that focuses on prototyping. You find that the provider has a team of engineers and developers ready to help you. At your first meeting with the provider, you explain your vision of the IoT solution you are trying to create. You show them the initial model for the device that your programmer created. Based on those conversations, the provider is able to build a prototype. Their engineers select a smaller microcontroller board instead of the SBC. They incorporated required sensors and communication modules for your intended IoT solution. At the end of the process, you are presented with a device that is custom built for your IoT solution and smaller than a deck of cards.

As you have seen, though, the device only constitutes a portion of the overall number of parts that need to be developed to have a full-functioned, scalable, secure IoT solution. You could continue to hire different integration providers to assist with individual aspects of the project. Instead, you may seek out a comprehensive IoT integration provider who is prepared to help you through the entire development cycle. Using this latter approach should result in a tighter integration between the all of the moving parts of your project.

2.8 Summary

At first glance, the IoT appears to be a simple platform upon which to design solutions. You have a group of little devices with sensors that collect data, actuators that respond to data, and are able to communicate independently via Internet protocols. It is just like sticking a small computer into an item such as a window shades and making them "smart." But as you have seen, there are many components that must be addressed when developing a full-featured IoT solution. With each of those components arise potential compatibility problems. Communication protocols between components must be compatible. The database has to accept the format of data captured by the IoT sensors. IoT systems integration is the key to a successful project. If an organization does not have the technical expertise in-house, IoT integration service providers can be hired for specific aspects of the project or to help with the entire product development life cycle.

2.9 Case Study

At the beginning of this chapter you were presented with a scenario in which you were going to prepare beef stroganoff. That scenario will be the basis for this chapter's case study. As you progressed through the different aspects of developing an IoT solution, the materials periodically referenced the opening scenario. For the purpose of this case study, you are asked to review the IoT solution presented in more detail based on your newly acquired understanding of the components that would be required to bring the service to market. Please review the Discussion Questions below to test your knowledge.

2.10 Discussion Questions

1. Identify possible sensors that would be required in the person's kitchen for the IoT solution to work.
2. Integration within the resources owned by a company is an expected challenge in the development and operations process. Identify points in the

scenario where interaction and integration with third-party products and services will be required.

3. The organization trying to create the IoT solution has decided that it needs help developing the physical hardware for the kitchen. Investigate three different IoT integration providers that could address this need. Prepare a report that identifies the services available from each provider. Which integration provider would you recommend and why?
4. There are many different wireless technologies adopted by IoT devices. Each wireless technology uses same or different electromagnetic spectrum. This raises the issue of Functionality and interoperability between IoT devices.
   a. How can we resolve this issue?
   b. Does this require using same wireless technology throughout out IoT infrastructure?
   c. Discuss the possible scenarios of having same wireless technology and having heterogeneous wireless technologies.
5. IoT devices require communication medium to interact between then and human interface devices. Such communication medium could be Wi-Fi, Zigbee, Bluetooth, Z-Wave, etc. Does this actually require connecting to the Internet or will it have its own dedicated local or wide area network?
6. Deployment and implementation of IoT infrastructure generates large amount of data. These data are usually stored on public cloud services which make it possible to retrieve data for monitoring and analysis at any time and any location.
   a. How are privacy and personal autonomy affected by these cloud services?
   b. Do we need to have private data repositories?
7. Privacy and Data Analytics are often at odds
   a. Describe some privacy concerns with regard to pervasive IoT spaces.
   b. How would their concerns be addressed?

References

Bojanova, Irena. 2015. "What Makes Up the Internet of Things?" Computing Now. March 31, 2015. www.computer.org/web/sensing-iot/content?g=53926943&type=article&urlTitle=what-are-the-components-of-iot-

Dixon, John. 2018. "An Introduction to IoT Components." 2018. www.avnet.com/wps/portal/us/resources/technical-articles/article/iot/nxp-intro-to-iot-components

Ray, Brian. 2016. "IoT Platforms: What They Are & How To Select One." LinkLabs. August 3, 2016. www.link-labs.com/blog/what-is-an-iot-platform

Srinivasan, Jay. 2018. "Top 8 Questions to Ask When Choosing a Data Storage Strategy for Your IoT Solution." January 10, 2018.

Chapter 3

IoT Positioning in the Verticals

Ronald J. Kovac
Ball State University

Contents

3.1 Introduction

The Internet of Things (IoT) is similar to a chameleon. It changes its colors to blend in with the surrounding environment. But also like a chameleon, its inner concepts stay the same. IoT exists, and is growing, due to a collective root goal: The efficiency and optimization gained by "linking" the various islands of information that exist within a corporation, the home, or the world. This process works through gathering information (through sensors), processing, deciding what to do, and acting on this information through actuators. The root causes and actions of IoT stay the same, but its skin changes within each environment or, in our case, each vertical.

Verticals are segments of the economy that are grouped together by common functions. Verticals include manufacturing, utilities, transportation, healthcare, telecommunication, and a host of others. This segmentation of the economy into verticals allows others to rationally make sense of the whole. In this chapter, we will see how IoT "changes it spots" within each of these verticals. We will also explore how IoT is currently operating within each vertical, and we will gaze into the crystal ball to see how IoT is projected to advance in each vertical within the next 5–10 years. Domains within each vertical will also be examined and specifically with use cases that fit the particular vertical. Additionally, we will identify the large companies within each vertical and give some examples of how they are employing IoT throughout their wheelhouse.

In this chapter, we will note the major players within the IoT community. These players, again, fall into segments such as Telecommunication (to connect all devices together), Analytics (to process the information), Sensors and Actuators manufacturers (as noted in Chapter 5), and other segments that make IoT function. This will give the reader a good breadth in understanding the various pieces and parts that make IoT work.

We will close the chapter with a case study and some insightful questions for you to think about. Please actively engage as you read along to fully understand the concepts and material.

3.2 The Verticals

The world economy is broken down into segments. These segments, called verticals, break down the whole into manageable pieces. We will approach this chapter with using these verticals in order to make more sense out of the data and make it simpler for the mind to understand. We will not look at all verticals but only the ones that IoT can play a big part in.

3.2.1 The Verticals—Healthcare

We are all somewhat familiar with the healthcare industry (Figure 3.1). It is your family physician that you go to for a broken bone or for flu medication. It is the hospital you go to when you break an arm playing ball and the pharmacy where

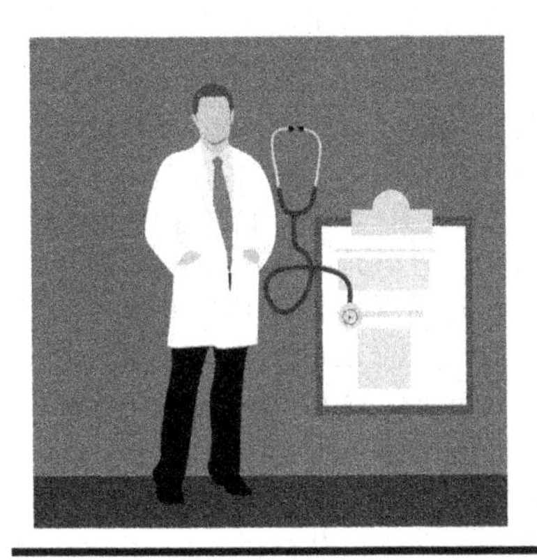

Figure 3.1 Healthcare vertical.

you go to fill prescriptions. But is it more than that behind the scenes; it is the economic system that provides goods and services to treat patients with curative, preventative, and rehabilitative and palliative care. As a whole, healthcare is one of the largest and fastest growing industries, and it consumes over 10% of the gross domestic product (GDP) of most developed nations. (Healthcare Industry 2018)

Typically, the healthcare vertical is broken down into three distinct categories:

a. Hospital activities
b. Medical and dental practice activities
c. Other human health activities

Listed in the diagram (Figure 3.2) are the ten largest healthcare companies. You may not have heard of many of these roles, as they are behind the scenes owners, or

Company	Gross Worth	Role
United Health Group	184 Billion	Payers
CVS Health Corporation	177 Billion	Drug Stores, PBM and Distributors
Amerisoruce Bergen	146 Billion	Drug Stores, PBM and Distributors
Cardinal Health	121 Billion	Drug Stores, PBM and Distributors
Walgreens Books All.	117 Billion	Drug Stores, PBM and Distributors
Express Scripts	100 Billion	Drug Stores, PBM and Distributors
Anthem	84 Billion	Drug Stores, PBM and Distributors
Johnson and Johnson	71 Billion	Pharmaceuticals
Aetna	63 Billion	Payers

Figure 3.2 Healthcare companies.

operators, of many of local companies and doctors' offices. These are measured by revenue and are certainly large organizations; although they seek economic advantages through technology and change, they are also large organizations that change slowly.

"Top 20 U.S Healthcare Companies by 2016 Revenues," Revenues and Profits, accessed February 19th, 2018, http://revenuesandprofits.com/top-20-u-s-healthcare-companies-by-2016-revenues/. (Revenues and Profits 2018) (Figure 3.2).

You will notice that the largest companies are in the drug store sector, pharmacy benefits manager (PBM) and distributors field, while the most revenue is earned by the pharmaceuticals and Biotech (where IoT can most productively live or at least be an incubator) companies. Certainly, more of the other industries within healthcare can be affected by IoT, and all can benefit as well (Figure 3.3).

Where can IoT most likely grow? The healthcare industry is large and continuously growing.

Due to the aging population of the world, the healthcare industry is growing at a higher rate than other verticals. Certainly, technology has been employed within the industry, as we see from any visit we take to a doctor's office or a hospital. Many companies have created innovative ways to use technology for the industry. A large problem with acceptance of technology within this vertical is due to its large size (big companies change slowly) and also the age of the vertical. The healthcare field has been around for a long time and has adopted older technology; these legacy systems are not compatible with today's technology. Conducting a massive change to implement new technology is both costly and time consuming (Zehel 2018).

> The healthcare vertical is expected to receive the largest IoT product development growth in the coming year; the projected growth is 163.2 billion by 2020 (O'Donnell 2016).

This is not to say change is impossible, but change will be difficult. What is the low hanging fruit for IoT within this vertical? The following provides some ideas and a few of the companies that are playing in that arena:

- Wearable devices—This is an arena that provides early warning, first-line, bio-sensing technology to the medical field. It allows all wearers the option to have their vital signs monitored in order to obtain early warnings of major medical changes. Currently companies that are important in this arena are iRhythm Technologies and Cardiac Insight.
- Artificial intelligence—Companies that utilize the emerging field of Artificial Intelligence to predict and/or make more accurate predictions on health-related issues. Many interesting approaches are being taken here including emotional artificial intelligence, cardiovascular imaging, and use of raw

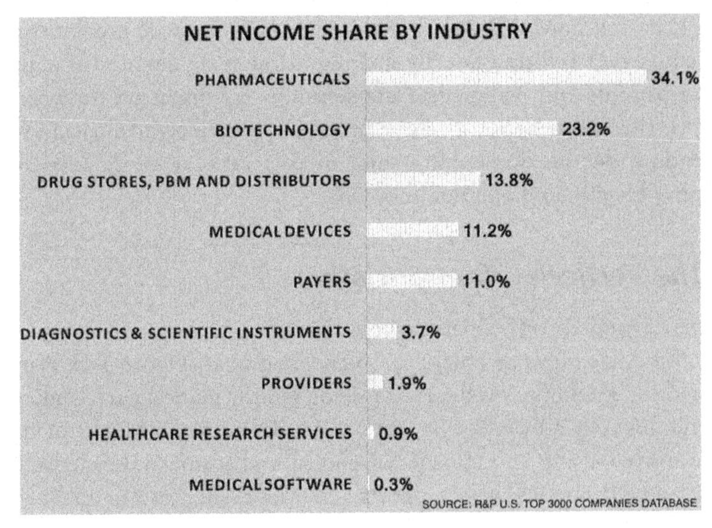

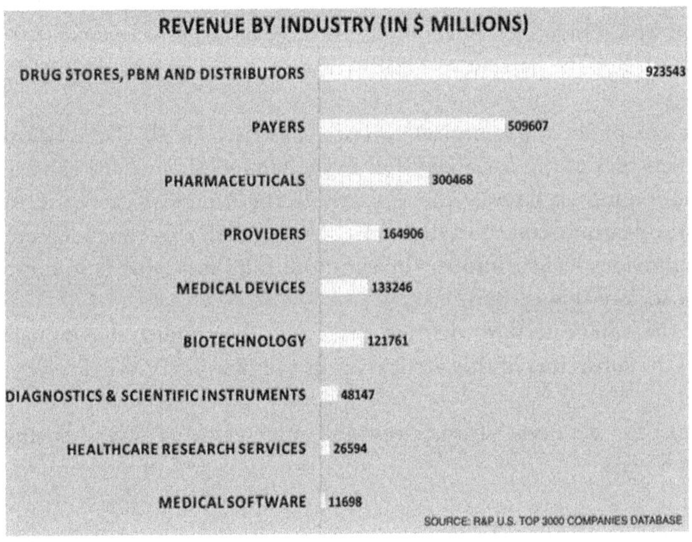

Figure 3.3 Healthcare revenue.

patient data. Some notable companies in this arena are BAYLABS, Imagen, and Affectiva.

- Imaging solutions—Imaging, whether in the visible spectrum or other areas, provides doctors a view of problem areas within a patient. The camera sensors, combined with data analytics, can help immensely in diagnostics. Some companies that are the forerunners in this arena include Medstreaming, iCad, and Medicalis.

- Big Data solutions—The goal of this arena is to provide predictive analytics based on past information. By looking at the large amount of data gleaned from patients and using Big Data solutions (as discussed elsewhere in this book), this arena hopes to make better or more insightful diagnosis to the attending doctor. Some companies in this vast arena are HBI Solutions, Zephyr Health, and Sandlot Solutions.

3.2.2 The Verticals—Transportation

A particular product, individual, or object is being transported every second of every day. Products move by truck, car, plane, and boat (Figure 3.4). People move by any of these methods. Products are made (ships, planes, cars, and trucks) to conduct the moving. Overall, this vertical provides transportation of passengers and cargo, warehousing and storage for goods, and scenic and sightseeing transportation; and this vertical also supports activities related to each of these modes of transportation.

As you would imagine, the numbers are large in terms of revenue (1,058 billion U.S. dollars), but the percent of GDP is also at a critical 6% (Plunkett Research Ltd. 2018).

The scope of the transportation vertical has risen in all of the key subsectors and parallels that of the national GDP in the United States. One can say that the transportation moves, literally and figuratively, the American economy. As one can see from the diagram, transportation only contracted during times of recession, and this data provides insight into the inventory-to-sales ratio that is so critical in our consumer society (Figure 3.5).

With the urgent need for delivery these days (overnight), this industry is only growing. The subsectors of this vertical are as follows:

- Logistics services—Transportation management, warehousing, order fulfillment

Figure 3.4 Transportation vertical.

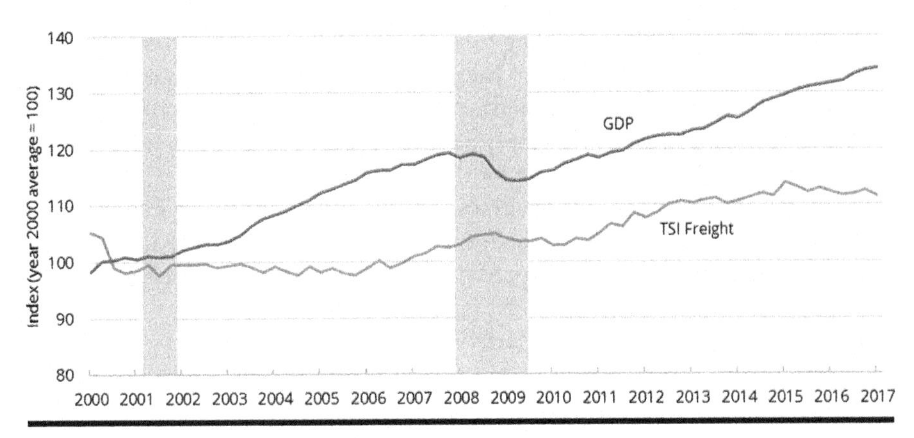

Figure 3.5 GDP and transportation services over time.

- Air and express delivery—Firms offer expedited time-sensitive end-to-end services
- Freight rail—Using rail network to transport people and products
- Maritime—Using ships, terminal for movement of cargo, and passengers by water
- Trucking—Over the road transportation of cargo over short and medium distances (Select USA 2018).

Similar to the healthcare vertical, this industry is large and has built up many systems over the years (some paper and pencil, others electronic) to meet its information management needs. Even though this is the case, the demands of modern industry are challenging these systems in accuracy, timeliness, and mass. It will take a broad overhaul of current systems to add the insights that modern America is asking for today (tracking shipments throughout, data disruption points, impacts of events, etc.).

The IoT applied to transportation is often called Smart Transportation or the Intelligent Transportation System (ITS). It is a key vertical application. These systems aim to provide services related to different modes of transportation and traffic management. This enables users to be better informed and make safe use of the transportation networks.

Where can IoT most likely grow? The following gives some examples of the useful cases for its use and some of the companies exploring these areas:

In 2010, the European Union had defined Intelligent Transportation Systems (ITS) as a system "in which information and communication technologies are applied in the field of road transport, including infrastructure, vehicles and users, and in traffic management and mobility management, as well as for interfaces with other modes of transport" (Kinney 2018).

- Navigation systems—Utilizing IoT; finds the best routes based on real-time conditions and alerts driver of hazardous situations to avoid crashes. A current application that fits this bill is Waze.
- Parking—Being guided to empty parking spaces by smart signs or navigation systems. A current application for this is SF Park.
- Traffic management and Routing—Detect and respond quickly to traffic incidents. A current application for this is created by Ericsson (Kinney 2018).
- Road adjustments—Adjust speed limits and signal timing based on current conditions.
- Freight tracking—Improve safety, inspection, and efficiency of freight traffic flow.
- Automatic road enforcement—Imagine a system that can search for stolen cars as they drive, automatically take tolls without stopping, and detect criminals as they drive around? This is automatic road enforcement.

The above are a small percentage of the applications that are currently being used and a fraction of the ones that can be developed. These applications can not totally cure all the problems within this vertical, but they can help alleviate some of the pressures on the transportation systems.

3.2.3 The Verticals—Cities

Smart cities are currently the general marketing term for this, so let's first explore the definition of a smart city (Figure 3.6). Generally, a smart city is an urban area that has electronic data collection sensors to collect information and actuators to effect positive change within the city. Data collected could be from devices (like traffic lights), people, and other assets. This data is further processed and analyzed to monitor and manage traffic, transportation systems, power distribution, law enforcement, schools and all the other elements of a city. As with our overarching goal of IoT, this vertical aim to enhance quality, efficiency, and performance of city service to reduce costs, help resource consumption, and to make the city more livable.

Due to the term being used widely and loosely, we have to put some parameters on it (found in the sidebar):

As one would imagine, there are many players in this arena. First, there are the many companies that are marketing initiatives for smart cities. These companies,

A smart city typically has four elements that contribute to the definition

- A broad range of electronic technologies in communities and cities
- Big Data analysis to enhance working environments within the regions
- Embedding Big Data into government policy
- Regional use that brings technology and people together to enhance innovation and knowledge.

Figure 3.6 City vertical.

IBM, Cisco, and Microsoft, have different approaches but use the products of the company to engage the city's population in the smart city realm. These companies make the compute, network, and storage of the ICT world and rely on senior developers to create the proper system.

Universities are also playing in this arena. Research laboratory studies are being conducted on intelligent cities to guide future action. Research consortiums are being created to focus on key areas, such as intelligence, planning systems, citizen participation, city platforms, and more.

Model cities are being formed. These cities have been actively pursuing a smart city strategy and have been noted for same. At the time of this writing, a few cites are noted as being models and we will mention a few here: Amsterdam, Barcelona, Columbus (Ohio), Madrid, Dublin, Singapore, and Stockholm. Each of these cities has a wealth of information available on their websites.

As one would expect, there is some criticism surrounding the smart cities movement. These are valid criticisms but are not being heard due to the loudness of the marketing engine of IoT. These criticisms focus around:

- Privacy. The large number of cameras and sensors will allot little privacy in the big cities. Coupled with Big Data analysis, this could lead to some serious impairment in personal privacy.
- The rush into IoT and smart cities is causing a focus that may overlook or underestimate the possible negative effects. This is especially apparent in the vendor push/pull within this arena.
- Historically cities have developed haphazardly. The smart city would be planned. This lack of serendipity could create a structure not desired in the end product.

Some areas that are popular for smart cities, at the time of this writing, are:

- Smart parking—The use of sensors within parking garages to guide you to the empty spots. This can also be built for city blocks to lower the amount of circling the block to find a parking spot.
- Traffic congestion—Monitoring city traffic and events in order to reroute traffic around accidents, congestion, and other events. Actuators would consist of the traffic light timing
- Smart lighting—Ability to control lighting based on pedestrian volume, weather, time of day, etc.
- Waste management—Monitoring rubbish level in containers in order to optimize trash collection routes.
- Smart grid—Monitoring and controlling energy consumption and management of same.

Of course, there are many more concepts being brought forward now and more importantly there are ideas and concepts not even thought of today to make life easier and better in the cities. The sky, or the smart sky, is truly the limit.

3.2.4 The Verticals—Manufacturing

Manufacturing is one of the verticals that have made our country great (Figure 3.7). Manufacturing, creating a finished product from raw materials, has created our economy and has defined our lifestyles. Manufacturing is a multiphase process that involves many elements such as the gathering of the raw material, processing of the raw materials into subsets, and then assembling all the subsets; we strongly suggest watching a few YouTube videos on this process that explains how the fundamental pieces fit together. Smart manufacturing

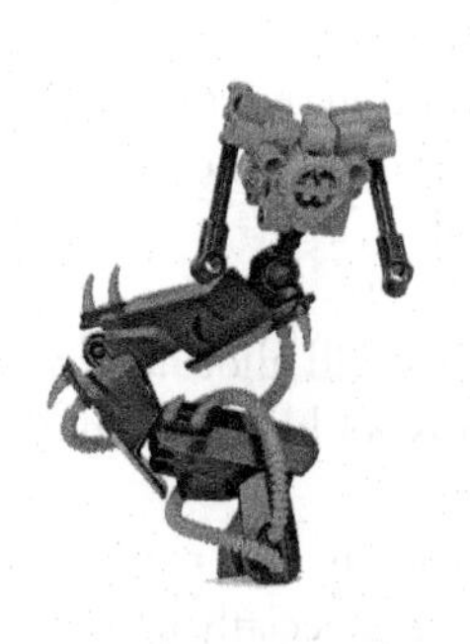

Figure 3.7 Manufacturing vertical.

starts from the multiple phases of the manufacturing process. The use of ICT makes the process more efficient and effective. As a broad example, this can include delivering parts to a factory minutes before they are required (so storage and inventory costs go down). It involves the integration of all steps of the product fabrication process. Analyzing the process to the nth degree smooths out all the wrinkles of the manufacturing process to make the product better and cheaper through increased efficiency. The goals revolve around creating a flexible manufacturing process that can respond quickly to changes in demand (at a low cost) without causing undo harm to other factors (such as workforce, environment). They look at the entire life cycle of a product to make predictions and therefore, control manufacturing.

Smart manufacturing involves many different technologies. We will discuss a few here that are different from other IoT permutations in order to give a better breadth of things (Figure 3.8).

- Robotics—It is becoming a key field of any manufacturing enterprise these days. The use of Robotics lowers long-term costs while also increasing the safety for workers. Advanced robots, or smart robots, can learn from their actions and therefore be better at reconfiguration and repurposing. This increase in rapidity of change is a key element in the smart manufacturing world.
- Big Data processing—Like the entire IoT field, there will be a lot of data to be processed and processed quickly in order to react in a timeframe of merit. The key elements of processing large data sets are the three Vs: *Velocity, Variety, and Volume.* Velocity refers to the amount and frequency of data that is collected, Variety describes the different types of data that have to be handled, and Volume represents the amount of data. Proper use of Big Data analytics allows an organization to be predictive in its actions rather than reactionary.
- Connectivity and services—Due to the high processing needs and "rough" environment of a typical factory, cloud services should be able to adapt quickly and effectively to the changing needs of the manufacturing industry. The cloud can provide the storage, compute, and network in an agile manner to companies.

Figure 3.8 Manufacturing life cycle.

Where can this be used effectively? The following are some currently effective uses of IoT.

- Utilization—The use of a piece of equipment is critical to efficiency. If a machine is broken, no products can be produced. Utilization is a measure of cumulative time of operation divided by total elapsed time. In order to keep utilization high, we must lower the down time for the machines. By using a multitude of sensors companies can predict and lower downtime maintenance and perform preventative maintenance therefore raising utilization time. Examining the broad life cycle of a device can give tremendous insight into its condition.
- Quality—Quality shifts across and along product manufacture. If it is in the acceptable range, the product is accepted and moved on. If not, the product is thrown away pending a fix of the equipment and/or the process. Monitoring quality in a real-time sense permits companies to lower the toss rate while also keeping the product specifications in a tighter line. This involves the measurement of many elements of a product and of the process, but IoT is capable of these requirements.
- Efficiency—By improving the efficiency of labor, energy, and material, more products can be made at a lower cost. Opportunities are there to improve across all dimensions of the manufacturing process (Parker 2015).

3.2.5 The Verticals—Home

As normal citizens in this world, we will see the home vertical of IoT as the most visible (Figure 3.9). This verticals output in IoT will be present in smart thermostats, smart lighting, smart heating ventilation and air conditioning (HVAC), and other areas that will make our home more adaptable to our lifestyles.

This concept, which of the smart home, has been around for a long time and has gone through much iteration before the IoT phase. The last iteration, circa 2005, came out with many products for the home in the concepts of:

- Lighting control
- Computer networks
- Security systems
- Home entertainment
- HVAC controls and monitoring
- Video, data, and voice
- System integration

Although many products were made, the area never gained traction due to a myriad of reasons. These reasons include the lack of standards and protocols due to each vendor's uniqueness, the lack of common layer one (physical) wiring, and the lack of attention to the user experience (UX). All in all, there were many innovative

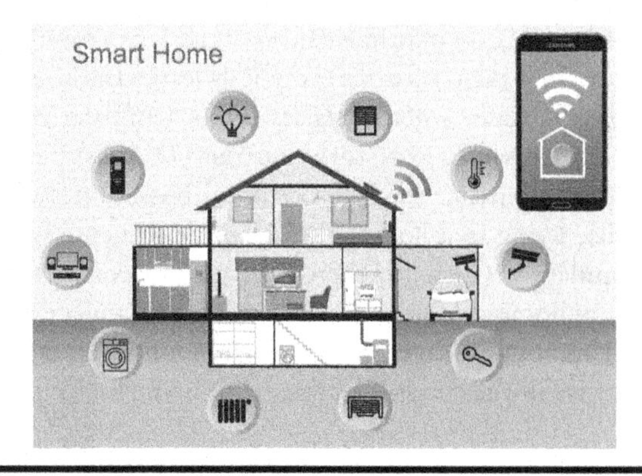

Figure 3.9 Home vertical.

products out there that the techno savvy people tried but eventually they were ignored. There were some movements within the construction industry to build all new houses with the wiring installed, but the user base never saw it as an area in which to be concerned; thus, the end devices were never purchased.

This time could be different! The concentration on software development and integration could bring the field to a new level, assuming the various vendors agree to some standards so end users do not have to fight with product integration. Additionally, the field appears to be using wireless as the method of communication between devices, and this alleviates the need to run wires in existing walls (a very difficult, expensive, and messy task). When you add this to the increased attention focused on the UX, the field can be capable of thriving. Additionally, the work that has been done in the voice activation field can really assist here as we would all like to control our homes without pulling a switch or pushing a button.

The following use cases are being heralded as useful and creative:

- Energy savings—With the rising cost of energy and the attention on making efficient use of our resources, this use case can really thrive. The ability of a house or room to sense your presence and adjust lighting accordingly can save electricity. Similarly, the ability of your home to adjust the temperature to your liking only when you are present can be positive. All of this requires smart sensors to sense a person's presence and a particular person's presence. With this, actuators can adjust the lighting and/or temperature system to our liking. Difficulty here will come in the form of actuators to control currently installed older systems, and integration to adjust to various user needs.
- Entertainment—The concept of the home theater has been growing, especially in more affluent areas. Using projectors, surround sound systems, motion actuated seating systems and environmental control all have the ability to bring the movie theater into the home. This home theater concept linked with the power to access any content you desire over the Internet can change the way we are entertained.

- Safety—This is a use case of prime concern, and it can save lives and property. Imagine a stove that turns off after everybody leaves the house, a safety system that turns off the water when it senses a leak (running water with nobody home), or a smoke sensor that turns off the HVAC system when smoke is detected in the basement. The list goes on of possibilities that can make life easier for those living in a home. This is especially acute when coupled with the aging population. Currently there is a large movement happening to allow people to live at home as long as possible before they must enter an assisted living facility. This not only saves tremendous amounts of money but also allows the elderly to live in the comfort of their own home for as long as possible.

As with all the verticals of IoT, there are elements that need to be solved before this really takes off. These include:

- Acceptance of standards, preferred solutions, and best practices that allow the consumer the freedom to mix and match products and integrate their home.
- Attention to the user need and habits, rather than producing technology without a defined and NEEDED purpose.

We suspect that this is again a young market and is still trying to figure out the best way of integrating this vertical into our lives. Certainly, there will be a lot of shake-up and consolidation in this arena. The lack of acceptable standards is really lacking. At the minimum, we need to accept some preferred solutions.

3.3 Estimates of Growth within the Verticals

There are various ways of gazing into the crystal ball to predict the future, and there are various crystal balls to gaze into. Both of these variables make the accurate prediction of the future difficult. To add to the assumptions that are always given (such as our economy will perform as it has, or that a war will not occur), we see the difficulty of predicting the future for IoT verticals or anything else for that matter. With that said, we will provide some generalized predictions.

Some beginning thoughts:

- Vendor predictions—We have reviewed most vendor predictions and consider them optimistic. Of course, this is to be expected as they are biased in their thoughts. Nonetheless, there are commonalities between their thoughts; these commonalities provide some measure of stability and predictive value.
- Research-based predictions—We have reviewed predictions from the major research firms (Gartner, Forrester, many more) and their research focused predictions. These often vary between each firm (due to different research base and sampling).

The following chart gives our predictions of growth of the vertical, as based on our intensive reading of all the previous mentioned materials, and of our interactions with the corporations (Figure 3.10).

Vertical	Key Use Cases	Size	Growth Potential
Health Care	Remote Diagnostics	Large	High
	Bio wearables		
	Equipment Monitoring		
	Elderly (at-home) Monitoring		
Home	Energy Savings	Small	Low
	Safety		
	Entertainment		
Transportation	Traffic Routing	Large	Medium
	Smart Car		
	Telematics		
	Car Sharing		
	Logistics (container tracking)		
Cities	Infrastructure Management	Medium	High
	Maintenance		
	Surveillance		
	Emergency Services		
	Traffic Management		
Manufacturing	Operational Efficiency	Large	High
	Energy Efficiency		
	Predictive Maintained		
Utilities	Predictive grid maintained	Medium	Medium
	Efficiency Gain		
Agriculture	Crop Management	Medium	Low
Retail	Store in a box – Parcel Delivery	Medium	High
	Automatic checkout		
	Personalized Experience		
	Wearables		

Figure 3.10 Vertical growth projection.

There is a current love affair with "anything technology". This is certainly driving our economy, our values, and our daily life! The IoT is just an extension of that love of technology. Remember, the internet is not just for human to human communication, but for human to machine, and machine to machine.

> IoT adoption reached some 43% of enterprises worldwide by the end of 2016, according to a Gartner Report (Rayome 2017).

Metcalfe's Law, created in 1980, states that, "The value of a network is proportional to the square of the number of compatible communicating devices." In other words, the more devices on the network, the greater the value. With IoT, the value is greatly enhanced by connecting devices to each other and gathering the data.

Given a stable economy and no wars, the IoT will grow and will assist us to make our world more effective and efficient; this in turn will hopefully simplify our lives as humans. At this point we are coexisting with technology at an awkward level. Possibly IoT will allow the relationship to flourish and be mutually beneficial and comfortable. With that, growth is imminent. How much? Which areas first? All educated guesses... But as the expression goes," a rising tide, rises all boats."

3.4 Other Ecosystem Players

Besides the verticals, which are the users of the IoT, there are many other players in the IoT arena. The verticals use the products of these multiple ecosystems' players, and these players listen, guide, influence, provide, and generally support the verticals if they wish for the IoT to be successful. We need a global partner ecosystem in order for all the moving pieces of IoT to work together.

These other players tend to be focused around these four areas of concern.

- Security—Provided for the general security of data and personal identification through authentication, accountability, and encryption. This is an area that is of HIGH concern these days.
- Standards—The development, through the standard organizations, noted below, govern communication, integration, etc. Imagine if every light bulb manufactured used a different socket to connect to? It would be chaos.
- Protocols—Based on these standards agreed upon above, common languages or protocols are selected and used to ensure that all devices, no matter what manufacturer, can talk to each other. Imagine if each country did not have a standard language protocol and you had to figure out what language to speak?
- Network and connectivity—To connect the network or the Internet, many considerations need to be determined from beginning to end. This is crucial for Internet integration.

But who are these players? The following diagram shows the various ecosystem players and how they interact (Figure 3.11). We will break these ecosystem players down into the following categories.

> Interoperability is key to unlocking the full business potential of IoT. McKinsey estimates that up to 40% of the value to be unlocked by IoT can be attributed to "cross-vertical" applications, i.e. requiring multiple IoT solutions to work together (Piet Vandaele)

3.4.1 *IoT Top Level Vertical Providers*

> According to Gartner, global semiconductor revenue totaled $333 billion in 2015, with an annual increase of 3%–4% with the help of IoT (Harper Partners LLC 2018).

- Chip manufactures—These are the companies that manufacture the integrated circuits, chips, that almost all electronic equipment uses today. The IoT will require a new set of chips in order for costs to be maintained and for size to be minimized. The large-scale integrated circuit chips, in the form of the microprocessor, will probably not have to change, but the IoT will require many ASICs (application-specific integrated circuits). These chip manufacturers stand at the core of all technology and can assist heavily in the rollout of IoT. Large-scale manufactures include Intel, Samsung, Taiwan Semiconductor, Qualcomm, and Broadcom.
- Network providers—In order to truly connect with the Internet, there must be a network provider integrated into the process; we would call this an Internet

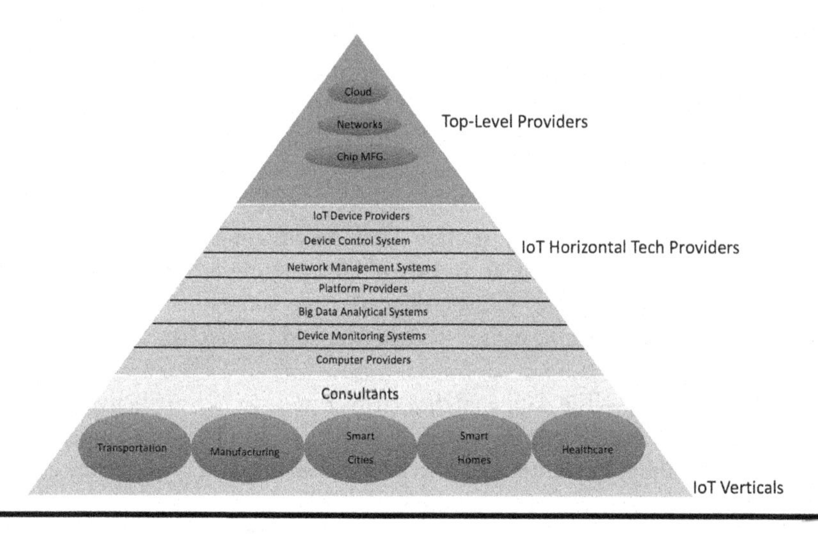

Figure 3.11 Ecosystem players.

Service Provider. These companies have developed and implemented circuits between any two points, and we lease bandwidth from the providers. They are integral players in the Internet, and therefore, they are integral in the IoT. Large-scale network providers include AT&T, Verizon, Comcast, Cox, Vodaphone, Time Warner, and others.

> Global Data Transmissions are expected to increase from 20%–25% annually to 50% per year, on average in the next 15 years in terms of IoT (IHS Market 2017).

- Cloud providers—For the IoT to operate there has to be some network/storage/compute of the data that is collected and eventually acted upon. These three elements network/storage/compute were traditionally completed in "On-Premise" computer centers. The current trend is to put these three elements in "the cloud." Here, we rent use of network/storage/compute, rather than buy equipment to complete tasks. This is similar to renting a car when we need it, rather than buying one to sit in parking lots and driveways. With the varied and distributed nature of data sensors, the Cloud seems like a logical place to store this data. Large scale providers of Cloud include AWS, Google, Azure, IBM Cloud, Rackspace, and many others.

3.4.2 IoT Horizontal Enablers

- Sensors and devices—These are the companies that manufacture sensors and actuators using the microchips mentioned above. These products can be low cost controllers and sensors, such as smartphones, drones, and automation equipment.
- Communication and gateways—In order to join the service providers network there must be an access method/on-ramp or gateway. Typically, these are wireless, such as Bluetooth, LTE, mobile networks, or edge networks, that provide the flexibility needed for IoT. The previously mentioned are protocols, and there are numerous companies that provide product sets for these protocols.
- IoT platforms, device registration, and provisioning—These IoT platforms provide an essential element to the IoT life cycle. Built upon microchips, sensors, and actuators, they provide these elements some intelligence so that they can enter the network, register themselves, and provide control over the data elements.
- Data analytics—The data analytics enabler provides for predictive analysis of the data collected. Based on current data and past trends, we try to predict, with some certainty, the future action to take. For example, if given a choice of running into a tree to the left, or a bush to the right, the choice to take would be the bush given it is a softer impact and less chance of physical danger.

- Cloud infrastructure and IT integration—Beyond the core of the cloud and network/storage/compute, there are many other elements needed for IoT to exist. These elements reside in the category of integration or making the pieces all play with each other. This is a lot harder than one might imagine and is a nightmare for IT personnel.

In terms of growth in the AWS sector: the company is continuing to aggressively invest in the cloud and, late last year, Amazon launched *AWS IoT*, a managed cloud platform that is says enables connected devices to connect securely to cloud applications and other devices.

3.4.3 Standards Development Organizations

The Internet is a strange entity: It is important to remember it has no governing body (Figure 3.12). There are laws that it abides by, but those are the laws of physics not the laws of man. It has no international regulation, although individual governments do impose restrictions on its use and reach. The Internet, and IoT work, only because groups of people develop "standards, best practices, and guidelines" that others' follow. Why do others follow? This is because it is in their best economic interest to do so. Imagine if you invented a great car that went fast and was economical but did not fit on the roads. No one would purchase the vehicle and your company would go out of business. Who are the bodies that create these "standards"? What do they do? Below are some of the standards development organizations that create order for the Internet and therefore IoT.

- IEEE—Started in 1963, it is the largest association of technical professionals with more than 420,000 members from over 160 countries. The Institute

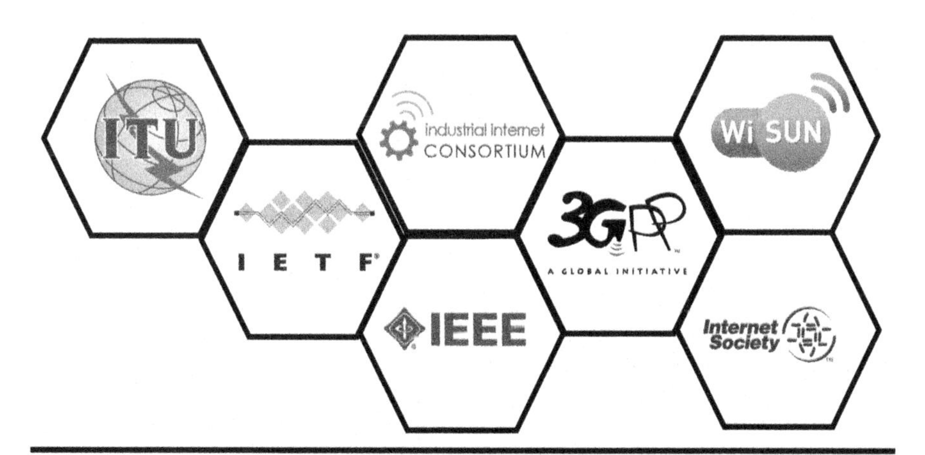

Figure 3.12 Standards development bodies.

of Electrical and Electronic Engineers has as its objective the educational and technical advancement of electrical and electronic engineering, telecommunications, computer engineering, and allied disciplines. The organization puts on numerous conferences, publications, and has membership levels for all involved. Many of the standards of the Internet started here.

- International Engineering Task Force (IETF)—Starting in 1993 as a standards development body, under the auspices of the Internet Society (ISOC), it develops and promotes voluntary Internet standards, in particular dealing with TCP/IP (the Internet Protocol Suite). It is an open organization with no formal membership or membership requirements. It is organized around working groups and informal discussion groups; this rough consensus is the primary basis for decision-making. There are no formal voting procedures.
- ITU—The ITU, coordinates the shared global use of the radio spectrum, promotes cooperation in satellite orbits and works to improve the Telecomm infrastructure in the developing world. The International Telecommunication Union is a member of the United Nations Development Group and is an intergovernmental public–private partnership. It has 193-member states and about 800 public and private sector companies and academic institutions.
- ISOC—The mission of ISOC is to "to promote the open development, evolution and use of the Internet for the benefit of all people throughout the world." It is a nonprofit company based in Virginia and started in 1992. Has a global membership base of more than 100,000 organizational and individual members. The ISOC provides a corporate environment to support many other working organizations such as IETF.
- Industry Alliances—There are also MANY other groups, alliances of people, who get together to try to solve the problems of interoperability and forward momentum of the ICT field. There are too many to name, but a few will be mentioned here that deal with the IoT realm: 3GPP, IIC, and Wi_Sun.

3.4.4 IoT World Forums and Innovation Centers

Besides the various standards organizations for the Internet, there are also organizations specifically targeted at making the IoT work better and smoother in our world. Some of these are:

Industrial Internet Consortium—This group was formed to accelerate the development, adoptions, and use of interconnected machines and devices. Founded by large corporations in March 2014, it coordinates the priorities of the IoT for manufacturing.

> According to its proponents, IoT also has the potential to save huge sums of money for all variety of organizations. For example, the city of Barcelona saved $37 million per year thanks to its new smart-lighting systems. Additionally, the city claims that new IoT initiatives alone have created 47,000 new jobs (Sagiss 2017).

Academic Institutions—Besides these organizations listed above, there are many innovation centers located within academic institutions (Ball State University has one named "INDi—The interconnected Device Institute"), and there are entire cities that have become laboratories for the IoT. The following are the cities that are most often associated with IoT projects and innovations centers: London, Rio de Janeiro, Toronto, and Barcelona.

3.5 Open Source IoT

Another factor that is always surrounding the technology field is the debate between open source versus proprietary. Most of the vendors and ecosystem players listed above would aim to create the killer protocol, application, and device that would secure their proprietary technology into infamy. All these players will fight it out to see who wins. On the outside looking in, are the open source players who want to keep the technology free for all and open to all.

The concept of open source is the culture of the Internet as TCP/IP and Ethernet, the hearts of the Internet, are open protocol and not owned by any one company. We all use them freely. The proprietary versions of these protocols, SNA, Decnet, Appletalk, and Token Ring have all thrown in the towel to these open protocols. Again, the struggle is on with IoT for each of the manufactures to create their proprietary technologies and make these the standards. We see the fight closely with Amazon Echo and Google Home. Each technology does not work well with the other and outside items are difficult, if not impossible to play in their sandbox. Apple IOS is a proprietary phone operation operating system, but Android, open source is growing quickly. Windows is a proprietary computer operating system, but Linux and UNIX are fast becoming important for corporations. Many current technologies are centered on free, "open-source" technology. Therefore, large companies are creating their businesses around new, community-based technology. This is a vast change from past actions of corporations (Figure 3.13).

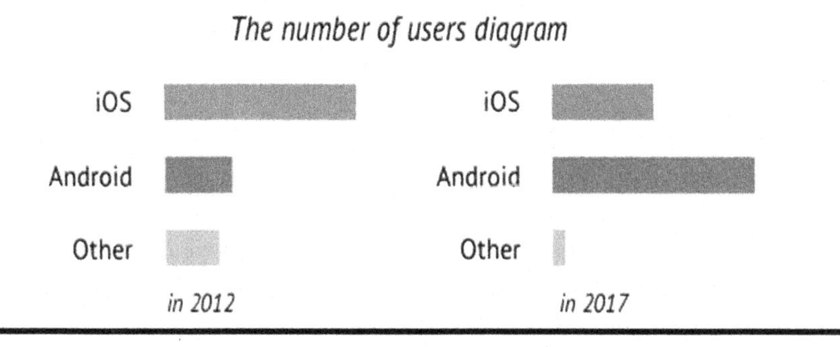

Figure 3.13 Growth of android versus IOS.

3.6 Summary

In this chapter we looked at the verticals and players within the IoT space in today's world. We started out by looking at the vertical industries that IoT can play a large part in. These include Healthcare, transportation, cities, manufacturing, and the home. We included use cases for each of these vertical.

Following we studied the growth estimates for IoT within each of these verticals. We moved on to look at other ecosystem players; including top level vertical providers (such as Chip and network providers), top level horizontal players (such as sensors and actuators providers and Management providers), and finally standard development bodies (such as IEEE, IETF, ad ITU). We closed this section with a look at IoT world forums and innovations centers located around the world.

We then looked at the field of Open Source software, such as Linux, and how it will and can affect the IoT world. Open source IS having a tremendous impact on the entire Information Technology field.

We close the chapter with a case study and some discussion questions to tie all this information together and to test your understanding of this information.

3.7 Case Study

The Experiment of Things (EoT) is a test laboratory that was formed by the Interconnected Network Device Institute (INDi). INDi has, as its purpose, research, awareness, and exploration of the IoT. INDi is based out of Ball State University in Indiana and is an academic initiative in the high-technology arena.

The IoT is a very broad area, and in order to form focused and tangible outcomes for INDi, we cut down our work into studying only one vertical, the home vertical. We realized that this was limiting our scope, but we were better able to maximize our expertise and our laboratory environment. We also limited our scope to studying the UX sector of IoT. We recognize that security and technology have broad reach within the IoT arena, but we strived to focus our research in an area that was manageable and controllable.

The EoT is a simulation of a modern apartment that we populated with the appropriate furniture and décor; likewise, it was populated with IoT equipment within the home vertical. A fair amount of funding was spent on the creation of this laboratory (Figure 3.14).

In regard to the IoT home vertical equipment purchased, we bought the most current editions of the Amazon suite of products, as well as the Google suite. We also purchased other ancillary equipment. An additional part of the laboratory was an experiment with the smart home, circa 2006. As we recognize the smart home/vertical, a concept that has come and gone the past few decades come to light. Our experiment of 2006 taught us a lot and was an excellent learning platform for both students and industrial concerns. One of our goals with the

Figure 3.14 Photo of the EoT Laboratory.

current EoT is to try to integrate the smart home movement of 2006 with the current equipment of 2018.

The following are our goals with the UX testing of the current IoT equipment:

1. Test the setup and operation of all the equipment by both the current generation and by people over 50 years of age.
2. Look at compatibility of equipment to keep vendors.
3. Observe the compatibly between vendors.
4. Look at compatibly between generations of smart home movements.

From our experimentation to date, 13 months, we have learned the following:

1. **Device compatibility:** Very few devices are compatible between the Amazon suite and the Google suite. Many separate devices are required to properly implement a smart home. Even though this is the case, some devices were compatible with one operating system. For example, the August Smart Lock was only compatible with IOS, not Android. This obviously caused some problems in terms of configuration and operation.
2. **Verbal commands:** Verbal commands had to be exact for the main smart home devices to correctly operate. To have a smart home hub response to a verbal command, the command must be clear, concise, and match the exact structure that the device requires. For example, the Amazon Alexa has a website page devoted to specific commands to be spoken. If the command is not clear and exact, the action will not be performed, thus causing the smart home hub to indicate they do not understand.
3. **Purchasing devices:** Various devices had to be purchased to achieve simple smart home tasks. It became clear very quickly that in order to have a simple

smart home established many other devices needed to be purchased. Light outlet sensors, buttons, locks, and various speakers had to be purchased and configured with the smart home hub(s). For example, in order to verbally turn on the lights, special light bulbs and outlet sensors were required. Over time, this cost increased with each new device.

4. **Application requirements:** Multiple tablets required corresponding device applications to configure. As stated above, only some smart home devices were compatible with a particular operating system. With this, two different tablets had to be purchased and configured with all of the smart home device applications. In order to connect the device to the smart hub(s), the corresponding application had to be downloaded and linked to both the device and the smart hub. Along with this, multiple device accounts had to be created and maintained.
5. **Wi-Fi connection:** This was pivotal in the operation of all smart home devices. The installation of a private Wi-Fi connection was implemented in order to link all devices onto one connection. It was important to make sure that the Wi-Fi connection remained successful throughout the entire process due to devices being constantly connected. For example, there were a couple instances where the Wi-Fi went down; this resulted in having to completely reconfigure the devices all over again.

Food for thought:

a. Why was compatibility so difficult within and between vendor equipment?
b. What standards would be advantageous to be created for home vertical equipment?
c. What skills are necessary because of the implementation of smart home devices?

3.8 Discussion Questions

1. There are many varying characteristics among the IoT verticals in today's society. Which one of the main verticals is growing at the fastest rate? Why could this be the case?
   a. Answer—Here, after reading through the first section, the student should have observed that healthcare is the fastest growing IoT sector. To support their answer, details and examples from the chapter paired with their own knowledge regarding healthcare should be applied.
2. As mentioned in the chapter, the transportation vertical consists of a variety of subsectors. By observing these five subsectors, which sector do you believe is the most relevant within today's transportation methods? How might this sector benefit directly from IoT advancement?

   a. Answer—Students have the opportunity to choose which transportation sector they think is the most relevant based off of the five provided. Along with this, students can support their choice by stating how IoT technology could further advance that specific sector.
3. Smart cities are making a major wave within today's IoT environment; new developments are being rapidly innovated to keep up with demands. Even though this area is facing major growth, what are some of the obstacles facing smart city creation?
   a. Answer—By reading the text, students are able to infer and provide context to possible obstacles that smart cities may face in the future. Any supporting knowledge is encouraged.
4. Manufacturing is a complex area of IoT development and innovation due to the multiple steps required within the process. What are some of the goals that the manufacturing vertical has to focus on in order to be successful?
   a. Answer—Students can provide various goals they see as priority within the manufacturing vertical. Students should utilize information provided in the text to aid them in their decision-making. Goals should illustrate how they could benefit the manufacturing sector as a whole.
5. IoT is expansive in its various uses. What are some of the primary areas within effective uses of IoT today? Where do you believe IoT is growing the most?
   a. Answer—We would expect the student to identify the major verticals that IoT is playing in, as defined in this chapter, and extrapolate on one of the examples uses that we provided. For the "where IoT is growing" part, the student can either do research for their answer, or use intelligent reasoning for their response.
6. There are many benefits surrounding the housing sector of IoT technology. What population/demographic has the potential to face large benefits from new IoT safety features?
   a. Answer—Students should respond with answer relating to one of the growing verticals above. For example, transportation would be a great answer due to enhancement of safety features for travelers/daily commuters.
7. IoT technology is defined by a multitude of perspectives and ideas. What law helps explain how IoT devices work together?
   a. Answer—Metcalfe's Law, created in 1980, states that, "The value of a network is proportional to the square of the number of compatible communicating devices."
8. IoT expands into various ecosystems players. What area of concern is centered around the governance of communication?
   a. Answer—As a response, students should answer this question with some insight gathered from the text centered on how IoT devices could face issues in the future concerning communication. For example, security of

these devices, as well as compatibility with each other (in terms of sharing information without the user's knowledge) could pose a challenge for regulation.

9. As we have learned, there are IoT vertical providers that exist to aid in the operation of IoT as a whole. What provider consists of three major elements? What are these elements and how to they benefit IoT?
   a. Answer—As covered in the text above, a provider that consists of three major elements could be cloud computing (*Cloud infrastructure and IT integration*—Beyond the core of the cloud and network/storage/compute, there are many other elements needed for IoT to exist. These elements reside in the category of integration or making the pieces all play with each other. This is a lot harder than one might imagine and is a nightmare for IT personnel). These three elements benefit IoT by providing the proper framework for future growth and integration.
10. What are the main standards bodies involved with IoT? Which body consists of the largest association of technical professionals?
    a. Answer—The main standards bodies associated with IoT consist of (but not limited to) IEEE, ISOC, IETF, and ITU. IEEE is the body with the largest association of technical professionals.

Bibliography

"20 Fascinating Facts about the Internet of Things", Sagiss, accessed August 23rd, 2017, www.sagiss.com/blog/20-fascinating-facts-about-the-internet-of-things.

Affectiva.com, accessed February 20th, 2018, www.affectiva.com/.

Automatic Road Enforcement, https://techcrunch.com/2016/07/13/how-iot-and-machine-learning-can-make-our-roads-safer/.

Baylabs.io, accessed February 20th, 2018, https://baylabs.io/.

Cardiacinsightinc.com, accessed February 20th, 2018, www.cardiacinsightinc.com/.

Dickson, Ben. "How IoT and Machine Learning Can Make Our Roads Safer", Tech Crunch, accessed July 13th, 2016, https://techcrunch.com/2016/07/13/how-iot-and-machine-learning-can-make-our-roads-safer/.

"Freight Tracking—Improve safety, inspection and efficiency of freight traffic flow". A current application for this is, www.freightquote.com/blog/how-the-internet-of-things-will-make-freight-shipping-more-efficient.

Hbisolutions.com, accessed February 20th, 2018, https://hbisolutions.com/.

"How the Internet of Things Will make Freight Shipping more Efficient", Freightquote, accessed October 29th, 2015, www.freightquote.com/blog/how-the-internet-of-things-will-make-freight-shipping-more-efficient.

Icamed.com, accessed February 20th, 2018, www.icadmed.com/.

Imagentechnologies.com, accessed February 20th, 2018, http://imagentechnologies.com/.

"The Internet of Things: A Movement, Not a Market", IHS Market, accessed January 1st, 2017, https://cdn.ihs.com/www/pdf/IoT_ebook.pdf.

Irhythmtech.com, accessed February 20th, 2018, http://irhythmtech.com/.

Kinney, Sean. "Dallas Investing in Smart Traffic Management with Ericsson", Interprise IoT Insights, accessed January 8th, 2018, https://enterpriseiotinsights.com/20180108/transportation/dallas-investing-in-smart-traffic-management-ericsson-tag17.

"The Logistics and Transportation Industry in the United States", Select USA, accessed February20th,2018,www.selectusa.gov/logistics-and-transportation-industry-united-states.

Medicalis.com, accessed February 20th, 2018, https://medicalis.com/.

Medstreaming.com, accessed February 20th, 2018, https://medstreaming.com/.

O'Donnell, Lindsey. "6 Red-Hot Verticals with The Biggest IoT Opportunities for Solution Providers," accessed October 11th, 2016, www.crn.com/slide-shows/internet-of-things/300082387/6-red-hot-verticals-with-the-biggest-iot-opportunities-for-solution-providers.htm.

Parker, Bob. "Making the Business Case for Smart Manufacturing", Industry Week, accessed February 27th, 2015, www.industryweek.com/smart-manufacturing.

Patterson, Steven Max. "The Open Source Internet of Things Has Some Big Aspirations", Network World, accessed June 14th, 2013, www.networkworld.com/article/2224803/software/the-open-source-internet-of-things-has-some-big-aspirations.html.

Rayome, Alison DeNisco. "The Five Industries Leading the IoT Revolution", ZD Net, accessed February 1st, 2017, www.zdnet.com/article/the-five-industries-leading-the-iot-revolution/.

Road Adjustments—Adjust speed limits and signal timing based on current conditions. Richard van Hooijdonk, "The Internet of Things: How the Problem of Traffic Congestion can be Solved", IoT Transportation, accessed May 3rd, 2015, www.richardvanhooijdonk.com/en/the-internet-of-things-how-the-problem-of-traffic-congestion-can-be-solved/.

Sandlotsolutions.com, accessed February 20th, 2018, http://sandlotsolutions.com/.

Top 20 U.S Healthcare Companies by 2016 Revenues," Revenues and Profits, accessed February 19th, 2018, http://revenuesandprofits.com/top-20-u-s-healthcare-companies-by-2016-revenues/.

"Transportation Industry Statistics and Market Size Overview, Business and Industry Statistics", Plunkett Research, Ltd., accessed February 20th, 2018, www.plunkettresearch.com/statistics/Industry-Statistics-Transportation-Industry-Statistics-and-Market-Size-Overview/.

Waze.com, accessed February 20th, 2018, www.waze.com/.

"What is the Internet of Things (IoT)? 10 Surprising Facts", Harper Partners, LLC, accessed January 1st, 2018, www.joinharper.com/10-surprising-facts-about-the-internet-of-things/.

Wikipedia, s.v. "Healthcare Industry", accessed February 19th, 2018, https://en.wikipedia.org/wiki/Healthcare_industry.

Wikipedia, s.v. "International Standard Industrial Classification", accessed February 19th, 2018, https://en.wikipedia.org/wiki/International_Standard_Industrial_Classification.

Zehel, Julia. "70 Companies Driving the Future of Healthcare Technology," July 27th, 2018, www.redoxengine.com/blog/70-health-tech-companies-disrupting-healthcare/.

Zephyrhealth.com, accessed February 20th, 2018, https://zephyrhealth.com/. www.ioti.com/strategy/20-most-imp;ortant-iot-firms-according-you

Chapter 4

Emerging Wireless Communication

Lauren McNally, Aaron Khoury,
Jerry Walker, and Stephan Jones

Ball State University

Contents

4.1 Introduction

As the world continues to turn, technologies come and go. The purpose of many of these technologies is to keep up with the need to provide faster speeds and better capabilities as consumers demand greater computing power and new products to improve all aspects of their lives. Many legacy technologies cannot handle the upgrades or improvements, and components can only advance to a level that the system was initially designed to accommodate—making the technology outdated and requiring system/power upgrades. As the energy and computing power needs of consumers and industries have increased, the manufacturers have advanced their capabilities to keep consumers connected. The wave of the Internet of Things (IoT) has led companies to reevaluate their wireless needs in order to keep connectivity active for their highly mobile constituencies—both employees and customers.

Wireless advancements have been fueled by the demands for greater data throughput (e.g., streaming video) and the constant push to stay competitive across numerous industries. Companies tend to state that their wireless connectivity improvements and technical advancements are occurring, but an understanding of the wireless capabilities is needed to advance areas that move products into the future. There are multiple dimensions that influence connections for IoT: future scalability, the ease of product deployment and management, cost to make and support the infrastructure, security, and constant reliability. Innovations for wireless technologies focus on some of these characteristics and provide a connection that will boost the use of IoT power. As there are many different sectors of wireless communications, engineering and technologies (WCET), understanding where WCET capabilities started and where wireless is going in the future related to IoT is important to define.

4.2 Wireless

IoT devices send and receive data information over a wireless network connection with the help of specific protocols. Wired connections can be used, but most products pursue a wireless approach to be able to connect in varied and untethered locations. A wireless connection over the most common radio frequency, is known as wireless fidelity (Wi-Fi), does not require wires or cables to be able to make a connection to the mobile user. When utilizing a wireless connection, the devices that encompass a network will not need physically wired cables to be able to send and receive data. These wireless connections are needed to provide devices with the capability to talk to similar devices with ease, accomplished by using an embedded chip set within the device or a specifically designed component that will allow communication to occur. The devices in the category of IoT host a multitude of abilities in diverse fields like automobiles, manufacturing, home appliances, and even fitness trackers. The movement from wired to a wireless connection has driven the industry to improve their connection capabilities, creating an exceedingly competitive marketplace of the high-end, highly reliable wireless products. While the realm of IoT allows for connectivity that appears to be available anywhere at any time, the underlying concept of their connections is simple, but these everyday objects are both helping and hindering the realms of business and communications [1].

Different varieties of IoT wireless network have created opportunities within diverse fields of innovation that will allow for communication on secure fields. As the number of individuals that have access to the internet increase, the amount of energy needed to connect these individuals' needs to increase as well. The IoT has been able to integrate wireless capabilities into everyday household items that are not specifically a smart phone. Smart phones are the most popular ways to send and receive data (which includes voice, streamed video, and data). This will change our perception of networks and "always on" in ways we have yet to imagine [2].

The power of a wireless connection depends on the amount of throughput and the spectral bandwidth of frequencies it may receive. Being considered a wireless network does not always imply one with the best connection. Wireless networking usually allows multiple devices like computers, cell phones, and printers to intermingle with network access. As the industry of wireless connectivity increases, it blends its capabilities with those of similar networks. Various wireless networks include: Wi-Fi, low power wide area networks (LPWANs), light fidelity (Li-Fi), and Narrowband Internet of Things (NB IoT) or otherwise known as personal area networks (PANs). All of these areas exhibit the unique characteristics of IoT and promote different capabilities for their needs.

4.2.1 Wireless Fidelity

The basics of Wi-Fi include allowing individuals the ability to obtain broadband network access on numerous devices with the help of wireless access points (APs) and specific unlicensed radio frequency signals. The commercial use of Wi-Fi did not begin until the end of the 20th century with the creation of the protocol standards now designated with the "IEEE 802.11 alpha-numeric designation" [3]. The popularity and accessibility of Wi-Fi have prompted the need to strengthen its frequencies and transmission capabilities. As a wireless low-power source of communication, Wi-Fi utilizes a main router hub to provide connections to many everyday devices. With an increase in connectivity, the increase in the number of participants on specific networks hinders the ability to communicate. Although there have been some drawbacks (e.g., security issues), Wi-Fi as a stand-alone technology is one of the most innovative and pervasive network access platforms ever created. Continuous innovations for Wi-Fi are crucial to the future of its networking capabilities, and it relies on unlicensed bands of spectrums. The standard Wi-Fi and its interaction with wireless devices create cases in which the technology will be able to help manage the backhaul issues and prevent traffic overloading. Wi-Fi was developed and deployed well in advance of IoT. The surge of IoT and the rapid increase of network access requirements has stimulated the need for rapid advancement in Wi-Fi throughput [3]. Wi-Fi and IoT have opened up new paths for internet access that are convenient to almost anyone.

Some of the benefits that are driving innovations with Wi-Fi relate to an increase of information technology (IT) support, the offloading of cellular/mobile traffic, and network densification. By expanding the spectrum of Wi-Fi access past the 2 and 5 GHz bands, the ability to adopt IoT devices has been greatly improved. The spectrum of connectivity for legacy devices, mixed with the new innovations (e.g., multiple input/multiple output antenna arrays, millimeter wave frequency use) have led to the ability to coexist on the same spectrum. Additional wireless protocols like Bluetooth and Zigbee help Wi-Fi to meet the requirements to work with IoT. With Wi-Fi's evolution, the proposed standards show the future support for the devices. As some IoT devices need a constant connection, the strain on wireless access increases to produce the amount of power that is required and expected of many devices. Wi-Fi and its role in the IoT are crucial as it creates many points of presence and supports mixed-vendor operations. The connections of IoT devices to Wi-Fi are still complicated at best, even though it seems simplistic and transparent in nature. As the playing field of IoT starts to increase, Wi-Fi networks are still the most efficient, popular, and widely used source for connections.

4.2.2 Low Power Wide Area Networks (LPWANs)

LPWANs allow for secure and scalable global coverage in a low-cost module. Unlike Wi-Fi, LPWAN is not a standard that was created by prominent IT committees. LPWAN branches off of wireless capabilities to help connect devices that operate in low-bandwidth spectrums so that their rates of power last over long periods of time. Although it is a lower cost wireless capability, it allows for the same, or greater, amount of power provisioning like traditional mobile networks. The coverage is improved over a variety of technologies like LTE-M and NB-IoT. LTE-M is the abbreviation for Long-Term Evolution of Machines and is a categorization for the evolution of technology in IoT devices that work directly with a 4G network [4]. LTE-M and NB-IoT work well with LPWANs to help strengthen the battery life of products, while reducing service costs. More devices are able to be supported in a LPWAN when linked to one another. When cellular connections are too expensive for some IoT applications, the use of LPWANs can come in handy to develop networks over the unlicensed mobile spectrum. Mobile technologies are the best way to integrate a solution with a LPWAN [5]. Many homes, schools, and work places have a form of a wireless AP. LPWAN can help IoT devices work in the home and in everyday lives through selecting desired speeds depending on data pertaining to varied areas. Due to the cost and nature of the technology, most opt for a low power option to be able to supply the wireless capabilities at limited quantities without high costs. The future of LPWAN and the IoT is unclear due to the rate of change and maturity curve. As IoT devices are constantly sending data, LPWAN is able to sustain device battery life by keeping power consumption low. Energy consumption is a major concern with IoT technologies and low power incorporates the efficient connection capabilities to the internet [2].

4.2.3 Light Fidelity

Li-Fi is a technique to help move to higher frequencies using light wave communications on the electromagnetic spectrum. Li-Fi as an emerging technology within the wireless space that will help to increase the speed of wireless communication through light. Li-Fi utilizes light-emitting diodes (LEDs) and its modulation techniques: single carrier (e.g., pulse amplitude modulation, pulse width modulation, and on–off keying) and multi carrier (e.g., orthogonal frequency division multiplexing), Li-Fi can cover more of the spectrum [6]. Being able to utilize LEDs instead of radio frequencies will cut down interference. LED lights are high speed transmitters of data over reduced transceiver units.

With the current evolutionary state of wireless and mobile communications, a change is necessary to keep up with the speed that customers expect. Li-Fi in conjunction with a Wi-Fi network could enhance the capabilities of networks in the areas of security and positioning. Li-Fi and its substitution for the use of Wi-Fi would lessen the congestion of frequencies and increase security [6]. Many consumers are nervous about their information being accessed through the transfer of IoT data information back to either data centers or other outsourced vendors. This simple act would help to decrease insecurities associated with IoT and the loss of information. Li-Fi is one of the newest technological advances for wireless networks, still in its development phases and only available in some hybrid states [7].

4.2.4 Narrowband Internet of Things or Personal Area Networks

Forging their way into the IoT realm, mobile providers launched themselves into the new world of NB or PANs by offering some of the first NB plans. NB-IoT is a branch off of LPWAN, having wireless capabilities and catering to low power networks while not utilizing LTE networks. Creating and offering this plan significantly reduces the cost and increases the efficiency to be able to connect items together. NB-IoT will work on kHz bands that are typically unused and independent of others. This allows for NB-IoTs to have their own block of stations to place their operating resources. As the number of devices trying to connect to networks increases, keeping the networks open and capable is achievable with NB-IoT [8].

As the fifth generation (5G) of wireless technologies evolve, the need to continuously improve becomes more crucial. NB-IoT software can release its own spectrum that has the capabilities of hosting more devices than ever before. Hosting static devices can lead to inactivity and is perfect with IoT as data transmission is constant. The new bandwidth capacity means that smaller data packets can be serviced and transferred for communication purposes in a quicker fashion [8]. NB-IoT data transfers serve the everyday intelligent device.

Processing powers of wireless capabilities can support IoT devices to different setting and needs. The sectors of Wi-Fi, LPWAN, Li-Fi, and finally NB-IoT contribute different levels of connectivity and bandwidth transmissions that can ease the strain that is put on the transference of data and information over the varying wavelengths. Each of these wireless protocols is constantly transforming and more innovations are being created. While these sectors are evolving there are many new service demands that create strains on the improvements associated with wireless capabilities.

4.3 Demands of Mobile IoT Devices

As the complexity of devices has increased, the need to make devices smaller and more compact has increased as well. There is a need to make devices smaller and hand held so the user can take them anywhere and utilize them at any time without creating inconvenience. But there is an unnoticeable strain on devices as they are miniaturized. Manufacturing companies must find a way to diminish the size of parts in devices but still make them functional enough to send and receive information with the same amount of power as before. Receiving products need to be modified to be able to connect with devices and wireless technology. Following are some of the technological aspects that must be considered. These considerations are not dependent on one another but intermingle in how they may help or hinder wireless mobile technologies.

4.3.1 Licensed/Unlicensed Spectrum

The spectrum of wavelengths available to providers is divided into licensed and unlicensed frequencies. Licensed spectrum is available to providers that pay for specific portions of the frequency bands. This stops unwanted traffic and interference as only license holders are allowed to transmit on the specific spectrum. Unlicensed frequency bands are open to anyone who can obtain the signal. Unlicensed spectrums allow for individual regulation of networks by the user, rather than relying on an agency (i.e., FCC) to do so. Utilizing a licensed bandwidth would theoretically be more reliable for IoT as the transmission of information to and from devices will have less interference, even if the distance between devices is great. With unlicensed spectrum there is an increased vulnerability that comes from interference by other users of the unlicensed frequency. Wi-Fi is the most common unlicensed option and carries distinct features for its use [9]. When operating on a licensed or unlicensed spectrum, the device capacities should be measured to ensure proper connectivity and transmission abilities. Overall, licensed spectrums offer more flexibility as they provide for higher power output and less interference from random device use of the frequencies within the Industrial, Science, and Medical (ISM) bands [10]. A calculation should be defined in the amount to charge for IoT device support when operating on licensed networks as the operator's costs to purchase the spectrums may go up with the increasing demand.

4.3.2 Signal Propagation

Dividing the spectrum that is available to IoT devices is challenging with the number of devices that are now available and trying to connect. As data needs to be transmitted, lower frequency waves are better at sending this information as it will be able to travel further and penetrate infrastructure easier. With signal propagation, the need to evenly distribute the scheduling functions and communication phases will allow neighboring devices to identify those devices surrounding them on the wireless network and can work cohesively to maximize the network services without a high packet/data loss created by interference and collisions [11].

4.3.3 Antenna Considerations

Having the ability to connect devices anywhere and anytime is made possible with differing types of antenna. Most systems require varying sizes of antenna to perform [12]. Antennas are not always visible on IoT devices; most commonly, antennas may be an embedded chip within the device (see Figure 4.1). Characteristics of antennas are crucial in the design and construction phases to make sure that their materials match the applications they are supporting [13]. Many IoT devices have forgone the addition of an antenna into their devices. This change to the features might hinder IoT in the future as the connections between devices grow in distance and the needs for communication grow.

4.3.4 Power Considerations and Battery Life

Increased need for wireless connectivity can create a power strain on devices. Changing activity periods of devices will help to alleviate the strain. IoT devices must provide enough battery life to host connectivity needs. Most are in a state where the device must be plugged in throughout the entire use. IoT device batteries are relatively simple in nature, being similar to most mobile devices. These batteries

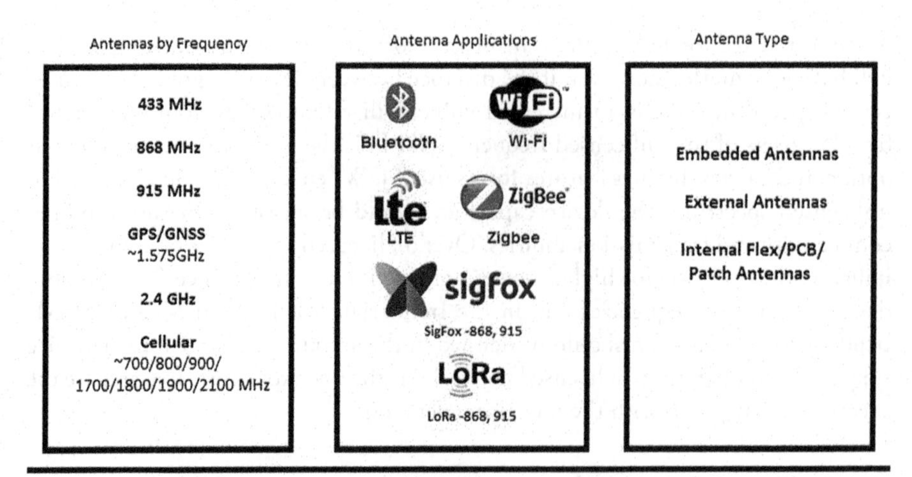

Figure 4.1 Antenna characteristics for IoT.

are not suited for the longevity of power needed by devices [14]. There are specific techniques that can be deployed to sustain the energy available for the batteries. Inserting additional wireless sensors or optimizing the skills of these sensors will help to optimize the battery needs without compromising the IoT device.

4.3.5 Proprietary versus Nonproprietary Solutions

Solutions related to wireless innovations are also debated by popular regulatory agencies as vendors/manufacturers withhold their information from competitors in order to keep their innovation creative and evolving. The debate on whether a specific vendor of a product must provide their information to others or keep it proprietary is resulting in a competition for not only products but the overarching characteristics of wireless connectivity. Information obtained from IoT devices can be controversial regarding privacy [15].

IoT has been working in a proprietary ecosystem of innovations; this means that the improvements have been kept secret from competitors. But, as the technologies change and adapt to the customers' needs, a nonproprietary system of innovations and wireless capabilities is needed to provide companies and customers with the device changes after tests and failures. Proprietary systems are suffocating the IoT space and preventing any scalable options and interoperability and need to be abandoned in order to keep a steady occurrence of technological innovation [16].

4.4 The 5G Network

One of the most important features of wireless IoT development is the ability for devices to be mobile and have nearly uninhibited network access. Historically, each generation of cellular devices has built upon its ability to offer more reliable, faster, and more widely available network access. The first generation of cellular communication was conceived in the 1960s and was developed in the 70s—however, the commercial use of such technologies did not become mainstream until the dawn of the 80s [17]. This network is referred to as the 1G network and, unlike the following 2G–4G networks, is no longer used. The 1G network used analog transmission, while the following network generations have all used digital transmission. The movement towards digitization of information (i.e., data) has allowed for greater bandwidth usage, more rapid exchange of information, and a safer and more reliable network [18]. The 5G network is currently being engineered and carries different burdens than many would have ever thought possible.

With the development of IoT devices in mind, the 5G network must support a vast number of devices. To accomplish this, 5G must accommodate Internet Protocol (IPv6) logical network addresses. The use of this rising protocol standard is important because of the rapidly diminishing number of IPv4 addresses. These addresses can be thought of like the address of your house—like a house, each device on a network is assigned a specific address that is used to guarantee the delivery and receipt of information. IPv6 guarantees—at least for many years to come—that each device on the world's networks will have its own address [19].

Further, the 5G network has greatly shifted its potential usage in comparison to prior generations. Where previous designs focused on simple voice calls, text messages, and, most recently, data—the 5G network will be especially designed for its ability to accommodate data delivery at speeds which triple current cellular speeds. Additionally, the availability of the network is expected to far exceed that of the current 4G LTE networks which are a current industry standard [17]. In keeping IoT devices in mind, it is important to focus on factors such as latency, energy consumption, and diversity of devices. Huawei, a company leading in the push for 5G, claims that the next-generation cellular network will allow for nearly real-time transmission, energy-per-bit reduction by a factor of 1,000, and a vast host of devices that will be able to utilize the network. The 5G network will undoubtedly lay a strong foundation for the future of wireless IoT [20].

4.5 Internet of Remote Things and Satellite Connectivity

It is undeniable that the rollout of the 5G network will allow access to many users who were previously unreachable. What about users/devices that still may face an inability to access the cellular network? It has been predicted that by 2020 there will be up to 20.6 billion connected devices in the world—certainly many of these will be located in hard to reach areas. These devices fall under the category of the Internet of Remote Things, or IoRT [21]. While it may not be viable to use satellite communication for all devices on future networks, satellites are able to offer 99.99% availability, which is very valuable to users and devices in areas where cellular penetration may be difficult to reach (e.g., rural communities). Many IoT devices are designed to transmit low volumes of data at a given moment. Additionally, sensors in an IoT-based system—such as the smart grid which powers millions of American homes—may function as a unit or act to accomplish similar goals. For these reasons, the use of satellite communication may be ideal. Existing satellites that support low bandwidth transmission can support IoT devices that require low data transmissions. Satellites can, by design, reach a wide geographic region. This will prove important for widely dispersed but interoperable devices—such as sensors in the smart grid. This presumably can all be done without significant changes to the current infrastructure.

To illustrate how satellite communication offers a real-life benefit to our economy, we will continue to use the smart grid as an example of IoRT. Hypothetically, temperature sensors could be scattered in hard-to-reach areas of the electrical grid. These sensors, via satellite, could relay significant changes in the grid which may cause a negative effect. The sensors might signal, based on temperature changes, allowing for a safer and more reliable network. This communication could be completed quickly because of the interoperability of IoT in areas that would otherwise take extensive labor to monitor. This methodology may also benefit the realms of emergency management and environmental monitoring [22].

4.6 Wireless Local Area Network

4.6.1 Security

In a 2017 study, IoT data management and new networking elements are two of three areas identified as key to successful enterprise level IoT use. There are exceptional difficulties in tracking high volumes of IoT devices in private networks—these high volumes of difficult-to-track devices pose issues in terms of security, stress on network infrastructure, and power consumption [23]. Since 2009, IoT device's offerings have continued to grow rapidly in a host of fields. As IoT devices become more widely used, so does the importance for standards in their regulations and implementations in wireless local area networks (WLANs)—especially with the aforementioned principles as guidelines.

In 2016, the dawn of a series of denial-of-service (DDoS) attacks began from a very unexpected source—more specifically, many small sources. The initial attack was led by a network of low compute power IoT devices interconnected in what was called the Mirai Botnet network. Such DDoS attacks typically flood a network or device with data rendering the device or network unable to function due to unmanageable amounts of data. This vast network of IoT devices was hijacked to direct attacks on companies with up to 600 Gbps volumes of data. The typical response to a DDoS attack is for an IT team to block the source from accessing the site or device; however, this proved difficult with the Mirai Botnet attacks —at its peak, there were 600,000 devices being implemented in the Mirai attacks. This attack was in part possible due to the low security standards implemented on IoT devices including DVRs, personal routers, and security cameras. These devices used universal default passwords which allowed them to be commandeered by unethical players. Developers are trying to combat the rise of other bot networks by making auto-updates an industry standard. Although it may seem bothersome, this practice has effectively reduced the spread of botnet infections [24]. For IT professionals, it is important to be able to track each device on a network—further, it is especially important to be aware of the typical traffic devices create on a network. Having insight on all devices allows for early identification of possible anomalies, breaches, or misuse of IoT devices.

4.6.2 Power Management

Another IoT concern is how power consumption issues may affect WLANs. When implementing IoT devices in a network, it is important to consider the effect they have on energy consumption; although IoT devices are typically designed to be low-power devices, the number of active devices can quickly increase power consumption. While there is no single solution to this issue, there are proactive measures being taken by industry to reduce the issue. In the IoT of healthcare field, there has been a review of the algorithms and means of optimizing power use and data transmission. Chen et al. [25] discuss the implications of focusing on how changing power consumption, data size, or both may benefit IoT. In healthcare and similar fields, this may be a sensitive topic as the accuracy and reliability of this data can have life or death implications.

Another workaround to power consumption issues is the design of devices that use radio frequency identification (RFID), Bluetooth, or Zigbee transmission—Zigbee seems to be the most promising of these transmission methods as it offers long range data transfer while still maintaining the ability to be a low power device and easy to use [26]. Zigbee offers a low power option for a wide range of IoT devices and has the potential to alleviate some congestion on WLANs and cellular networks. Unfortunately, industry adoption of this technology has not been widespread or implemented in many consumer devices.

One last solution under evaluation is the design of wake-up features for certain IoT devices. While IoT devices can be thought of as "always on" in many cases, they often only interact when triggered to "wake-up." Designing devices with proper sensitivity and the ability to ignore false alarms is a key to being able to bring about cost-effective IoT implementations. This task will prove difficult as the device must still maintain reliability and performance while trying to reduce power consumption.

4.7 IoT Disruption

As mentioned, the proliferation of IoT devices and solutions has spread to many industries. This section offers some insight to what is being done in just a few industries;

it is important to consider where these technologies have growth opportunities and how they may improve quality of life and efficiencies in business throughout the world.

4.7.1 Agriculture

There is major waste with agriculture. It is estimated that 30% of the total food produced is lost globally each year. This is estimated to be $1 trillion U.S. dollars [27]. It is hoped that IoT can be used to collect real-time data allowing farmers to better tend crops and cattle. This technology could also transform supply chain management reducing the amount of waste seen before delivery to consumers. Soil moisture sensors are currently being used to allow for constant nourishment—this practice has shown to increase yields by 1%–2% and save resources by 20% [28].

4.7.2 Automotive

One highly visible IoT technology in the automotive industry is the use of autonomous vehicles. The launch of self-driving cars has some very clear implications for people who work as commercial drivers. However, the launch of these vehicles will not be a clear-cut path—this was made evident when a woman was struck and killed by an autonomous car being tested in the United States in March 2018. Other areas of automotive IoT being explored are traffic management and insurance policies. The use of modules for tracking driving patterns is becoming increasingly more popular among drivers. These modules can send data directly to insurers who may adjust policies in "pay-how-you-drive" plans. Additionally, across the world, cities are becoming "smarter" by implementing real-time traffic monitors in densely populated areas. These monitors can remotely deal out traffic tickets and provide data for designing more effective traffic engineering [29].

4.7.3 Healthcare

Implementation of IoT in the healthcare field can be very challenging because of data sensitivity and privacy. In the United States, the Health Information Privacy and Portability Act (HIPPA) compliance is a very serious regulation which protects accessibility, integrity, and confidentiality of health records. It is predicted that by 2022, healthcare IoT will more than double in valuation reaching nearly $500 billion. One current example is the use of wearable devices, allowing doctors to remotely monitor patients. This is partially possible due to the refinement of RFID sensors which can function in conjunction with Wi-Fi and Bluetooth transmissions. By setting up varying RFID chips/sensors, data can be transmitted to a central node where processing occurs. This method allows for sensors to be designed for discreteness and replaceability. While this adds a degree of luxury or convenience, it also opens patient data to security risks such as tampering or interception of medical data; security will certainly remain a driving factor in wireless IoT application [30].

4.7.4 Supply Chain and Manufacturing

One of the most widely used practices of IoT is through RFID and supply chain management. The practice of tagging and scanning RFID sensors during the process of product delivery has become a common act in the movement of resources and logistics handling practices. As these technologies and business processes become more interconnected and automated with databases, there becomes a larger pool of information for business analysis. Big Data can also be produced in factories by implementing machine to machine (M2M) and IoT technology. A demonstration of this factory application is with robotic machines and a smart conveyor belt. These devices are designed to interact with one another and RFID tagged products. With M2M technology, each device in the production line can provide feedback which controls speed, destination, and other factors associated with the manufacturing process. As these machines become more automated and with the potential of artificial intelligence (AI) solutions, this process could be further enhanced. If M2M and IoT can collect data and AI systems interpret and adjust settings for maximized profitability, factories could become a nearly automated back end of many businesses.

4.8 Summary

In this chapter, we looked at the current and emerging forms of wireless technologies that are and will be used with IoT. We started out looking at the popular current wireless technologies such as Wi-Fi and LPWAN. We discussed the different characteristics and needs of IoT devices versus regular devices. These demands manifested themselves in spectrum, propagation, antenna considerations, power consumption, and solution ownership.

We next looked at the 5G network, the next generation of the cell phone system, and its capabilities and promises. Following we discussed WLAN technology and the remote monitoring of devices via IoRT.

We closed with looking how IoT can disrupt various verticals.

4.9 Case Study

This case study describes the IoT innovations in parking services made in St. Louis under the direction of a new city treasurer. The St. Louis Treasurer's Office controls and monitors the bank accounts of the City. Through daily contact and detailed reconciliations of these accounts with the Comptroller's Office, this office provides a check and balance for the Comptroller's Office. In addition, this office is by law the depository for all receipts of the City and provides a means for departments to make daily deposits.

The Treasurer's Office issues all payroll checks, deposits funds for federal and state taxes, and funds for savings bonds and other payroll deductions. The Treasurer is also responsible for making all investments for the City. This includes purchasing, selling, and auditing the earnings on these investments as well as ensuring that City funds are safe and secure.

Since 1951 the Treasurer, by state statute, also serves as the Supervisor of Parking and Chairperson of the Parking Commission for the City of St. Louis. As the Supervisor of Parking for the City of St. Louis, the Treasurer oversees all aspects of parking, both on street and off street. This includes on-street meter inventory, maintenance, collections, enforcement, policies, and parking management systems.

Under prior administrations, on-street parking was done the same since fees were implemented. During that time, coins were the only payment method accepted for parking meters. However, collections were performed daily across the city, and each meter had to be checked manually. Maintenance of the meters was either checked manually or an issue was reported. This created a loss of time, revenue, and work hours which ultimately cost money. Off-street parking was similar in that the only payment method accepted was cash or checks for monthly parkers. Monthly parkers were tracked but with difficulty. All event parking was handled with cash as well, which made it difficult to track payments and reconciliations.

Parking enforcement was done with no real-time data and was connectionless. Parking tickets were posted a day after the ticket was written. Ticket disputes were left to one's word against another, with no real way to check if the violation actually occurred. This would often cause the dispute to go to arbitration with a judge and, again, cost money.

However, as credit cards, debit cards, and more versatile payment services became the norm, using coins as means of payment became outdated as they were aggravating to the consumers and harder to control with little to no accountability. With the rise of IoT, the overarching theme became "work smarter not harder."

In 2013, a new Treasurer, Tishaura Jones, took office and immediately made this issue related to parking revenue one of her high priorities. Treasurer Jones wanted to know how to leverage technology and use it to raise revenue, lower tickets, add transparency, and implement full accountability while not having an internal IT department.

First, a parking study was conducted to assess the current state of city parking with recommendations for change as an outcome. The study discovered that there were approximately 3,000 m on city streets that didn't collect more than a quarter per year. The study compared St. Louis to other cities and revealed that it was the lowest cost per hour for parking. During the parking project, credit cards were implemented in off-street garages. This caused a spike in revenue and also generated the need for an internal IT department in the Treasurer's office.

With the completion of the parking study, Treasurer Jones did two very important tasks. First, she created an internal team to research parking technology. Then she created a team of community stakeholders to test, rate, and recommend the parking technology that the first team researched. The Treasurer wanted to make sure that, with this radical change for the city, the community would have a voice and buy in for the new technology. After a public request for proposal (RFP) was made to bid on delivering a new parking system, three new technologies were selected for on-street parking: Parkeon, IPS, and Parkmobile—along with updated Parking Enforcement Officer (PEO) handhelds.

Parkeon uses pay stations for payments of parking spaces. When a consumer parks in a space, the payment is taken at the kiosk, which is usually a few feet from the actual space. At the pay station, the consumer types in the space number, adds the amount of time, and the station gives the amount. The payment can be either cash, credit, or debit. The pay station then sends all of the information in real time to cloud services. Each station is equipped with cellular services and linked to the cloud service using virtual private network (VPN) security. The VPN connection is always "on" for each pay station. This allows for the station to self-report alerts. Each pay station sends alerts, when needed, such as collections, error messages such as out of receipt paper, and coin jams. Likewise, each pay station will send alerts when a collection or maintenance on the pay station has occurred.

IPS meters are single meters for each parking space. These meters resemble old style coin meters with the ability to do more than just coins. Like Parkeon pay stations, IPS meters can take coin, debit and credit payments. These meters are linked back to the IPS cloud service in the same manner as Parkeon through a secured VPN connection. Like the Parkeon meters, all data collected and all alerts are sent real time to the IPS cloud service.

Parkmobile works differently than traditional parking pay stations and meters. Parkmobile is a pay-by-cell application (app) for parking in almost every major city. Parkmobile allows the consumer to pay for parking with an app on an Android or iPhone regardless of meter manufacturer. This allows the consumer to pay as you go using a debit or credit card. The app alerts the consumer when time is going to run out and allows the consumer the opportunity to add more time without the hassle of actually walking to the car or pay station. Parkmobile gives the consumer more insight on the location of available parking and cost in real time. The Parkmobile app then sends all of the data to the Parkmobile cloud.

The added feature about all three of the parking services technologies is that the data is always accessible in real time. Parking Enforcement Officers (PEOs) now have the ability to write tickets directly from what they see on their handhelds. Before a PEO issues a ticket, the handheld checks all of the parking systems for payment. At the same time, PEOs use the handheld to take pictures of the violation with a clear view of the car and license plate. The issued ticket along with pictures are uploaded immediately to the parking enforcement system via cell service with secured VPN.

The Treasurer created an IT department within her office before any of the technology was implemented. This was a crucial step for the office to further Treasurer Jones' vision and mission with technology. The IT department was responsible for overseeing everything dealing with technology for the Treasury and Parking

Division. This included policy and procedures, administration, security compliance, risk analysis, infrastructure, and vendor management.

The implemented parking IoT technology created a profound and positive change in parking operations for the City of St. Louis and the Treasurer's office. After credit and debit card payments were implemented, there was an immediate uptick in revenue. Cash usage started a downward trend while credit and debit cards usage along with Parkmobile app usage started going up. Over 3 years, this changed even more with the app becoming the preferred method of payment.

With the ability to access parking data real time, consumers are able to confirm and pay tickets online without having to walk into the Parking Violations Bureau (PVB). If a ticket was believed to be given in error, the consumer can dispute it immediately. PVB can bring up the violation and either confirm or deny the user by showing the actual time of ticket compared to time that was paid for by the consumer. This, along with pictures, is communicated via a web portal eliminating debate on the topic. Providing the citizens of St. Louis with this new parking technology caused the parking revenue to go up, over 100,000 less tickets to be written and in turn caused a significant drop in disputed tickets in court, allowed for better use of personnel's time, and sparked regional change as other nearby cities sought the Treasurer's advice on updating parking technology.

With the amount of data being collected, we were able to leverage Microsoft Power BI (business intelligence), Azure, and SharePoint to do some data analytics. Using PowerBI, we are now able to pull in all parking transactional data to better understand our current parking environment. Using one dashboard, we can view data from year to year, month to month, year over month, return on investment (ROI), credit/debit versus cash versus app, revenue generation by zones, meter usage heat map and more. The dashboard allows anyone with proper access the ability to drill down even more to get granular details such as viewing areas that are using more cash than credit or app. If the goal is to get higher adoption for the app, then the prior information allows us to pinpoint areas where we need to concentrate our app marketing. To attract more businesses to an area, we can drill down to provide data showing parking space turnover rates that could lead to higher foot traffic for restaurants or stores, which in turn leads to revenue for the business.

Ultimately, the new technology provided the city with more money into the general fund. Prior to Treasurer Jones' innovation project, the most the city general fund received from parking enforcement was less than $400,000 a year. To date, the revenue stream has grown to $2.1 million on a yearly basis.

4.10 Discussion Questions

Here are a few issues that the City of St. Louis Treasurer wanted to solve by leveraging IoT. Did they do it? And how did they do it? How was IoT employed to solve the problems? Please answer each of these for the following questions.

a. How can we increase revenue with new technology?
b. How can we make it easier for citizens to pay for parking by providing multiple means of payment?
c. How can we deter cash theft?
d. How to provide real time data for parking enforcement?
e. How can we add transparency between the citizens and parking administration?

References

[1] Mayordomo, Iker, and Peter Spies. 2011. "Emerging Technologies and Challenges for the Internet of Things." *IEEE* 1–4.
[2] Elkhodr, Mahmoud, Seyed Shahrestani, and Hon Cheung. 2016. "Emerging Wireless Technologies in the Internet of Things: A Comparative Study." *International Journal of Wireless & Mobile Networks* 67–82.
[3] Kosek-Szott, Katarzyna, Janusz Gozdecki, Krzystof Loziak, Marek Natkaniec, Lukasz Prasnal, Szymon Szott, and Michal Wagrowski. 2017. "Coexistence Issues in Future WiFi Networks." *IEEE* 86–95.
[4] GSMA. 2018. *Long Term Evolution for Machines: LTE-M*. February. Accessed June 1, 2018. www.gsma.com/iot/long-term-evolution-machine-type-communication-lte-mtc-cat-m1/.
[5] Purnell, Lauren. 2018. Mobile Technology—Ideal Solution for Low Power Wide Area (LPWA). Presentation, GSMA.
[6] Haas, Harald, Liang Yin, Yunlu Wang, and Cheng Chen. 2016. "What is LiFi?" *Journal of Lightwave Technology* 1533–1544.
[7] Ayyash, Moussa, Hany Elgala, Abdallah Khreishah, Volker Jungnickel, Thomas Little, Sihau Shao, MIchael Rahaim, Dominic Schulz, Jonas Hilt, and Ronald Freund. 2016. "Coexistence of WiFi and LiFi Toward 5G: Concepts, Opportunities, and Challenges." *Optical Communications of IEEE* 64–71.
[8] Militano, Leonardo, Antonino Orsino, Giuseppe Araniti, and Antonio Iera. 2017. "NB-IoT for D2D-Enhances Content Uploading with Social Trustworthiness in 5G System." *Future Internet* 1–14.
[9] Borges Neto, Joao, Thiago Silva, Renato Assuncao, Raqual Mini, and Antonio Loureiro. 2015. "Sensing in the Collaborative Internet of Things." *mdpi* 6607–6632.
[10] Bayhan, Suzan, Gurken Gur, and Anatolij Zubow. 2018. "The Future is Unlicensed: Coexistence in the Unlicensed Spectrum for 5G." 1–7.
[11] Cluney, Chrissie. 2017. "RF Fundamentals for the Internet of Things." *IoT Evolution* 1–2.
[12] Sohn, Illsoo, Sang Won Yoon, and Sang Hyun Lee. 2016. "Distributed scheduling using belief propagation for internet-of-things (IoT) networks." *Springer* 152–161.
[13] Lizzi, Leonardo, and Fabien Ferrero. 2015. "Use of Ultra-Narrow Band Miniature Antennas for Internet-of-Things Applications." *Electronics Letters* 1964–1966.
[14] Keysight Technologies. 2017. "Battery Life Testing Chalenges in IoT Wireless Sensors." *EE Publishers* 1–5.
[15] Maple, Carsten. 2017. "Security and Privacy in the Internet of Things." *Journal of Cyber Policy* 154–184.
[16] Javed, Bilal, Mian Iqbal, and Haider Abbas. 2017. "Internet of Things (IoT) Design Considerations for Developers and Manufacturers." *IEEE* 1–6.

[17] Bhandari, Nikhil, Shivinder Devra, and Karamdeep Singh. 2017. "Evolution of Cellular Network: From 1G to 5G." *International Journal of Engineering and Techniques* 98–105.

[18] Goleniewski, Lillian. 2006. *Telecommunications Essentials, Second Edition: The Complete Global Source.* Boston, MA: Addison-Wesley Professional.

[19] Naqvi, Izza Fatima, Adeel Khalid Siddiqui, and Awais Farooq. 2016. "IPv6 adoption rate and performance in the 5G wireless internets." *Region 10 Conference (TENCON), 2016 IEEE.* Singapore: IEEE. 3850–3858.

[20] Vision. 2013. *5G: A Technology Vision-Huawei.* Shenzhen, China: Huawei Technologies.

[21] Gupta, Jyoti Radheshyam. 2017. "Significance of Satellites in IoT." *International Research Journal of Engineering and Technology* 2690–2695.

[22] Sanctis, Mauro De, Ernestina Cianca, Giuseppe Araniti, Igor Bisio, and Ramjee Prasad. 2016. "Satellite Communications Supporting Internet of Remote Things." *IEEE Internet of Things Journal* 113–123.

[23] IDC Corporate USA. 2017. *IDC Survey Reveals Significant Impact of Internet of Things Initiatives on IT Infrastructure.* February 23. www.idc.com/getdoc.jsp?containerId=prUS42328117.

[24] Antonakakis, Manos, Tim April, Michael Bailey, Matthew Bernhard, Elie Bursztein, Jaime Cochran, Zakir Durumeric, et al. 2017. *Understanding the Mirai Botnet.* Vancouver,BC: Usinex.

[25] Chen, Xi, Ming Ma, and Anfeng Liu. 2017. "Dynamic Power Management and Adaptive Packet Size Selection for IoT in e-Healthcare." *Computers and Electrical Engineering* 357–375.

[26] Harris, Albert F., Vansh Khanna, Güliz Tuncay, Roy Want, and Robin Kravets. 2016. "Bluetooth Low Energy in Dense IoT Environments." *IEEE Communications Magazine* 30–36.

[27] Chandler, Leon. 2015. The IOT: Impact on Waste Reduction (The Internet of Things—pt. 2). December 7.

[28] Zhang, Xueyan, Jianwu Zhang, Lin Li, Yuzhu Zhang, and Guocai Yang. 2017. "Monitoring Citrus Soil Moisture and Nutrients Using an IoT Based System." *Sensors* 17(3):1–10.

[29] Wakabayashi, Daisuke. 2018. *Self-Driving Uber Car Kills Pedestrian in Arizona, Where Robots Roam.* March 19. www.nytimes.com/2018/03/19/technology/uber-driverless-fatality.html.

[30] Amendola, Sara, Rossella Lodato, Sabina Manzari, Cecilia Occhiuzzi, and Gaetano Marrocco. 2014. "RFID Technology for IoT-Based Personal Healthcare in Smart Spaces." *IEEE Internet of Things Journal* 144–152.

IoT ELEMENTS AND ISSUES

Chapter 5

IoT End Devices

Ronald J. Kovac
Ball State University

Contents

5.1 Introduction

Sensors are the start of the Internet of Things (IoT). Actuators are the end of the IoT. Without cost-effective, accurate, reliable, and secure sensors and actuators, S/A, the IoT could not exist. This is similar to our human existence. We see a blockage on the sidewalk with our eyes (sensors), and we adjust our body to walk around it. The IoT takes this beyond our five senses (sight, smell, sound, taste, and tactile) to things we cannot sense (infrared and ultrasonic) and with accuracy that we can only imagine (98.6°F body temperature as measured by a thermometer).

In this chapter, we will explore various types of S/A that are and will be used in the IoT space. We will also look at the various criteria that we must measure and affect and, what is coming down the pike as far as smart sensors go.

We, of course, will close with a case study that looks at the S/A in use in today's world and then some probing questions to discuss and consider.

5.2 Major Sections and Subsections of IoT

Depending on whom you listen to, IoT can be broken down into numerous parts and sections. For the subject here, we will break IoT into four major sections:

a. Input—The sensors and their telemetry
b. Decision-Making—The compute, storage, and networking that must take place in order for decisions to be made
c. Output—The activators
d. Services—The logic and systems underlying IoT (Figure 5.1)

5.2.1 Sensors—Input

The starting data that permits IoT to exist comes from sensors. These sensors can be standard, they can be smart, or they can be manual. A standard sensor, like

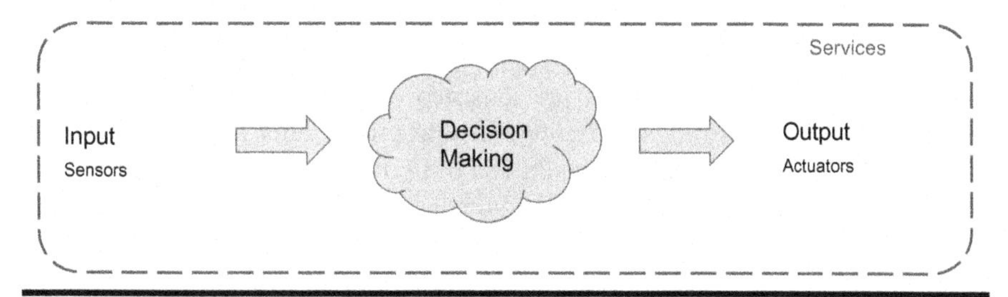

Figure 5.1 Major sections and subsections of IoT.

your car's gas gauge, senses one variable. A smart sensor detects multiple variables and makes decisions based on its readings. Finally, a manual sensor is like a light switch—human intervention is necessary. Whatever method, these are the start of the IoT cycle. In this chapter, we will look at various sensors and what they measure in order to create the starting point of IoT. Elements here are often called *data sources* or *IoT things*. The data transmitted from sensors is often called telemetry, meaning remote monitoring. Here we create and start the IoT world.

5.2.2 Compute/Storage/Networking—The Decision-Making

Based on the input from the sensors, decisions get made. The major parts of this decision-making consist of compute, storage, and networking elements. The compute elements do the actual processing of data, similar to your personal computer, and do so with the same architecture as common computers (Intel microprocessors, etc.). The storage elements hold the data before and after processing. For example, we store all the traffic flows for a week and then compute base-predictive scenarios to turn the traffic lights green or red to optimize traffic flow. The networking elements allow all the various devices to talk among themselves within a local system. Included in this decision-making is an area which is coming into prominence, called Big Data analysis. Essentially, this involves analyzing very large data sets to derive patterns and trends. This field is up-and-coming and is nowhere near as simple as it sounds. An in-depth look by the reader is strongly suggested (Figure 5.2).

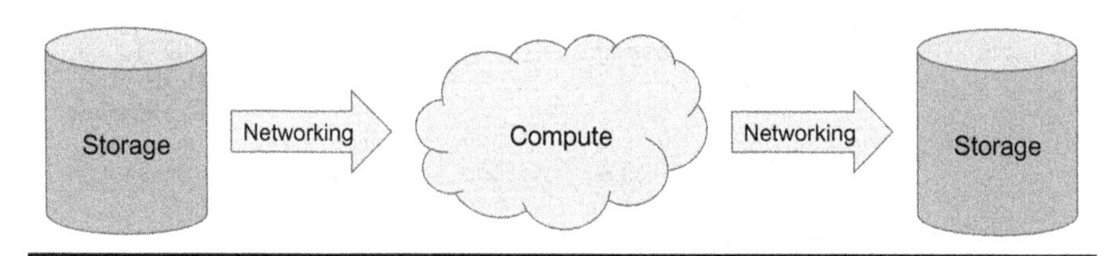

Figure 5.2 Major sections of the decision-making.

5.2.3 Output—Actuators

Based on the decision made in the previous step, an actuator changes the state of the environment (by turning the heat on, turning a traffic light red, etc.). Be aware, the output of the decision-making could also be in the form of a notification: a note or flag to a human to do something. The human element is often brought into this to assure the proper decisions are being made and to provide a failsafe for the IoT system. Here we act as the end of the IoT system.

5.2.4 Services—Logic Underlying All of This

Services are the software side of IoT that manage the infrastructure; guide the decision-making; and provide scalability, adaptability, portability, and automation. This section also includes the globally interconnected network of the IoT world and any applications that remove the raw processing from the user.

Figure 5.3 is a more complex build out of the IoT world that one would typically see in magazines and books.

5.3 Sensors and Telemetry

Sensors are the start of the IoT world. They measure characteristics for the rest of the IoT systems that think and act upon this data. They detect types of input from the physical environment and are the root of the IoT solution. The signals that the sensors create are often called *telemetry*. Sensors must comply with the following rules:

- They must be sensitive enough to reliably and accurately measure the noted property (for example—temperature).
- They cannot influence the measured property (such as a temperature sensor that puts out heat).

Input	Decision Making			Output	Services
Create & Capture	Compute	Storage	Networking	Act	Analytics
Sensors	Complex event processing			Actuators	MQTT, Push Notifications
	Batch ingest	Analytics & discovery	Hadoop clusters	Visualization	Aggregation

Figure 5.3 Complex IoT form.

- They must not be sensitive to other properties that can alter the measured property (such as a thermometer that is affected by wind chill).

Most sensors put out a linear function of telemetry, meaning that when temperature changes, they put out a correspondingly lower or higher voltage in a linear fashion (like a line). Sensors adjust the voltage according to the input temperature, or they can change capacitance, resistance or any other electrical property. If a sensor puts out an analog signal, as most natively do, this signal must be digitized in order to be used in digital computers. This function usually happens at or near the sensor itself.

5.3.1 Criteria for Sensors

There are many characteristics of sensors that must be considered before use or application. A few selected characteristics are discussed below:

- Range—Usually a sensor can measure properties only within a predefined and limited range. For example, a thermometer goes up to 110°F and down to −10°F. The selection of the sensor range must be qualified to the use of the sensor.
- Sensitivity—How precisely can the sensor read. Can it read to the 1°F level or to the ¼°F level? This is often called resolution.
- Nonlinearity—Often sensors measure differently throughout their range of measurement, and some type of correction must be put in to adjust the sensor output. For example, as a thermometer gets hotter, the scale of the output becomes nonlinear. (Refer to Figure 5.4).
- Drift—As the sensor measures its property over time, its characteristics change, and therefore, the measurement characteristics become wrong.
- Reliability—Drift, as discussed above, often happens over time. Reliability is the incorrect nature of a signal that happens during manufacturing of a device or over a period of time. Calibration (comparing the output of the sensor to a *known source*) can often correct issues with reliability and drift.
- Quantization Error—If the output of the sensor is analog, as it often is, the signal is usually digitized to work in modern digital systems. The translation of analog signals to digital signals is an approximation of the analog signal and therefore prone to errors.

The global market for sensors reached $110.4 and $123.5 billion in 2015 and 2016, respectively. (Rajaram 2017)

The sensor market is expected to increase from nearly $138.8 billion in 2017 to nearly $240.3 billion in 2022 at a compound annual growth rate (CAGR) of 11.8% for 2017–2022. (Rajaram 2017)

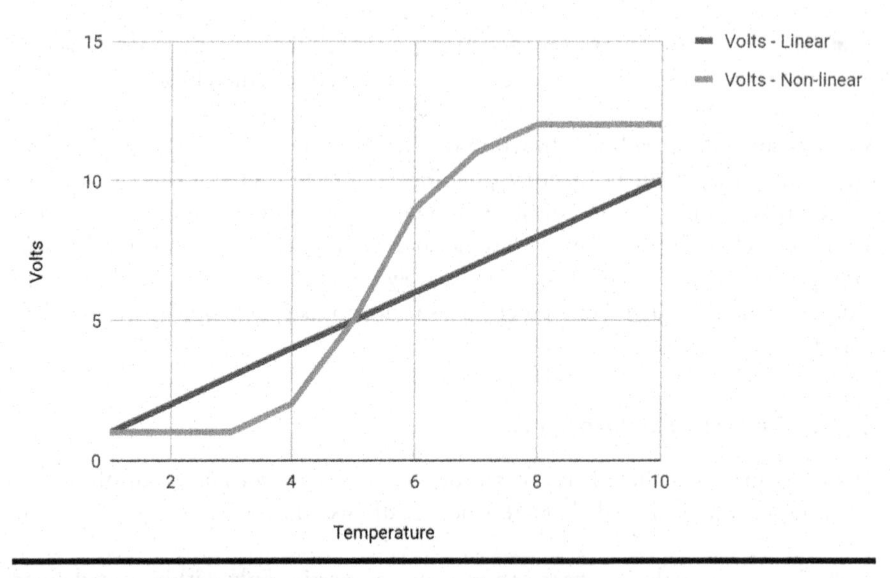

Figure 5.4 Linear output of a sensor and digitization.

5.3.2 Management and Control of Telemetry and Telecommand Messaging

To control and manage the stream of information coming from sensors and going to actuators, there are messaging protocols. These messaging protocols control the pace of information flow, the syntax of the flow, and generally manage and provide structure to all the information flow there is for the IoT world. There are many types of messaging protocols for sensors and actuators, and here, we will introduce you to a few of the common ones found these days. These protocols guide the information flow from the sensors.

> "When I developed MQTT, with my collaborator Arlen Nipper, 17 years ago, I had what I used to call my 'modest plan for world domination', which was that one day all devices would connect to TCP/IP, and all of them would talk MQTT."—Andy Stanford-Clark
>
> IBM Distinguished Engineer for the Internet of Things (Lewis 2016)

- MQTT—Message Queuing Telemetry Transport is an ISO Standard. It works on top of the TCP/IP protocol and is often used for remote connections that do not have a lot of information (such as a temperature probe). Therefore, it requires limited network bandwidth which is appreciated. Facebook Messenger used MQTT. https://en.wikipedia.org/wiki/MQTT
- Advanced Message Queuing Protocol (AmQP)—creates a standard of interoperability between various

pieces, parts, and systems. It is open standard based and has a high degree of reliability and security. This is often used to connect servers to servers.

- Extensible Messaging and Presence Protocol (XMPP)—is often used to connect devices to people. Extensible Messaging and Presence Protocol is based on the very popular markup language XML. Designed in 1999 for near real-time communications, it has been used for publish–subscribe systems, signaling systems, gaming, file transfer and, of course, the IoT. It has an extensive set of middleware programs to talk with any other protocol that exists.
- Streaming Text-Oriented Messaging Protocol (STOMP)—Formerly known as TTMP is a text-based protocol designed to be interoperable and therefore talk with any message broker. Like the other protocols listed above, it works above TCP/IP in the Open Systems Interconnection (OSI) model. It is easily implemented and therefore very well liked.
- Application Program Interface (API)—You may have heard of this term as it is popular these days as it acts like a universal translator. It allows different systems to talk to each other. APIs' make it easier for people to use programs and developers to use technologies in building applications, by abstracting the underlying implementation, and only using the elements needed. This allows for easy integration between the various types of systems that are prevalent in today's world. Like a universal translator seen in past episodes of Star Trek, it abstracts the conversation into ideas and not specific words. API, unlike the aforementioned messaging protocols, is a concept and not necessarily a messaging protocol.

The communication realm and especially messaging protocols within IoT are fractured and organic. Until some of these standards win out over others, the field will struggle. There are dozens of alliances and coalitions formed, and forming, in hopes of doing just this. Against them and the current products out there is the constant invention of the "Best" protocol.

One must also say here that although many things of the standard Internet are being capitalized on for IoT, many others are being modified for the IoT world, and these others exist at all levels and layers of the Internet: infrastructure (ipV6 Modified), authentication (URLs), transport layer (LPWAN), discovery (DNS-SD), device management, semantics, and many others. The IoT is truly inventing and reinventing new elements.

5.3.3 Sensor Types

Acoustic sound vibration—The first type of sensor we will cover is one that measures sound or acoustic vibration. A pickup on a guitar is a prime example of this, as is a microphone, but these are simplistic and common examples. Let's take a look at some others and their use in IoT.

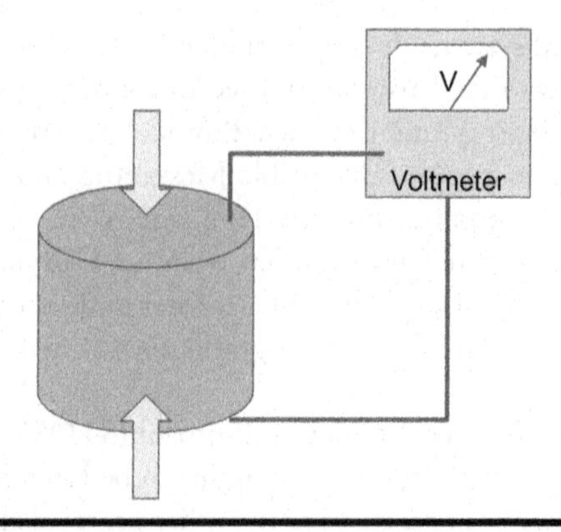

Figure 5.5 Piezoelectric effect.

Hydrophone—A hydrophone is a microphone that is specifically designed to measure sound underwater. With these we can listen to whales talking or sounds from around the world, as sound travels better in water than in air. Most hydrophones are based on a *piezoelectric* transducer, which generate electricity when exposed to sound (pressure) hitting them. The piezoelectric effect is simple: when certain materials (crystals and some ceramics) are subject to the pressure of sound hitting them, they emit electricity proportionally to the pressure. This effect was discovered in the Mid-18th century by Carl Linneaus and Franz Aepinus (APS News 2014) (Figure 5.5).

Microphone—A microphone is similar to a hydrophone but is used to change sound in air, rather than water, into telemetry. Of course, microphones are very popular and used in sound recording, megaphones, telephones, and various other uses. Although sometimes using the piezoelectric effect of hydrophones, most microphones are of the *dynamic* type. These use a moving coil and work via electromagnetic induction. This concept, credited to Michael Faraday in the early 1800s, involves moving a magnet through a coil of wire and therefore producing electricity proportionally to the movement of the magnet (Magnet Academy 2014) (Figure 5.6).

Sound locator—The last item in this category we will look at is the sound locator. Taking the concepts of the microphone further, we can also use sound to determine the distance and direction of an object. This localization can be done actively or passively. Passively we listen, actively we send out a signal and wait for it to bounce back to determine where the object is (this is commonly called radar for the air and sonar when used underwater). Using microphones and mathematical techniques for triangulation, we can determine an object's distance and direction from the sensor.

Temperature—Another category of sensor is the thermal sensor which measures heat or temperature. This is typically used for detecting and measuring the temperature of a flame. We will explore some of these technologies and uses.

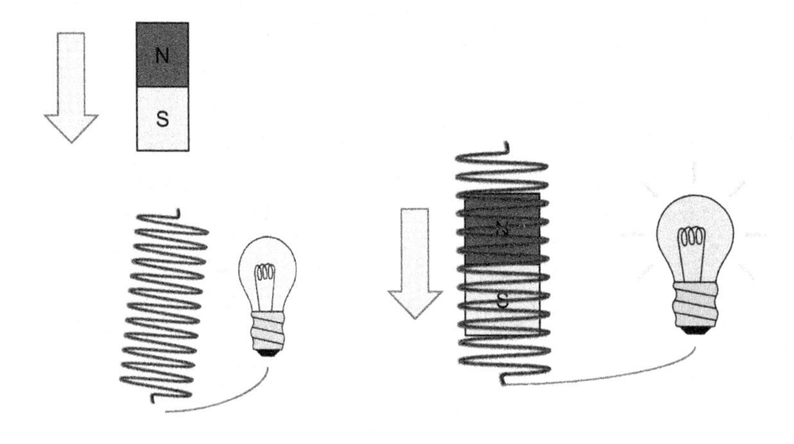

Figure 5.6 Moving coil microphone.

Thermometer—A thermometer is a device to measure temperature. The concept of a thermometer was known to the ancient Greeks when they saw elements expand when exposed to heat. With the addition of scale and the use of linear materials, such as mercury, the modern thermometer has become a valuable sensor. Currently, digital sensors are available to replace the mercury in a thermometer.

Resistance thermometers—These are sensors that measure temperature and have high accuracy and repeatability. It was discovered that the resistance of metals changes when exposed to changes in temperature. After determining the best metal to use and the best calibration instruments, this device is becoming very popular in the industrial setting. These are also known as resistance temperature detectors (RTDs).

Thermistor—These are similar to the above category (RTDs); however, thermistors use a ceramic or polymer material rather than a pure metal. Thermistors typically have a greater precision than RTDs but within a more limited temperature range.

Thermocouple—A thermocouple is based on a different concept from that of the thermometer. When two dissimilar conductors are put together and exposed to heat changes, a voltage is produced. The voltage is interpreted to measure temperature. Although widely used, thermocouples accuracy of less than 1°C is hard to obtain. This concept was discovered in the early 1800s by Thomas Johann Seebeck (California Institute of Technology 2018) (Figure 5.7).

Pyrometers—A pyrometer is a thermometer that can determine the temperature of a surface from afar (remote sensing). It determines the temperature of a remote surface by the thermal radiation spectrum (color) it emits. This process is known as pyrometry. First discovered to measure kiln temperature, the field has evolved into a critical remote sensing field use that is accurate and complex in its function.

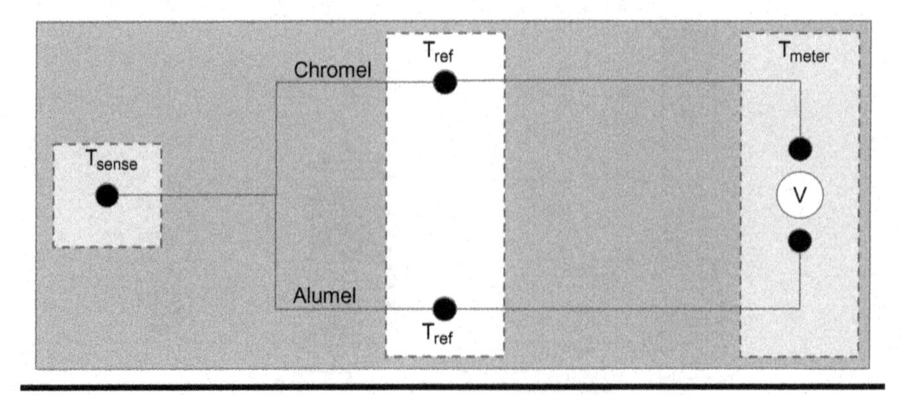

Figure 5.7 Thermocouple.

Proximity, presence—This third type of sensor measures the proximity, or closeness, and presence of things. It is used in a wide variety of settings. Let's explore some of the uses and the technologies behind it in the IoT world.

Motion detectors—Commonly used in alarm systems, home, and lighting control, motion detectors have become vital elements of the IoT world. Most sensors are passive, and they contain either an optical, microwave, or acoustic sensor that when interrupted, signal the motion or proximity of an object. Typically, they work within distances of 15 ft (Safewise 2018). They are used for lighting systems to turn on/off lights when people are present, to signal an alarm when somebody is entering a home or trigger a camera for possible intrusion.

Doppler radar—A specialized type of radar that measures the velocity of objects at a distance. This is based on the Doppler Effect, which notes that signals and waveforms change their frequency when they bounce off of different objects. We all know of these when used in a meteorological sense, but they are also used for radar guns and major league baseball speed pitching systems (Figure 5.8).

Reed switch—Invented at Bell Labs in 1936, this type of sensor measures the presence of a magnet near to it (Engineer's Edge 2018a,b). Once the magnet is removed, the switch goes back to normal. Used extensively in proximity switches, these are reliable, effective, and cheap devices that can be used in many IoT applications (Figure 5.9).

Figure 5.8 Doppler effect.

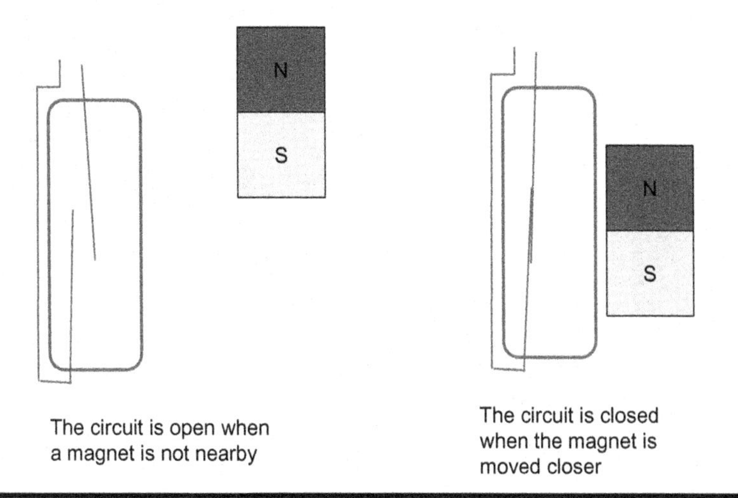

Figure 5.9 Reed switch.

Pressure sensors—Sensors in this category measure the increase or decrease in pressure on a sensor. Extensively used in many current systems, they will continue to hold value in the IoT world. Some typical use cases and technologies involved in these are:

Barometer—Typically used in meteorology, they measure the pressure of the atmosphere. This in turn can forecast short-term changes in weather. Barometers are very common devices, but like all sensors, there must be an offset factor for variables we do not want to measure. Typically, temperature and altitude are the two variables that must be kept constant for a barometer to provide accurate data.

Piezometer—A piezometer uses static pressures to measure liquid pressure. It does this by measuring height of a liquid column as it rises against gravity. Used to measure groundwater, a variant of this is the Pitot tube used on airplanes to measure air pressure.

Tactile sensor—These measure data arising from physical interaction in the environment. Your touchscreen on a mobile phone is an example of this. Typically, they work from piezo-resistive, piezoelectric, capacitive, and elastoresistive methods. The various sensors measure the resistance, electricity, capacitance of the touch, but to varying degrees of accuracy and cost. Used extensively in manufacturing and medical imaging, they are becoming common in robots (Figure 5.10).

Chemical—Sensors that measure the presence of certain chemicals in an environment (air, water, etc.). These sensors are typically used in manufacturing environments and are becoming more popular in the healthcare environment.

Breathalyzer—A device most have heard of that measures blood alcohol content (BAC) from your breath. The device is based on chemical transformation in which 1. ethanol present in breath is oxidized to acetic acid at the *anode*, 2. atmospheric oxygen is reduced at the *cathode*, and 3. the overall oxidation of ethanol to acetic

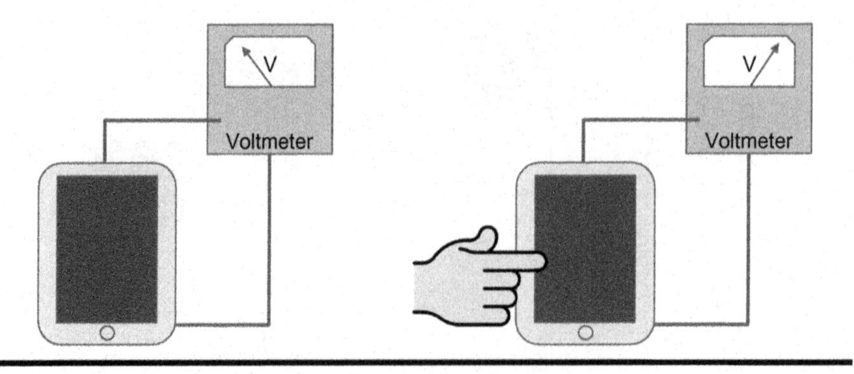

Figure 5.10 Tactile sensors.

anode and water is measured by a *microcontroller* and displayed as an approximation of BAC.

Carbon monoxide detector—Used to detect an abnormal level of carbon monoxide, a poisonous gas, in home or industrial environments. Typically, one of four different types of sensors are used in these: opto-chemical, biomimetic, electrochemical, and semiconductor (Engineer's Edge 2018a,b). The electrochemical is the dominant technology in the United States and Europe for these detectors and is the lowest cost option.

Chemoresistor—A class of chemical sensors that change their electrical resistance in response to changes in the chemical environment. Simplistic in form, but very accurate, the sensing material is the heart of the system but can only measure the presence of certain chemicals (Banica 2012) (Figure 5.11).

Electrical—Measures any parameter in the electrical world. These are voltage, resistance, current, capacitance, and inductance. A few of these case studies and principals of electrical sensors follow.

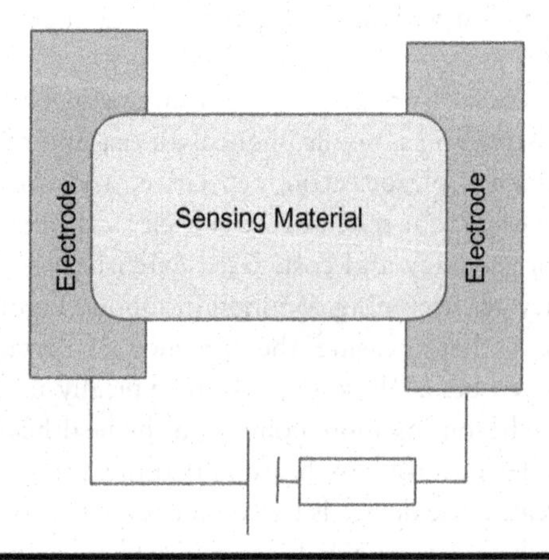

Figure 5.11 Chemoresistor.

Current sensor—Measures electrical current, amperes, in a wire by sensing the size of the electrical field produced by the current. The Hall Effect, and the Hall Coefficient, is the ratio of induced electric field to the product of the current density and therefore magnetic field produced by the current. This all is an offshoot of electromagnetism, which describes the electrical magnetic field produced around electrical voltage or current.

Magnetometer—Devices that measure magnetism direction, strength, or change in a magnetic field. A typical compass would be an example of a magnetometer. These magnetometers can be used to detect metal objects at a far greater range than the typical metal detector, which operates based on a different principle. Magnetometers have become miniaturized and can be incorporated in integrated circuits at a very low cost (MEMS magnetic field sensors).

Other Sensors—Whole books could be, and have been, written on sensors. Above is only a sampling of today's sensors. Only book length prohibited us from exploring the other types and technologies of sensors. Certainly, just studying the technologies and principals of these sensors is fascinating. What was attempted here was to introduce you to the popular sensors and principals that will be used in the IoT field. If you wish to learn more, plenty has been published in articles, books, and other sources.

Today, the bulk of revenue booked from MEMS sensors comes from the smartphone segment. There are more than a billion smartphones sold each year with each containing at least one MEMS sensor (Gemeli 2017)

"Sensors are experiencing a renaissance of sorts as microelectromechanical systems (MEMS) technology becomes less expensive and further miniaturized, in turn fueling penetration of sensors into new applications and creating new potential for the Sensor market."—Kaivan Karimi (Karimi 2013)

5.3.4 Smart Sensors

In this day and age everything has to be smart or have an "i" in front of it. For us older folks, it's difficult owning things smarter than we are. But now designers and manufacturers are developing smart sensors and actuators. What exactly is a smart sensor or actuator? To make a sensor or actuator smart, processing power and a communication interface are added to the device (Figure 5.12).

As mentioned before, the volume of data generated in IoT would be almost impossible to deal with using today's technology. In order to help alleviate that, some conditioning/processing of the data is done locally at the sensor level and maybe even stored locally. This reduces the amount of data to be transmitted and assists in data correction and accuracy. Data to be transmitted upstream would only be necessary for data analysis sake (for example, sending one packet summarizing the day's temperature… rather than every hour). Smart sensors also help with fault tolerance and conditioning of the signal.

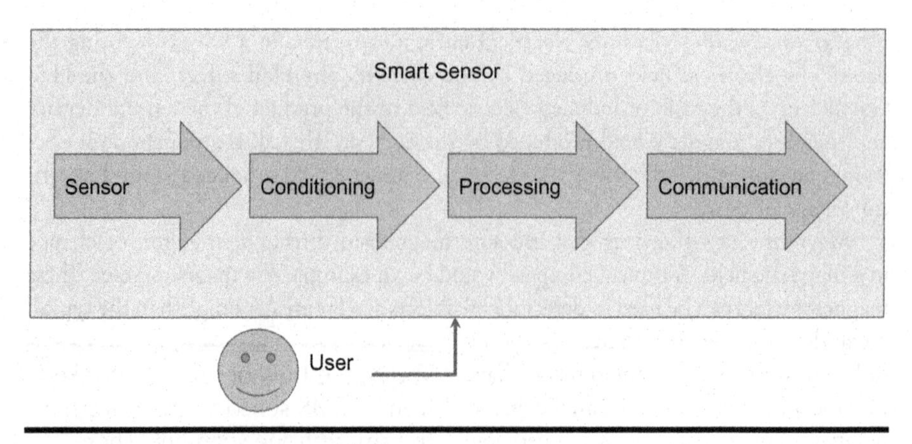

Figure 5.12 Smart sensor components.

A smart sensor may also include a number of other components besides the ones listed above. These components can include *transducers*, *amplifiers*, *analog filters*, and compensation. A smart sensor incorporates software-defined elements that provide functions such as data conversion, digital processing, and communication to external devices. Smart sensors are a heavy enabler for the IoT arena as they make data collection simpler. RFID tags would be an example of a semi-smart sensor.

A smart sensor offers many advantages such as:

- Ease of setup and use
- Higher reliability
- Size
- Cost (Total Cost of Ownership (TCO), of system)
- Reduce network load

There are smart actuators also, which include more than the standard actuator. Smart actuators include processing power and communications interface. For these actuators, there are standards for these devices as laid down in the IEEE 1451 standard.

IEEE 1451 is a rule that is laid down by the International Electronic and Electrical Engineering Society. If you wish to read more on IEEE, look further in this chapter for a section dealing with the same.

We must always keep in mind that just making devices able to sense and be connected is not enough to fulfill the promise of IoT. It will be successful if it follows a user-centric approach and solves real-life challenges of humans: making life easier!

The smart sensors market was estimated to be worth USD 3.90 BN in 2015 and is expected to grow at a CAGR of 24.22% to reach USD 11.53 BN by the end of 2020. (Mordor Intelligence 2018)

5.3.5 Sensor Fusion

The concept of sensor fusion involves combining sensory input from a multitude of sensors to gather an output that exceeds the output of the individual parts. The information from the various sensors can be combined, correlated, or related to infer conclusions. For example, you could monitor temperature and humidity and correlate their data in order to predict failure rates of food supplies. Sometimes these two sensors are available in one "fusion device," and other times their outputs are combined in software. The key concept is creating better, more reliable output so uncertainty of decisions is reduced. Sensor fusion creates a situation in which the whole is much greater than the sum of its parts. Sensor fusion is sometimes called data or information fusion.

Fusing data can be done in a decentralized way, where the individual data is put together locally and most likely processed there for a conclusion. Another way is centralized fusion, where the individual sensor data is fused together at a central site therefore allowing for more software flexibility. As you would expect, there is also a combination platform where some data is fused locally and other data centrally.

There are various "levels" of fusion that can take place (Damiani 2009). The levels are:

- Level 0—Data alignment
- Level 1—Entity assessment (e.g. signal/feature/object)
- Level 2—Situation assessment
- Level 3—Impact assessment
- Level 4—Process refinement (i.e. sensor management)
- Level 5—User refinement

The concept of fusion of data from sensors is fascinating and the possibilities are great. For example, we can use already existing data from roadside noise sensors, visual images, and other sensor data to predict/note road conditions (traffic jams). An issue that is holding down the concept of fusion is the various number and types of OSs that are out there and their limited ability (via drivers) to request sensor information. This has not been a priority for OS companies. This is changing. Sensor fusion is a part of the current Microsoft strategy. Starting with the Windows 8 OS, Microsoft supports sensors in a cohesive manner, using sensor-class drivers based on industry standards developed in collaboration with other ecosystem partners. The Windows Runtime programming module allows for lightweight executive calls that enable sensor processing at the hardware level.

The human body works with sensor fusion. We sense things from the environment in many ways (our five senses) and our brain puts these together to form a reaction. For example, we smell smoke, feel the heat from a fire, and see the flames, and this causes our brain to react (call for help, put out the fire, or flee). The fusion causes the brain to react because the composite information is greater than the sum of the disparate sensory inputs.

5.4 Actuators

5.4.1 Actuators Overview

We will now switch to the other end of IoT, the output, as embodied by the actuator. Measurement and processing are useless without action. It is the actuator that does the action: that turns the lights on, that shuts the water off, that moves the car though the city, that performs any action as deemed necessary by the controller.

Actuators are sometimes called telecommand devices, as they serve the command function from a distance. They are also sometimes called "movers," as they are tasked with moving a mechanism or system. Actuators require two major elements:

"First, we recognize that although IoT sensors provide useful data, it's only IoT actuators that provide the *So What* that the market is seeking. And second, we see a major challenge in IoT today: there are many more sensors than actuators. As we work to make IoT more relevant—to provide compelling answers the *So What*? Question—we need more devices like Nest and Hue that actuate."—Mike Vladimer (Vladimer 2015)

5.4.2 Actuator Control Signal

Tells the actuator when and how much to move or to act on the environment. The signal can be simple, software based, human, or any other input. A light switch is an actuator as it turns on a light bulb in response to a human control signal. A printer is an actuator that acts on control signals from your computer and prints out on paper.

5.4.3 Energy Sources

The energy source provides the power to move the world. Main energy sources can be electrical, thermal, hydraulic, or pneumatic. The actuator takes this energy source and turns it into the movement we want in the environment. Typically, this involves energy transformation, such as turning electrical energy into motion. When we turn on the light bulb, the energy source of electricity gets transformed into light. When the computer tells the printer to print, the energy source of electricity gets transformed into characters/text on a page.

The energy source is a prime consideration in the selection of an actuator, as energy must be supplied in adequate capacity to do the required movement of the actuator. Also, the energy must be conveniently supplied at the proper time. Four main types of energy sources are used with actuators.

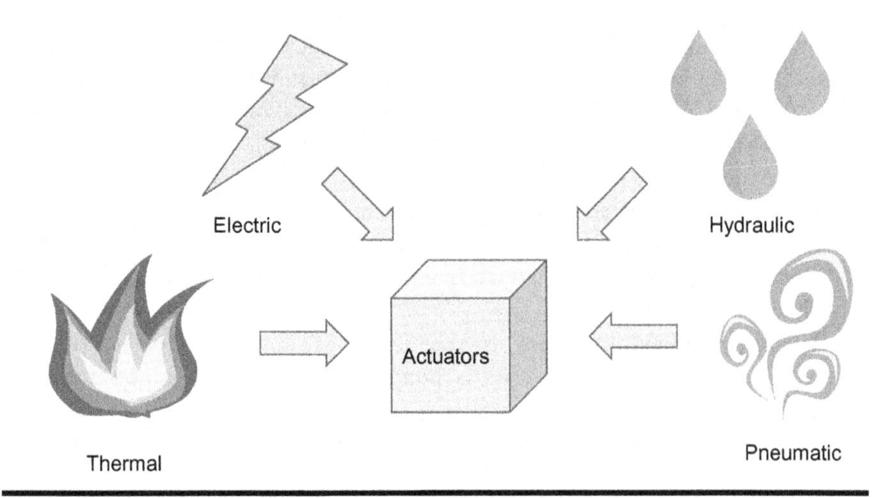

Figure 5.13 Energy sources.

- Pneumatic—Uses a gas in either a vacuum or compressed air form for either linear or rotary motion. The air pressure, or lack thereof (vacuum), is used to move the actuator based on the control signal.
- Hydraulic—Uses the power of a fluid to facilitate the mechanical operation. The mechanical motion provides a linear, rotary, or oscillatory motion. Liquids are used as they are not subject to compression, and the pressure of the liquid forces a required movement. Hydraulic movement is often slow, but very powerful. The brakes on your car use hydraulics to stop the car, and hydraulics are also used for elevator movement.
- Electrical—Uses electric current to induce a movement, either rotary or linear, desired by the actuator. It is one of the most common energy sources due to its availability and is also the cleanest as it does not directly involve oil or other fossil fuels.
- Thermal or magnetic—Although not a common energy source, we can use heat or thermal energy to move actuators. This is sometimes used in industrial settings and is also even used in residential settings for geothermal power (Figure 5.13).

5.4.4 Actuator Considerations

When we are selecting an actuator for a purpose, we need to consider many variables. The categories of variables follow:

Force—How much power or force is needed to move the environment? Can the actuator do it? When considering force, we must look at two different sides: static and dynamic loads. Static loads are considered when the actuator is at rest, not moving. Dynamic loads are dependent on the actuator load while it is in motion. The great differences in static and dynamic loads are a topic beyond the scope of this book but are a fascinating concept to be researched.

Speed—How quickly do we want the movement to happen? Does it have to happen instantaneously or with great force? Or, can it happen over a few minutes with little load? When applying the brakes to a car, the slowing down does not have to be, and should not be, instantaneous. It must be gradual. This parameter dictates the type of actuator to be used.

Operating environment—What temperature, humidity, and environment will the actuator be operating in? Can it be used in an industrial environment? With toxic or explosive gases? These are all questions that need to be considered in the selection of an actuator.

Reliability—How important is the actuator's movement? When it does fail, what must be done to fix it? Is human life dependent on the actuator's movement?

Movement—Does the actuator require a rotary motion (such as a motor), a linear motion (to turn on a switch)? There are mechanisms to convert a rotary motion to linear motion, and vice versa, but these are additional elements that are not native to the systems.

5.4.5 Actuator Types

In an industrial sense, actuators are put into categories. These categories are based on the energy source and their movement type (linear or rotary).

> Based on type, the linear actuator segment is estimated to account for the largest Actuators Market share in 2017 (PRNewsWire 2018).

- Electrical linear—Electrical energy source that converts electrical energy into linear displacement through mechanical transmission to provide straight line push/pull motion.
- Electrical rotary—Electrical energy source that converts the energy into rotational motion. What we would commonly call a motor.
- Fluid power linear—Hydraulic energy source that produces linear displacement by means of the hydraulic fluid.
- Fluid power rotary—Hydraulic energy source, turning the hydraulic power into rotary motion. Basically, a hydraulic motor.
- Manual linear—Converts manual energy into linear displacement. Used for precise positioning.
- Manual rotary—Converts manual energy into rotary action.

5.4.6 Actuator Examples in the IoT World

Electrical motor—Using electricity as the energy source, and a switch or computer signal as the control signal, creates a rotary motion. Motors operate through the interaction between a magnetic field and winding coils to generate rotary motion.

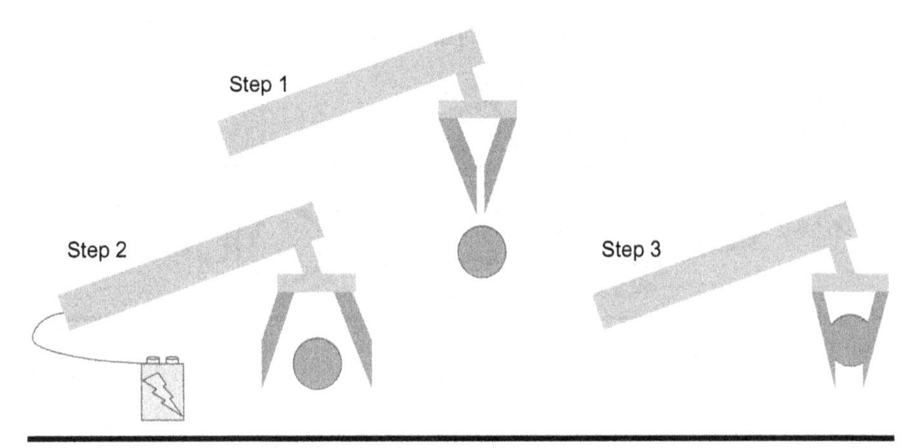

Figure 5.14 Electroactive polymers.

This rotary motion can be converted into linear motion. A motor can be used as an actuator in IoT heating systems. The sensor tells the controller that the room is too cool, and the motor kicks in to blow hot air out the furnace.

Electroactive polymers—Using electricity as the energy source, and a switch or computer signal as the control signal, creates linear motion. Electroactive polymers exhibit a change in size or shape when stimulated by an electric field. One common application of these is in robotics in the development of artificial muscles (Figure 5.14).

Hydraulic cylinder—Converts hydraulic power into linear motion. These are typically double-acting cylinders and can force movement in two directions. Hydraulic cylinders are very heavily used in industrial equipment. When you enter an elevator and press a button, this creates a control signal that forces hydraulic fluid into a cylinder and moves the elevator up to the next floor.

Servomechanism—Actually, an actuator and sensor combined. It uses error-sensing negative feedback to correct the action of an actuator. This controls position, speed, and other parameters. Your car cruise control mechanism is a form of this, as it keeps the speed set at what you desire by monitoring current speed and comparing it to set speed (Figure 5.15).

5.5 Key Criteria for Sensors and Actuators

There must be many sensors and actuators for IoT to work effectively. These will allow us a comprehensive view into our world and give us the ability to control it. These S/A will be placed "in the field" and therefore must be agile and tough enough to withstand the characteristics of the field. Most electronic equipment is placed in environmentally controlled rooms with consistent temperature, humidity,

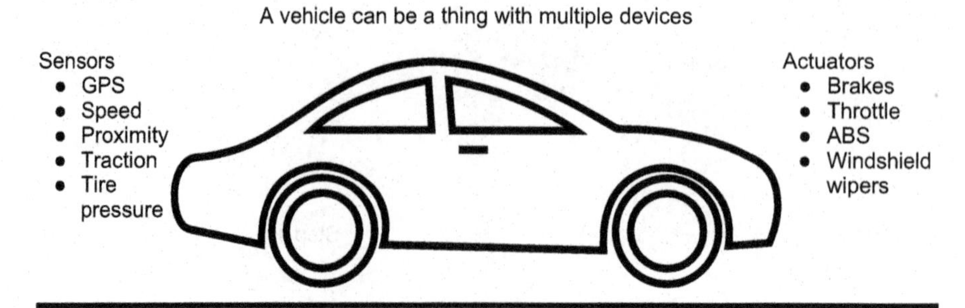

Figure 5.15 Servomechanism.

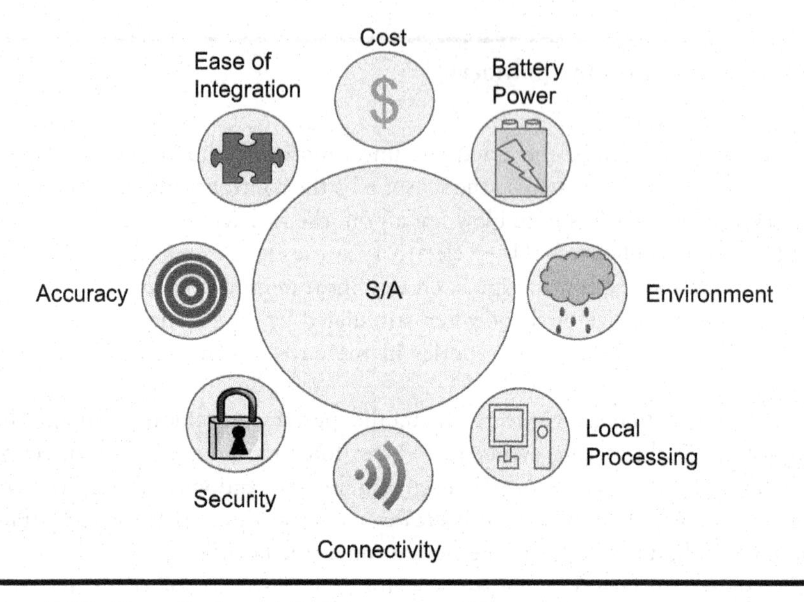

Figure 5.16 Key criteria and their impact on S/A.

and clean conditions. The field is harsh, with wide temperature variations and little control over the environment. Therefore, S/A must be made to endure these surroundings. Specifically, what are these conditions? Below are mentioned the key criteria for S/A and the conditions they must endure (Figure 5.16).

5.5.1 Battery Power

Most sensors, and many actuators, will not be able, effectively nor efficiently, to be powered by normal current. They must be independent and run on battery power or low power. This limits the power that they can draw from the source current and therefore their capabilities. We see this everyday with our cellphones, the more you

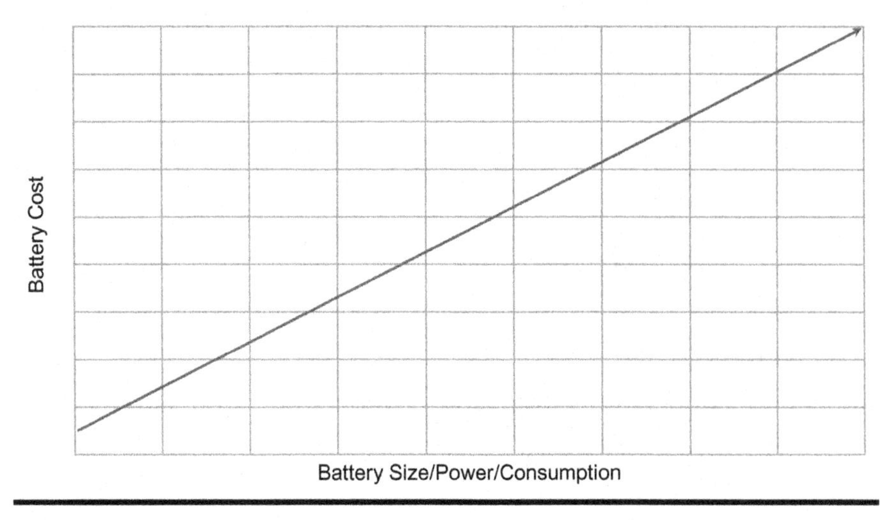

Figure 5.17 Battery consumption/power/size/cost.

have them turned on, the faster the battery is depleted and therefore the more often we have to charge them. Unfortunately, we don't have easy access to S/A in the field to charge them (Figure 5.17).

5.5.2 Environment

The field is sometimes a harsh place. A factory floor is dirty, cold, and has other conditions that are often damaging to electronics. If we place S/A in the outside world, we add humidity, water and other factors that could be damaging to electronics. Weather-resistant cases are often necessary to shield the electronics. Often times S/A are put in vulnerable situations where the general public can steal or destroy the device. These factors add bulk, weight, and cost to the device.

5.5.3 Local Processing

It would seem logical to provide some local storage for the S/A so that is does not have to transmit constantly to a central site and to provide some local compute so that raw data can be transformed to useful information to the central site (such as an average of temperature reading taken every minute but transmitted every hour). But both add power consumption and size, therefore working against two of the elements that we want S/A to be: low power consumption and small in size. Adding to this, sensors and actuators are typically analog devices, which means that an analog-to-digital/digital-to-analog converter must be placed locally with the S/A. Sensors and actuators with storage, compute power, and conversion capabilities are often called smart sensors and actuators.

Wired	Wireless
Category 5 Connection	PSTN Connection - 4G, 5G
Tap Into Cable Connection	WiFi Connection - 802.11 X
Power Line	Near Field - Bluetooth, Other

Figure 5.18 Connectivity solutions.

5.5.4 Connectivity

Whether a sensor or actuator, in order to be effective in the IoT world, a device must connect with a central site. For a sensor, it must provide information to the mother ship so that the data can be acted upon. For an actuator, the decision is made by the central site and sent to the actuator to make it actionable. All require connectivity to a network. Connectivity can come in two broad forms: wired or wireless (Figure 5.18).

But within each on these broad categories, there are a multiple of options. In general, wired connection is better in performance, security, and reliability but far less agile and may not be available due to the lack of a close wired infrastructure. Wireless offers agility and ease of access but at the sacrifice of performance, security, and reliability. As for the sub options, each has its advantages and disadvantages, with the first level of decision-making being "is it available to me."

5.5.5 Security

There are two levels of security that must be thought of: physical and cyber. For physical security you want to assure that the S/A cannot be stolen or removed and that it cannot be subject to vandalism by employees or the general public. For *cybersecurity*, you want to assure that the device cannot be hacked (broken into electronically) and that the transmission/reception of signals are confidential and not subject to inspection/modification. This is an area that is currently being vastly overlooked! In the race for product and low cost, security is being sacrificed. This is generating a lot of security holes into networks and electronics and providing many a sleepless night for security professionals (Figure 5.19).

5.5.6 Cost

Cost of S/A is always a factor. We can hope/rely on the costs of production coming down with the mass of volume expected but that is a good guess and should not be considered a fact. When you add on costs of security, environmental shielding,

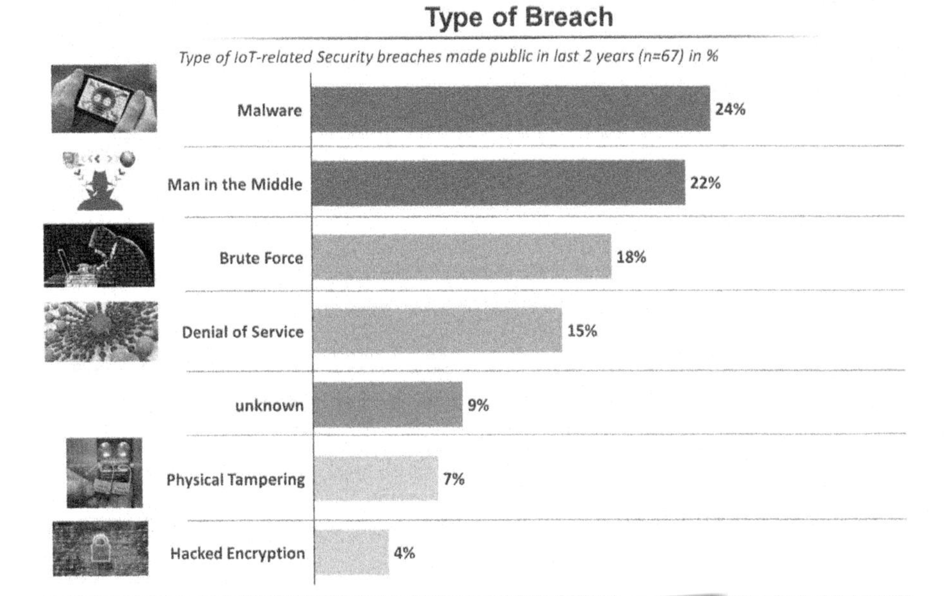

Figure 5.19 IoT security.

and connectivity options, costs may actually increase. The judgment about excess cost will always come down to what the IoT system saves/offers in real money (Figure 5.20).

> The market for actuators is forecast to reach $49.28 billion by 2022 from $38.80 billion in 2017 at a CAGR of 4.90% during (2017–2022) driven by technological advancements in actuators, increased investments in process automation across end-use verticals, increase in the number of aircraft deliveries, the establishment and growth of new smart cities, and the growth of infrastructure, such as smart buildings, smart healthcare, smart mobility, smart infrastructure, smart transportation, and smart security. (PRNewsWire 2018)

5.5.7 Ease of Integration

The ability to play in the sandbox with other components is called ease of integration. When you buy a whole IoT system from a particular vendor, they assure things go together (at the price of vendor lock in). When you do it yourself, you risk parts not integrating with each other (Figure 5.21).

5.5.8 Accuracy

If the restaurant we are seeking is 50 ft ahead and not 40, as reported by your phone, it's a minor inconvenience and maybe an unnoticeable one. If your self-driving car

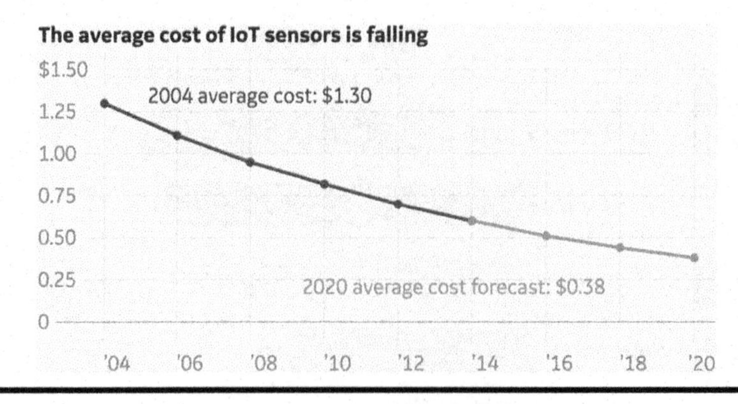

Figure 5.20 Projections about cost of sensors and actuators.

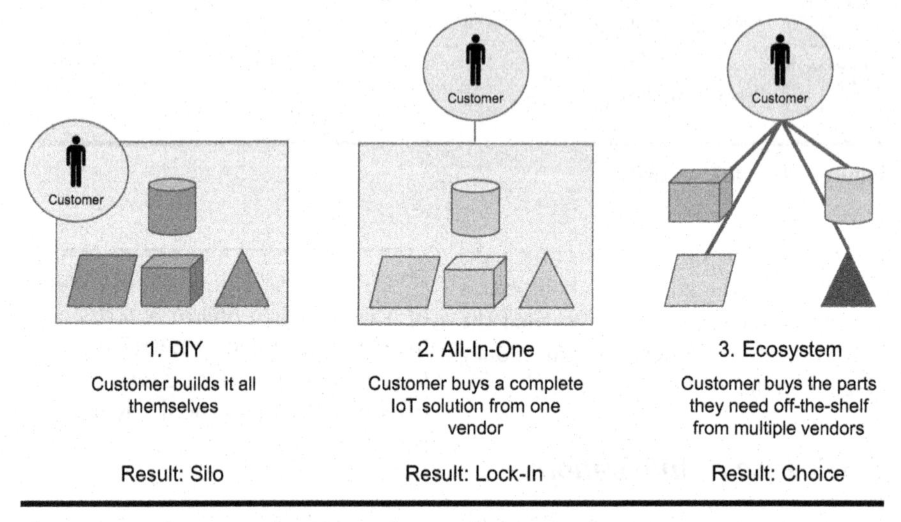

Figure 5.21 Sensor and activator integration strategies.

missed your garage by 10 ft, it's a catastrophe. These errors depend on the accuracy and calibration of the sensor, the accuracy of the underlying compute infrastructure and its algorithms, and also the precision of control of the actuator. As you would imagine, as things get more precise and accurate, they cost more and then need increased maintenance to keep them accurate.

5.6 Summary

In this chapter we looked at sensors and actuators, the beginning and end of the IoT life cycle. We studied the important criteria for sensors and actuators (range, sensitivity, nonlinearity, drift and others, and then the management and control of telemetry that sensors produce and that actuators act on.

Moving on, we looked at various sensor types in the acoustic sound vibration mode, the temperature mode, the proximity and presence sense, pressure, chemical, and finally electrical. Other sensors were introduced. We then looked at actuators and the control signals and energy sources (pneumatic, hydraulic, electric, thermal, or magnetic) that actuators use to change the world.

Finally, we looked at key criteria for sensors and actuators such as power, environmental concerns, local processing necessary, connectivity requirements, security concerns, costs, ease of integration, and accuracy.

We closed the chapter with a case study and discussion questions for you to review to tie in all together and to test your understanding.

5.7 Case Study

Autonomous cars are an intriguing piece of emerging IoT technology. Some sources suggest widespread adoption of autonomous vehicles would reduce traffic accidents by upward of 90%. While that kind of mass adoption might seem like a pipe dream, a recent Forbes article predicts that due to how rapidly the technology has been advancing, autonomous cars will be the norm in as little as 10 years.

Waymo, the self-driving car unit of Alphabet Inc., is arguably leading the pack in this emerging market. The long-term valuation of Waymo is estimated by some to be in the range of $41–$125 billion. As of the time of writing, the business plan for the Waymo autonomous vehicle is understood to be to package the enabling hardware and software and sell that technology to existing automotive manufacturers. Chrysler is currently Waymo's flagship partner in automotive manufacturing.

While much of the technical details appear to be held somewhat under wraps, what is known about Google's autonomous vehicle project, Waymo, is that it uses LiDar sensors, radar sensors, and 360° video-capturing sensors to construct a comprehensive map of its surroundings—both near and far. This data is then supplemented with a pre-generated, extremely detailed map of that particular stretch of road as well as GPS data in order to pinpoint the position of the car within centimeters. Integrating all of this data via sensor fusion allows the vehicle to perform all of the calculations necessary to smoothly drive.

LiDar sensors are one particularly interesting piece of hardware contained within Waymo's autonomous vehicles. A Waymo car actually includes three LiDar sensors optimized for near, medium, and long-range readings. The main advantages of LiDar as opposed to radar or traditional video sensors include high performance in estimating distances and the ability to construct 3D images of sensed targets. Outsourced in early Waymo models, these units are now manufactured in-house for significant cost savings.

One key service for supporting the IoT aspect of autonomous cars is the SLAM algorithm. SLAM stands for simultaneous location and mapping and describes the underlying process used in the software of a self-driving car to construct a map of the car's surroundings and to determine where the car itself is with respect to those surroundings. Neural networking is also projected to be important to the processing architecture working behind the scenes in autonomous cars.

The CAN bus is an important actuator enabling autonomous cars. Essentially, a CAN bus creates a localized network for the car's disparate components—from A/C to throttle and braking. The information and control from these components can then be centralized into one area. In the case of the autonomous car, the CAN bus also interacts with external sensors such as LiDar, enabling rapid, centralized decision-making, and action for the vehicle.

In a significant first step towards mass-market introduction, Waymo is set to deploy a fleet of self-driving taxis to Phoenix, AZ in late 2018. The vehicles will be available for public ride hailing. The company has been testing its vehicles in Phoenix since October of 2017.

5.8 Discussion Questions

1. Pick a few devices you use in your everyday life and list what sorts of sensors and actuators enable them.
   a. For example, if they picked the refrigerator, you would expect them to talk of the sensors of the door/temperature and actuators like light. For examples listed, you would expect them to list subcomponents.
2. All sensors are not created equal. Looking ahead to the IoT future is there one type of sensor (proximity, temperature, tactile) that you believe will see significantly more use than the others? Why?
   a. A good answer to this question would reference the characteristics of sensors (range, reliability) as well as perhaps cost. The student might also tie the sensor in with common devices that rely upon that sensor to support their argument.
3. As discussed in this chapter, sensors can be described as standard, smart, or manual. List a few examples of standard and/or manual sensors and how they are used in common devices. Would these devices be better served by smart sensors?
   a. Student should list three or more devices/sensors and should support their argument as to whether or not a smart sensor would help. Ideally, the student would talk about a smart sensor helping to solve a problem. They might also reference the characteristics of sensors and how smart sensors could improve them.

4. Actuators are described as relying on one of four primary energy sources: hydraulic, pneumatic, electrical, and thermal/magnetic. Which of these do you think is the most common? The most powerful? The most reliable?
   a. The student should demonstrate an understanding of each of the energy types and be able to speak on their pros and cons. Bridging in real-world examples would also be good.
5. Security is a major concern for the future of IoT. How might you go about ensuring the physical security of IoT devices? The cybersecurity?
   a. Students should demonstrate a clear understanding of the difference between physical and cybersecurity. The focus is not on how much they know about risk mitigation in cybersecurity but on what sorts of weaknesses they might be able to see in the physical and digital architecture of IoT.
6. Do you think that security concerns will slow down or even halt the popularizing of IoT, or do you think these issues will go largely overlooked?
   a. The student should be able to recognize the value proposition of IoT, as well as have a basic understanding of what sorts of risks are involved.
7. Do you think that standards will be put into place to help ensure the interoperability of IoT devices? Who/where do you think such standards would or should come from?
   a. The student should demonstrate an understanding of the principle of ease of integration and how it relates to IoT. Examples of where the standards could come from could include government, an industry consortium, or a dominant industry player.
8. Only a select number of sensors were discussed in this chapter. Can you think of any kinds of sensors that were not discussed? Can you think of any that may not even exist but would be useful?
   a. The student should be able to come up with a few sensors that were not covered. Examples could include gyroscopes, cameras/image sensors, or a creative idea.
9. Connectivity was discussed as key enabling criteria for actuators. List a few cases when wired connectivity might be better than wireless and vice versa.
   a. Students should list three or more examples and use, at least in part, the key criteria of actuators to support their argument. An advantage of wired connectivity might be reliability or throughput. An advantage of wireless connectivity might be mobility.
10. The characteristics of sensors (range, sensitivity drift, etc.) were discussed in this chapter. Consider the devices you use in your everyday life—do these devices struggle with regard to any of the listed sensor characteristics?
    a. Students should list three or more examples. Common examples might be the GPS on mobile phones not being precise enough or smart home hubs like Amazon's Alexa not having enough range.

References

APS News. 2014. *March 1880: The Curie Brothers Discover Piezoelectricity.* March. Accessed April 19, 2018. www.aps.org/publications/apsnews/201403/physicshistory.cfm.

Banica, Florinel-Gabriel. 2012. *Chemical Sensors and Biosensors: Fundamentals and Applications.* Chinchester, UK: John Wiley & Sons.

California Institute of Technology. 2018. *Brief History of Thermoelectrics.* Accessed April 19, 2018. www.thermoelectrics.caltech.edu/thermoelectrics/history.html.

Damiani, Ernesto, Jechang Jeong, Robert J. Howlett, Lakhmi C. Jain. 2009. *New Directions in Intelligent Interactive Multimedia Systems and Services—2.* Berlin, Germany: Springer.

Engineer's Edge. 2018a. *Carbon Monoxide Sensor Detector Review.* Accessed April 19, 2018. www.engineersedge.com/instrumentation/carbon_monoxide_sensor_detector_review_9776.htm.

Engineer's Edge. 2018b. *Reed Switch Review.* Accessed April 19, 2018. www.engineersedge.com/instrumentation/reed_switches.html.

Gemeli, Marcellino. 2017. *Smart Sensors Fulfilling The Promise of The IoT.* October 13. Accessed April 19, 2018. www.sensorsmag.com/components/smart-sensors-fulfilling-promise-iot.

Karimi, Kaivan. 2013. *The Role of Sensor Fusion and Remote Emotive Computing (REC) in the Internet of Things.* June. Accessed April 19, 2018. https://cache.freescale.com/files/32bit/doc/white_paper/SENFEIOTLFWP.pdf.

Lewis, Karen. 2016. *Meet Master Inventor & IoT Evangelist, Andy Stanford-Clark.* September 1. Accessed April 19, 2018. www.ibm.com/blogs/internet-of-things/andy-stanford-clark/.

Magnet Academy. 2014. *Electromagnetic Induction.* December 10. Accessed April 19, 2018. https://nationalmaglab.org/education/magnet-academy/watch-play/interactive/electromagnetic-induction.

Mordor Intelligence. 2018. *Global Smart Sensors Market.* March. Accessed April 19, 2018. www.mordorintelligence.com/industry-reports/global-smart-sensors-market-industry.

PRNewsWire. 2018. *Actuators Market Growth Steady at 4.90% CAGR to 2022.* February 13. Accessed April 19, 2018. www.prnewswire.com/news-releases/actuators-market-growth-steady-at-490-cagr-to-2022-673898083.html.

Rajaram, Srinivasa. 2017. *Global Markets and Technologies for Sensors.* March. Accessed April 19, 2018. www.bccresearch.com/market-research/instrumentation-an.

Safewise. 2018. *The Beginner's Guide to Motion Sensors.* Accessed April 19, 2018. www.safewise.com/resources/motion-sensor-guide.

Vladimer, Mike. 2015. *Sensors, Actuators and IoT.* June 4. Accessed April 19, 2018. https://medium.com/@mikevladimer/sensors-actuators-and-iot-ca3361a9fc71.

Additional Sources

Blasch, Erik P. and Susan Plano. 2003. "Level 5: user refinement to aid the fusion process." *SPIE 5099.* Orlando, FL.

Bojanova, Irena. 2015. *What Makes Up the Internet of Things?* March 31. Accessed April 19, 2018. www.computer.org/web/sensing-iot/content?g=53926943&type=article&urlTitle=what-are-the-components-of-iot-.

Francis, Sam. 2018. *How Sensors and Actuators Are Being Used to Create Self-Driven Vehicles.* January 3. Accessed June 4, 2018. https://roboticsandautomationnews.com/2018/01/03/how-sensors-and-actuators-are-being-used-to-create-self-driven-vehicles/15537/.

Greig, Jonathon. 2018. *You Can Catch a Self-Driving Taxi in 2018, If You're Traveling to Phoenix, AZ.* Accessed May 9, 2018. www.techrepublic.com/article/business-travelers-self-driving-waymo-service-will-be-available-in-phoenix-in-2018/.

Guizzo, Erico. 2011. *How Google's Self-Driving Car Works.* October 18. Accessed June 4, 2018. https://spectrum.ieee.org/automaton/robotics/artificial-intelligence/how-google-self-driving-car-works.

Haines, Osian. 2016. *An Introduction to Simultaneous Localisation and Mapping.* May 13. Accessed June 4, 2018. www.kudan.eu/kudan-news/an-introduction-to-slam/.

Korosec, Kirsten. 2017. *5 Things to Know about the Future of Google's Self-Driving Car Company: Waymo.* January 8. Accessed June 4, 2018. http://fortune.com/2017/01/08/waymo-detroit-future/.

Krause, Reinhardt. 2018. *How Google Can Race Ahead of The Pack In Self-Driving Cars.* April 16. Accessed June 4, 2018. www.investors.com/news/technology/self-driving-car-waymo/.

Lang, Nikolaus. 2017. *Making Autonomous Vehicles a Reality: Lessons from Boston and Beyobnd.* October 17. Accessed June 4, 2018. www.bcg.com/en-us/publications/2017/automotive-making-autonomous-vehicles-a-reality.aspx.

Nuñez, Ariel. 2017. *An Introduction to the CAN Bus: How to Programmatically Control a Car.* June 4. Accessed June 4, 2018. https://news.voyage.auto/an-introduction-to-the-can-bus-how-to-programmatically-control-a-car-f1b18be4f377.

Thomas Publishing Company. 2018. *What are Actuators? Learn about Actuators.* April. Accessed April 19, 2018. www.thomasnet.com/about/actuators-301168.html.

Waymo Team. 2017. *Introducing Waymo's Suite of Custom-Built, Self-Driving Hardware.* February 15. Accessed June 4, 2018. https://medium.com/waymo/introducing-waymos-suite-of-custom-built-self-driving-hardware-c47d1714563.

Wikimedia Commons Contributors. 2013. *File:EAP-example2.png.* December 27. Accessed April 19, 2018. https://commons.wikimedia.org/w/index.php?title=File:EAP-example2.png&oldid=112677418.

Chapter 6

IoT Architecture and Compatibility with Current Infrastructure

Frank Groom
Ball State University

Contents

6.1 Introduction

With the advent of the fifth generation of cellular networking expected in 2020 and the rapid introduction of Internet connected devices, the local access, metropolitan, and national backbone networks are being reengineered to meet the expected connectivity demand from such devices. Network functions are becoming virtualized with the Control Layer which manages the devices being separated from the Transport Layer which performs the switching and routing. The Wide Area Backbone is operating as a Software Defined Service (SDS). And the complete network is being logically sliced with vertical slices of services at the edge and horizontal slices of isolated services across the country. Ultimately, the customer services over the completed network will become customer configurable in the fashion of a cloud computing-like network.

6.2 Some Classifications of IoT Devices

Millions of devices are being created with communication capabilities and many can be imbedded within industrial, healthcare, home, and office instrumentation, such elements have been dubbed "the Internet of Things" or IoT. There are many ways of classifying IoT devices. Among these classifications are:

1. According to what duties the devices perform
2. According to whether they are Constrained or Gateway devices
3. The Internet Engineering Task Force (IETF) has identified a Class structure composed of: Class 0, Class 1, and Class 2.

However, another approach addresses how these devices communicate and in particular how they interface with the local access network facilities and downstream with the metropolitan and national backbone networks. To handle the packets of information transmitted and received by these IoT devices, the access, metro, and national networks are undergoing serious modification both to endure the traffic load itself and deal with the unique and diverse characteristics of these devices. This evolving traffic load will soon be upon us as the billions of IoT devices forecast to be connected in the early 2020s become active.

The major changes in networks to satisfy the transmission demands of IoT devices are resulting in the rapid deployment and upgrade of the Access networks to the fifth generation or 5G, particularly 5G Cellular Access networks. The established 4G Long-Term Evolution (LTE) Cellular network with its Evolved Packet Core (EPC) network and set of Gateways is being seriously modified with new radio, new and closely spaced antennas, and high bandwidth backhaul to the edge of the metro Internet and national backbone transport.

A gross level of network affecting IoT devices is those which are:

1. **Mobile Broadband** (smartphones and UHD devices (Ultra High Definition)—IoT devices demanding high network capacity and precise video streaming and video caching capability.
2. **Mission-critical** IoT devices—Motion control, autonomous driving, automated factory control, and control of smart electric grids with low latency and high reliability requirements.
3. **Massive number** (IoT devices)—Sensors and sensor networks for electric, gas, and water metering; agriculture component sensors; buildings sensors and cameras; logistics location; and availability, individual home devices, city facilities, street and highway cameras, and recording devices.

   Further classification of these IoT devices can be by:
4. **High Bit Rate devices** (smartphones, tablets, and laptops)—These devices can further be classified as Fast Mobility devices moving over significant distances by walking, auto, train, and airplane.
5. **Sensors**—Industrial sensors, medical sensing, and relaying devices need to provide communication of high reliability and low latency while consuming little energy as they transmit only a few bits per episode.

In summary, we have devices which are mobile, have high bit rate transmission, high usage, and others which use infrequent connection but require high reliability and low latency when they do communicate. Figure 6.1 supplies the usage categories, data rates, latency, and mobility that the diverse IoT elements will demand from the network (NGMN, 2017).

Categories of Use	User Data Rates Up and Down	Latency	Mobility
Broadband Access Dense Area	Up 50 Mbps Down 300 Mbps	10 ms	On-demand 0 - 100 km
Indoor Ultra high BB	Ul 500 Mbps DL 1 Gbps	10 ms	Walking
Broadband Crowd	UL 50 Ms DL 25 Mbps	10 ms	Walking
Mobile Car Train	UL 25 Mbps DL 50 Mbps	10 ms	0 - 120 km
Massive low cost Long range	1-100 kbps	seconds – hours	0 - 500 km
Ultra- low latency	UL 25 Mbps DL 50 Mbps	< 1 ms	Walking
Surge Traffic /resilient	UL 1 Mbps DL 1 Mbps	Not Critical	0 - 120 km
Ultra-high reliability Ultra Low Latency	UL 10 Mbps DL 10 Mbps	1 ms	0 - 500 km
Ultra- High Available Ultra-high Reliable	UL 10 Mbps DL 10 Mbps	10 ms	0 - 500 km
Broadcast Svc Quality	UL 100 kbps DL 200 Mbps	< 100 ms	0 - 500 km

Figure 6.1 Usage categories, data rates, latency, and mobility that IoT elements require.

Beyond requiring a new radio connection and utilizing increased bandwidth, the major change to the physical access, metropolitan, and national backbone networks will be a requirement to allow customers and devices to demand an elastic service from the transport network. They will require the network to appear as an elastic facility, one that can expand and contract to meet each user's dynamically changing set of requirements. The desire of the national carriers is to recreate the national network as a "Cloud Network" along the guidelines of the Cloud Data Centers created by Amazon, Google, Microsoft, IBM, and Oracle. To achieve this, the network vendors intend to provide virtual logical "slices" of the network facilities which will isolate traffic according to the demands required by each particular IoT device set and user need and will guarantee a specific quality of service (QoS) required by the device and user (GSMA, 2017).

6.3 Access to the Network, Particular Network Access Technologies, and IoT Devices

We have become a mobile world. It is estimated by AT&T that 70% of all dataflows by cellular phones and tablets will be on the move.

Other devices, although moved from place to place, but not while in use, such as laptops, employ access by short-distance Wi-Fi wireless connection to access points and then on by wire to routers. The router then connects by private line to the Central Office access point of the local and then Metro Network. From there connection is made across the Metro network to other sites, to the public Internet, or across the metro backbone to the Wide Area Backbone Network to other business or residence locations across the country.

Still other devices are fixed in location but wirelessly connect over very short range to a controller and then by wire line to computers or by a router to the wide area communication network to ultimately connect to remote controllers. As a result, we shall first look at what is being done to acclimate the cellular 5G LTE (Long Term Evolution) network for the IoT devices, then the business connections to handle the increased volume and transfer speeds required by growth of business augmented by supporting IoT devices.

6.3.1 The Cellular Network

The ubiquitous deployment of smartphones was supported by the upgrade of the cellular network. A new radio access was introduced with the introduction of the fourth-generation 4G LTE (Long Term Evolution) process. The antennas are placed closer to each other and each equipped with its own computer to control the interface with the cellular phones. This radio connection network was termed the Evolved UMTS (Universal Mobile Telecommunications Service) Terrestrial

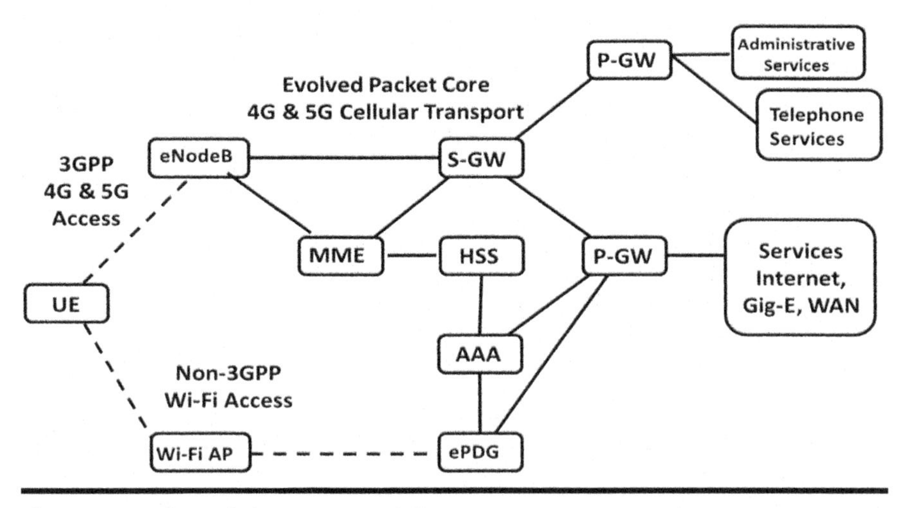

Figure 6.2 The cellular access and downstream EPC network.

Radio Access Network or E-UTRAN which handles all the communication from the cell phone to and through the computerized tower antenna, now called an eNodeB. Figure 6.2 presents the radio connection between the user smartphone (UE) and the new radio antenna eNodeB, as well as a short distance Wi-Fi wireless connection. These packet transmissions are then transmitted over backhaul transmission links, increasingly fiber, to the Central Office housed authentication and transmission control database systems—Mobility Management Entity (MME), Home Subscriber Server (HSS), and the associated Gateways to the Metropolitan Backbone—the Evolved Packet Core (EPC) network.

The modern eNodeB antennas are provisioned with a computer service (at their base or on the antenna) which allows them to completely enable and control traffic to and from user devices. The Evolved Packet Core (EPC) network carries user information, authentication, and location information as needed to the eNodeB (Base Station) antenna and provides a high-speed backhaul connection for user packet traffic to and from the Central Office and on to the Metropolitan Backbone and Wide Area Networks.

Among these servers and gateways are:

The Mobility Management Entity (MME)—Supports user equipment context, identity, authentication; manages session states; and authenticates and tracks a user as the move locality and as a result across the network.

The Home Subscriber Server (HSS)—contains a database of user-related and subscriber-related information.

TheServing Gateway (SGW)—Receives packets from the eNodeB antenna and sends them on to the core network routing data packets through the appropriate access network. This may be off to the traditional telephone switched network or to the various packet routing facilities.

The Packet Data Network Gateway (PGW)—Connects the EPC with external packet carrying networks for transmitting data downstream to the public Internet or private data networks.

The current fourth-generation cellular network in place over the past decade has evolved to become a principally packet-oriented network while simultaneously continuing to support transmission from the older 3G (third generation) cellphone devices from the beginning of the 21st century.

However, this 4G network is inadequate to provide the connectivity demanded by 5G cellular devices with their data end streaming entertainment transmission. To support these demands, the 5G cellular network is expected to deliver 20 Mbps bandwidth and support the ubiquitous set of services customers have become accustomed to (RAS, 2017).

The Radio Access Network (RAN) needs to be upgraded and possibly replaced with a new RAN facility employing the application of Carrier Aggregation, massive multiple input and multiple output (MIMO) parallel transmission, quadrature amplitude modulation (QAM) of the wireless signal, and orthogonal frequency division multiplexing (OFDM) of the transmissions (Ijaz et al, 2016).

Furthermore, other Wi-Fi-based transmissions providing connection for some of the less demanding IoT devices will also be supported. The connection spectrum was augmented by using a portion of the unlicensed spectrum. The 5 GHZ spectrum is now being added to the previous 3.4 GHz spectrum for handling both LTE traffic in the unlicensed spectrum (U-LTE) and the new Wi-Fi traffic.

6.3.1.1 New Phased Array Antennas

To handle the new radio transmission technology and expanded spectrum usage from Wi-Fi traffic, the existing antennas 4G LTE eNodeBs (generically called Base Station antennas) are being upgraded to transport in four carrier wireless bandwidths—first in the 5 MHz spectrum transmitting at 17 Mbps, then in the 10 MHz spectrum transmitting at 37.6 Mbps, then in the 156 MHz spectrum at 56 Mbps, and finally at the 20 MHz spectrum transmitting at 75 Mbps. Much of this employs special multiplexing 4×4 antennas operating at up to 300 Mbps.

6.3.1.2 Expanded Frequencies

In addition of the currently used access frequencies below 6 GHz, an expanded set of frequencies termed Millimeter Wave or MmWave ranging from 28 to 100 GHz are being utilized. Some of these lie in the unlicensed spectrum. The range of existing frequencies below 6 GHz and those newly being deployed above 28 GHz range all the way to 100 GHz and even beyond. These frequencies are presented in

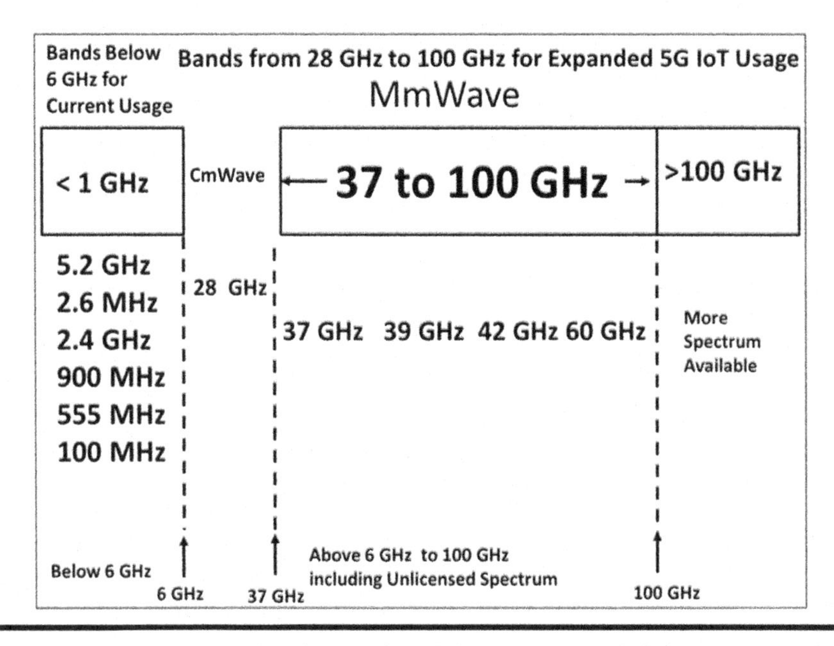

Figure 6.3 Frequency bands employed in current and future IoT wireless transmission.

Figure 6.3. (Millimeter wave spectrum is the band of spectrum between 30 and 300 GHz which is located between microwave and infrared waves and will be used for high-speed wireless communications with the recently released 802.11ad Wi-Fi standard which operates at 60 GHz (Rappaport et al, 2013)

With the new Radio Access Technology (RAT), the addition of the unlicensed spectrum for carrying lower demand traffic, and the short-range Wi-Fi connections, a new generation of phased array antennas is being installed. These new antennas have advanced digital signal processors with intergraded circuit units installed to handle the new high-frequency wavelengths used by small special service devices. The new antennas will have up to 128 receiving and transmission elements compared to the common set of four to six elements installed for today's 4G LTE service. A number of small-size antenna elements are deployed which will compensate for signal loss and distortion.

Along highways and in rural areas, a significant increase in tower placement is anticipated to handle cellular traffic increase and rural business requirements for such businesses as farms with expanded IoT sensors and other devices as well as significant information transmission for operating the business.

Within cities it is anticipated that we will need radio tower devices which are closely packed, placed every 200 m, one per block in the city or industrial areas, usually on the roof of buildings. In the rural areas—towns, highways, and countryside, fewer will be required due to fewer devices, but with the limited transmission range of many new IoT devices, they may need connection from device to smartphones to piggyback on their smartphone transmission capability.

For each user, a deliverable transmission speed will be increased to up to 20 Gbps through antennas, across the EPC network, across the Gateways, through the Central Offices and on through the Metro and national backbone networks (Donovan and Prabhu, 2017)

6.3.2 Backhaul Connection Networks

To connect the multitude of new smart IoT devices through enhanced phased antennas with hundreds of canisters receiving units and connect them to a distant Central Office which is 10–39 miles distant, a new backhaul network must be deployed employing a new architecture. Currently, T-1 copper circuits are deployed to connect a tower to the Central Office, and in special locations, fiber has been deployed. This must all be replaced with fiber employing minimal optical transport technology (termed General Passive Optical Network technology or GPON over fiber), which will use less expensive multiplexing and less expensive lasers or LEDs. These GPON Service (GPONS) optical connection links currently are limited to 12 miles for reliable transmission but will need to be extended.

Another architecture will be employed in flat rural Midwestern areas. Instead of backhauling traffic by copper or fiber, the signals may be retransmitted wirelessly from the receiving antenna to a centralized master antenna. This antenna which has active optical multiplexing components will then forward all received traffic over fiber to the closest Central Office which serves as the connection point and information center for all the numerous antennas which service a particular area.

In rural areas, the carrier is utilizing this new transmission capability to forward all received traffic. This includes both the high-speed cellular traffic plus new low-bandwidth traffic, which will then be all transmitted wirelessly to the centrally located relay antenna placed appropriately among the group of distributed peripheral antennas.

This antenna will serve as a relay station which utilizes a high-speed backhaul link employing the more expensive Active Optical Network capability of Dense Wavelength Division Multiplexing components. This relay antenna will transmit parallel traffic flows over the separate wavelengths, or "colors," that make up the white light beam (currently up to 90 in the wide area and usually 30 in the metro area). Figure 6.4 presents how such Relay Radio Towers are deployed.

From the Relay Radio Towers, even newer fiber backhaul links from the advanced 5G eNodeB Antennas to the Intelligent Servers in the renovated Central Offices (15–30 miles away) are being deployed. These are advanced Gigabit Passive Optical Network (GPON) fiber links (up to 12 miles) and augmented by newer and longer distance 10G-Passive Optical Network (XGPON) fiber links are being deployed for longer distance transmission to the Central Offices.

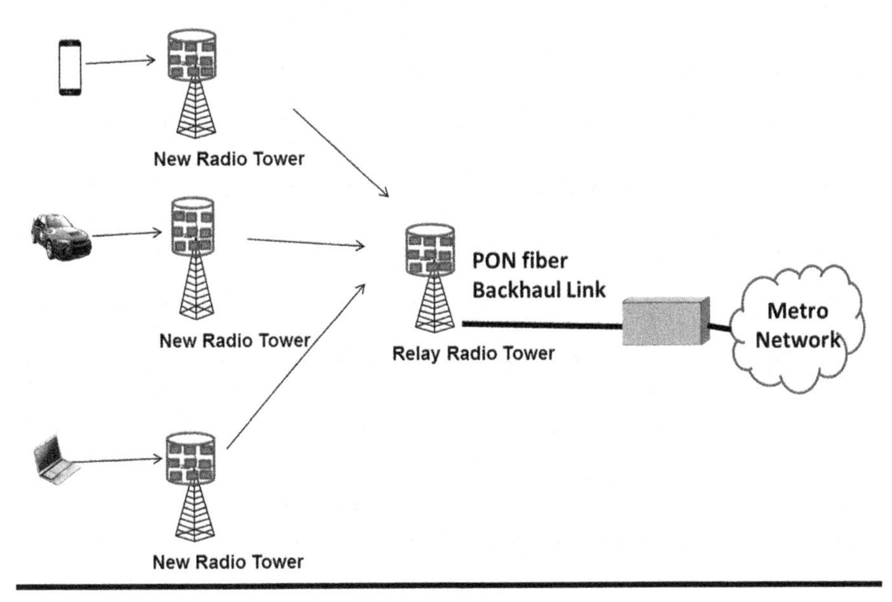

Figure 6.4 Centralized relaying radio tower and passive optical backhaul network.

However, Relay Radio Towers are only one component employed in the suburban and rural areas as well as along major highways. Distance becomes a major issue. One approach to minimize the number of required fiber trenches and long-distance fiber back to the serving Central Office is to create a short passive fiber links to the closest Radio Relay antenna. They can use individual short-haul GPONS fiber links up to 12 miles or short-haul radio transmission to carry traffic to the closest Relay Tower (Donovan and Prabhu, 2017).

6.3.2.1 Common Backhaul

A Central Backhaul Fiber is provisioned with more expensive Active Optical Elements to transmit multiple channels over each "color" of the laser beams. Optical Add Drop Multiplexors directly connect the traffic from each Relay Tower onto (and off) an optical channel of the Central Optical Backbone fiber. Every 12 miles, another Relay Tower can add optical traffic on to this common fiber link back to the Central Office.

Furthermore, each fiber has up to 16 or 30 lasers focused on each of 16 or 30 isolated "colors" or wavelength channels of the beam of white light which travels down the fiber. Each wavelength channel carries isolated traffic at the transmission speed of the fiber. Bundles of such fiber can be placed within the pipe placed in trenches extending back to the controlling Central Office. The deployment of a Common Backhaul Fiber from distributed rural antennas to their Controlling

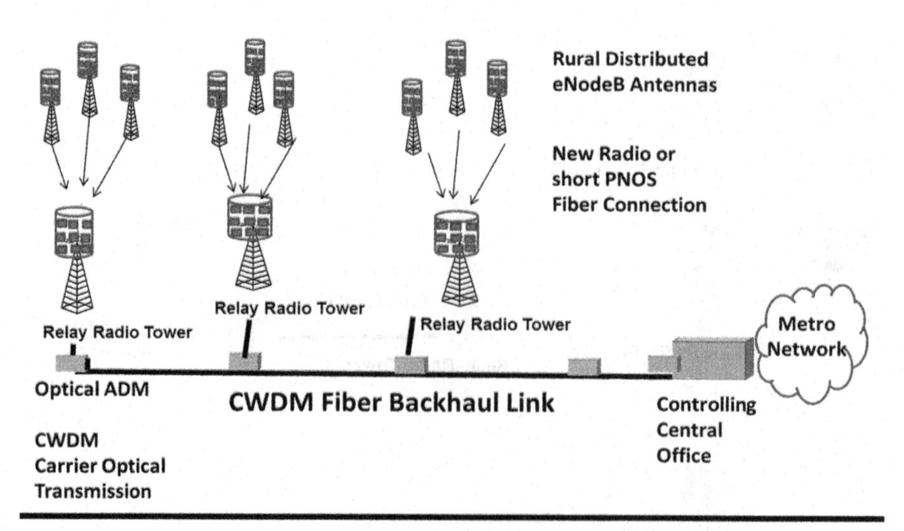

Figure 6.5 Crude WDM optical fiber backhaul as rural antenna backhaul relay link.

Central Office is portrayed in Figure 6.5. Such a Common Backhaul fiber can be placed along highways connecting a set of distributed antennas back the closest Central Office.

6.3.3 Central Office Data Centers

The Local Telephone Central Offices are being upgraded to operate as local "Edge" Data Centers containing the computer servers to support the 5G EPC, Cellular, and Wi-Fi networks with the appropriate MMS, HSS, and Gateway Servers and devices. Furthermore, a number of content-specific services can be stored and delivered from that location including information, entertainment, and emergency services. There is the additional possibility that local businesses may desire to provision local data processing services in these local networks connected centers within close range to the business and many of its users.

6.4 Virtualizing the Carrier Networks

Services offered over each carrier's optically switched Wide Area Backbone is created as a software-defined and software-controlled service. Traditional MPLS (Multi-Protocol Label Switching) provides for the routing of packet traffic, while the underlying optically switched core of the network carries traffic as a set of individually switchable wavelengths. In order to provision the connectivity for the vast number of newly conceived IoT devices and the elastic demand other users will require from a "Cloud-like" elastic network service, centralized

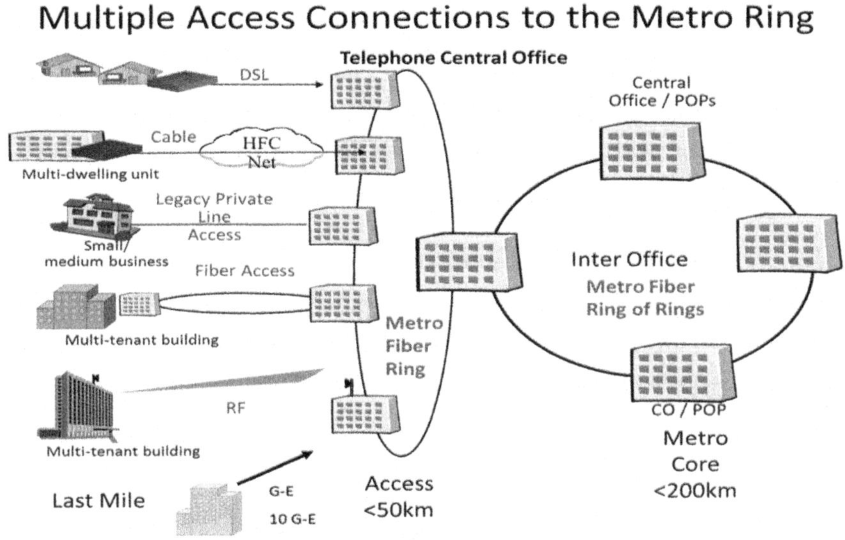

Figure 6.6 A variety of connection alternatives for businesses and residences.

software control of the complete network is being created with software tools that allow the connection facilities to be logically divided into horizontal slices of isolated services which span regions and the entire country. Simultaneously, the edge access facilities are expanded as a set of vertical slices of services from the existing variety of offered services. The intent is to allow customers to demand from the national network the same nature of "elastic services" as the Cloud Data Center providers. It is anticipated that customers will eventually be able to request services across the network when and at what speed and capacity they require and contract or cancel that service when no longer needed (Donovan and Prabhu, 2017) (Figure 6.6).

6.4.1 The Edge Central Office Becomes an Edge Data Center

Traditionally, all traffic destined for local connection, across metropolitan backbone facilities, or through distant connection over the Wide Area Backbone must first connect by means of the local Central Office. The telephone Central Office serves as the unique access point to distant connection. Generally, Central Offices are place every 30–50 miles. All traffic whether telephone, DSL, cable, private line, or cellular/radio must first connect to the closest Central Office. These Central Offices have expanded their traditional function, housing all the serving components of the cellular EPC, the DSL Modem's bank and routers, and all private line connection facilities. Now those local Central Offices are being converted into

Content Storage and Delivery Centers as well as providing the traditional connection functionality. With the emerging set of IoT devices, many such devices will require local processing services. The distributed Central Office through which all traffic must flow is an ideal place to store and provide such functionality. Otherwise that Central Office will connect the device to a public or private Data Center which is reachable across the local Metropolitan Network.

6.4.2 The Metropolitan Backbone

The Metropolitan Backbone since the 1990s has been a set of Sonet-based rings forming an Optical Backbone. These optical rings have more recently been augmented by a parallel switched Gigabit Ethernet Mesh Network carrying high-speed 1, 10, and soon 40 Gbps business data traffic (Groom, 2001a). Work is being done on standardizing an even higher speed 100 Gbps Ethernet specification.

6.4.3 The National and Regional Wide Area Backbone Network

To satisfy the additional distant transport required by IoT devices, the Wide Area Network will need to be significantly expanded. Currently, the transport is provided by an all optically switched network with all traffic being optically connected to the network and optically dropped off from the network. The fiber links between optical switches to and from cities are deployed in pipes placed underground. Multiple-shielded fibers are bundled together, and multiple bundles of these fibers fill up the containing pipes.

Each individual fiber has a beam of "white" light from a laser focused down the shielded fiber. Every 80–100 miles that light beam must be amplified by specialized Raman Amplifiers. From New York to Chicago this requires between eight and ten such amplifiers for a direct link. Fewer amplifiers may be required for links that are sequenced from city to city to span from New York to Chicago.

Following Isaac Newton's discovery that light is composed of many wavelengths or "colors," a separate laser is focused on each wavelength of the light beam. Currently 88–90 wavelengths are isolated in the light beamed down each fiber in the national network while up to 30 wavelengths are utilized within Metropolitan networks. Each wavelength transports a separate 100 Mbps payload. In order to handle the traffic load of a massive number of IoT devices and the 20 Mbps traffic of cellular devices, special modification are being made to increase the transmission speed on each "color" (wavelength) on each fiber from 100 to 400 Gbps.

In the future, the number of "colors" of wavelengths contained in a beam of white light transmitted down each fiber is expected to be increased from 90 traffic carrying wavelengths on each fiber to up to 180 wavelengths as other bands of the spectrum are employed.

For the Metropolitan areas where IoT devices will gain access to the network through local Central Offices, these access points are spaced about every 30 miles. Each Central Office connects traffic from telephone subscribers, cable subscribers, private line business subscribers, or connected cellular and Wi-Fi users and connects them to others by means of the interconnecting-backbone of the city or metropolitan area. Due to the density of these facilities and connection points, hundreds and sometime thousands of optical components are required, from add drop multiplexors, laser devices, optical receivers, and all-optical switches which are deployed. Due to the vast number of such optical devices required to isolate the 90+wavelengths per fiber in the national network, a cheaper, smaller number of optical components are deployed. This optical technology employed in the Metropolitan backbone is termed Crude Wavelength Division Multiplexing using fewer (16–30) wavelengths and require fewer lasers and wavelength receivers. Many of these Crude Wavelength Division Multiplexing (WDM) devices will be upgraded to handle the IoT traffic load, but the cost to carry that traffic will be dramatically higher due to the sizeable number of optical devices required to serve a metropolitan area with its sizable number of connection requirements to the various Central Offices in that city.

The complete end-to-end network transporting traffic from each fixed or mobile device across the country is portrayed in Figure 6.7. Traffic flows across the Access Network to the Central Office, across the Metropolitan Backbone in each city, across the country over Carrier Network such as AT&T, Verizon, or Level 3's national backbone network, to the appropriate destination here identified as a Cloud Data Center located in a destination city (Groom, 2001a).

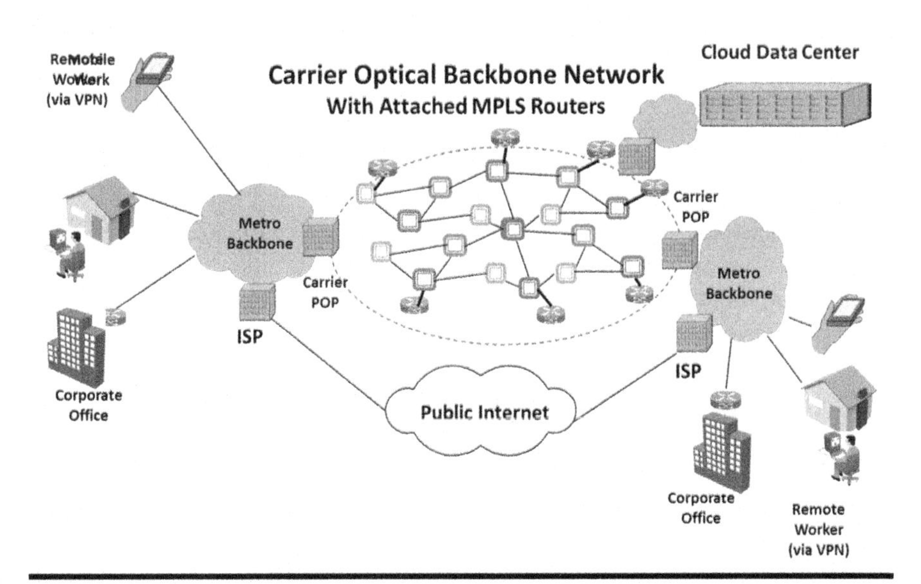

Figure 6.7 The complete end-to-end national network including the public internet.

6.5 MPLS and the Creation of Network Slices

All traffic is carried by a set of fibers packaged in bundles, with a set of bundles buried underground in a pipe. Many of the fibers remain "dark" and reserved for future need. Each fiber has a set of lasers, currently 90, each focused on one of a set of wavelengths or "colors" of a single beam of white light. Guaranteed transmission currently is 100 Gbps on the total network although transmission is at the speed of light across these connected sets wavelength channels. This guaranteed national transmission is in the process of being upgraded to 400 Gbps to handle the needs of the 5G and IoT traffic. The number of carrying wavelengths for each fiber can currently be upgraded to 180 channels and potentially even more as more bands of wavelengths are employed (in the 1,400 and 1,600 nm bands). Traffic is placed on and dropped off of this optically switched network by employing reconfigurable Optical Add Drop Multiplexers which avoid the slowing of traffic due to the reduced speed of employing electronic Add Drop Multiplexers and Electronic switches (Li and Yates, 2005).

Furthermore, AT&T and Verizon have deployed a set of specialized MPLS router/switches at various offices across the country. These MPLS Routers allow precontracted and guaranteed QoS private line services to be established across the national fiber network. These can be semi-permanent channels or dynamically established on-need channels.

MPLS service can carry any of the possible Layer 2 (Point-to-Point, Ethernet, Frame Relay, and ATM) and Layer 3 IP traffic across the optically switched network (Griffin, 2002). As previously mentioned, customer traffic transport across the network is currently guaranteed at 100 Gbps with upgrade to 400 Gbps planned in the near future to handle the increased 5G traffic loads.

MPLS can provide the initial ability to create isolated traffic "slices" across the country for different traffic classes where each require their unique transmission service and transport speed and latency. More sophisticated traffic isolation into virtual slices will be enabled by "virtualizing" the functions, specifications, and policies for each device employed in the Metropolitan and national Wide Area Backbone networks. The individual network services and slices are intended to be created as software-defined network services and centrally allocated and managed (Donovan and Prabhu, 2017).

6.6 Virtualizing the Local, Metro, and National Network

The basic idea of virtualization is to abstract the fundamental specifications of each component in the network and the items used to control the execution of the functions performed by each network element. These specifications are then placed in a central controlling computer center. However, the original specifications still remain in each network element since they are required for that element's performance of basic network functions. When changes to the performance of the function are required, the central site downloads those changes to the appropriate component. Since the overall control has been removed to the control center and only execution remains under the specifications that are at the moment downloaded, the network is operating under the virtual control of that control center. This process is termed "Network Function Virtualization" (NFV).

When new services are offered, such as a new Wide Area Network Transport Service, the software controlling the network elements that provide the new service can be placed in the central Control Center and downloaded as appropriate to the switching and routing components that provide that service. This is termed "Software-Defined Networking' (SAN) with the "Software-Defined Wide Area Network" service (SD-WAN) as a common example as well as SD-MPLS as another.

To minimize the sizable investment in converting the network to Software-Defined Virtualized Network and allow it to be operated from a Central Control Center, standard "off-the-shelf" switches, routers, optical apparatus, digital cross connects, tunable lasers, Dense and Crude Wavelength Division Multiplexers, Optical Add Drop Multiplexers, and MPLS switches and routers are being deployed to seriously reduce the cost. Furthermore, standard "Open-Source" software to operate and control these devices is being deployed. And the complete set of specifications, controlling policies, and open-source software are being placed under the control of an overall "Orchestrator" software system. The control technicians will use the "Orchestrator" system to observe, manage, and execute from the Control Center through the physical network elements spread across the country. The "Orchestrator" must also have access to the traditional and still functioning Operations Support Systems (OSSs) which maintain all the detained records of every network element, component, file, order, inventory record, and person involved in the network. Figure 6.8 depicts such a central control site "over-looking" the physical network (Ohlen et al, 2017)

The complete Controller and Orchestration function is portrayed as existing on a Control Plane, separate from the physically functioning Transport Layer. The actual functioning switches, routers, optical appliances such as Add Drop Multiplexers, DWDM Multiplexers, and lasers all existing on a physical network transport plane interconnected by fiber. Figure 6.9 isolates the depiction of these two layers—the Control Layer with the abstracted device and connectivity information as well as Orchestrator software and the Transport Layer of switches and routers connected by fiber links (Donovan and Prabhu, 2017)

With the advent of the 5G network and the expected onslaught of billions of IoT devices demanding connectivity through this national network, the carriers are in the process of using the capabilities of centralized network control, function virtualization, off-the-shelf inexpensive hardware and Open-Source software to carve out Logical Horizontal Slices of network functionality. They will then make the appropriate slices available "on demand" to certain

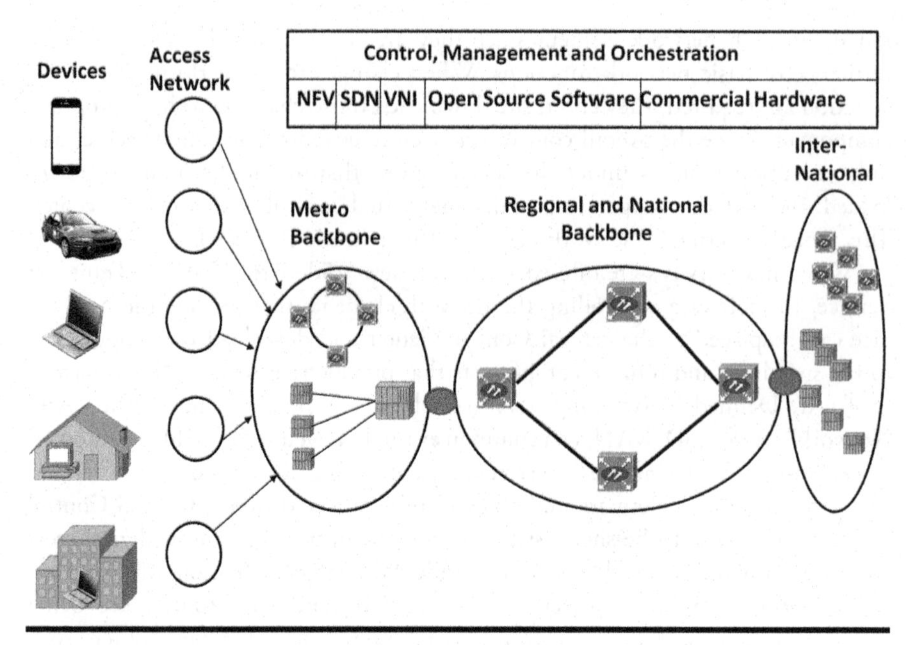

Figure 6.8 The control software managing a set of virtualized network components.

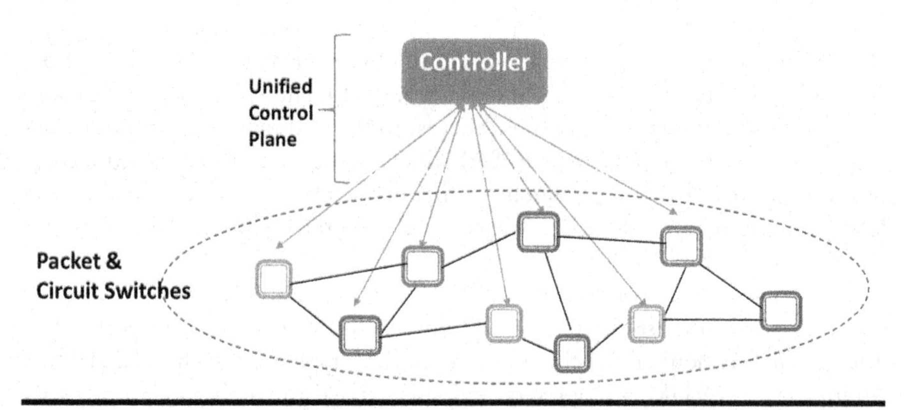

Figure 6.9 A central control plane and functioning elements of a physical transport plane.

classes of IoT devices for their transport services. The key word is "on demand." When not demanded, those components are available to support other transmission requirements for all customers.

At the Access point, all the various access services are considered "vertically slides" to support the separate vertical placement of all such devices in association with the appropriate connection whether it be cellular, public Wi-Fi, within building Wi-Fi, GPONS, T-1 or DS-3 trunks, or other access technology.

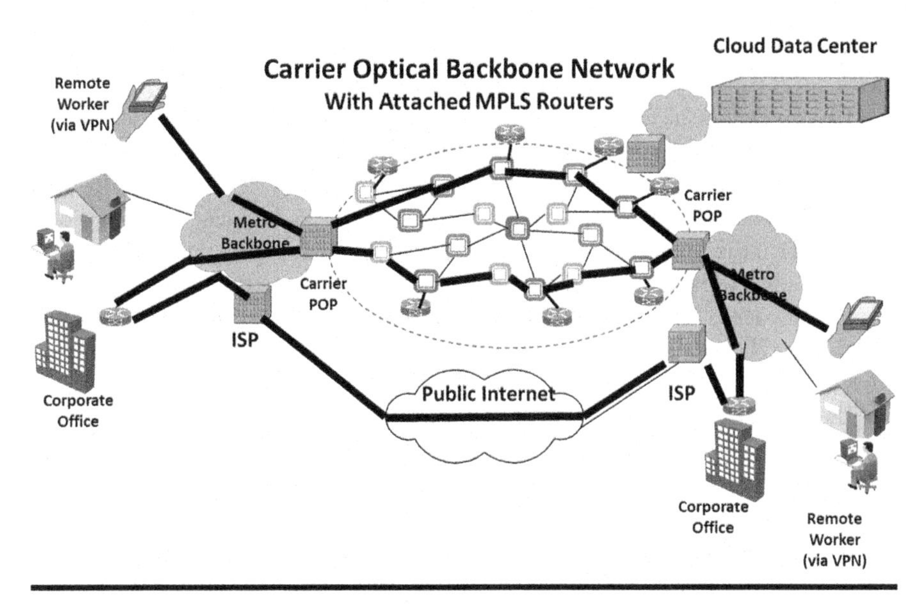

Figure 6.10 Slicing the national network into one-to-end transport services as demanded by select IoT devices.

Figure 6.10 displays a conceptual view of separate network slices across the local, metro, and national networks appropriate to distinct IoT classifications of devices and their transmission requirements.

Many devices will also transport their information across the public Internet, which is not expected to have the services of network slicing available but will considered to be another link across the county available to the Access and Metro slices at cities at both ends of a national connection.

6.7 Control of Network Slicing across the National and Local Networks

To control this network and provide the elastic capacity desired by users appropriate for each type of IoT device instance, the specifications, functions, and policies associated with each device are abstracted and centralized in a Control Center. These are provided for each device in the network whether electronic switches, routers, hubs, add-drop-multiplexers, optical switches, digital cross connects, Sonet components, or dense wavelength division multiplexing and laser components. In the Control Center, a set of software will manage all the components in the Metro and National Backbone Network. Meanwhile, the set of individual physical components will continue to perform the basic transfer functions of routing and switching while the centralized Control Layer decides the basic rules which these functions will follow in performing these standard functions. Furthermore,

Software-Defined Metro and Wide Area services can be "virtualized" and carved out as paths through this centrally managed and controlled network.

This collection of all the policies, rules, specifications, and direction of the offered services is centralized under the direction of a General "Orchestrator" software package operated by network technicians.

Traditionally, the collection of network hardware has been engineered specifically for the individual requirements of AT&T and Verizon's proprietary Metro and Wide Area Network deploying components supplied by such companies as Siemens, Cisco, and Ciena. Work is underway to replace these function-specific components with off-the shelf switches and routers which can be centrally controlled and operated with Open Source, off-the shelf software (Figure 6.11).

The architecture of the overall Control Center Software is displayed in Figure 6.12. This architecture follows the basic layered model provided by the IETF. Various network application software routines and algorithms for network management are clustered in the top layer. The moment-to-moment interactions between the Orchestrator software and the events and requirements of the network components are placed in the middle Control Plane. And the connection to the executing individual element "instances" exist in the physical network which perform the transmission of data. Attached to this model are the connection to the traditional Operations and Support Systems of the Telephone Company and the Policy and Control specifications used at the moment by each individual network device (Donovan and Prabhu, 2017) and (5G PPP Architecture Working Group, 2016).

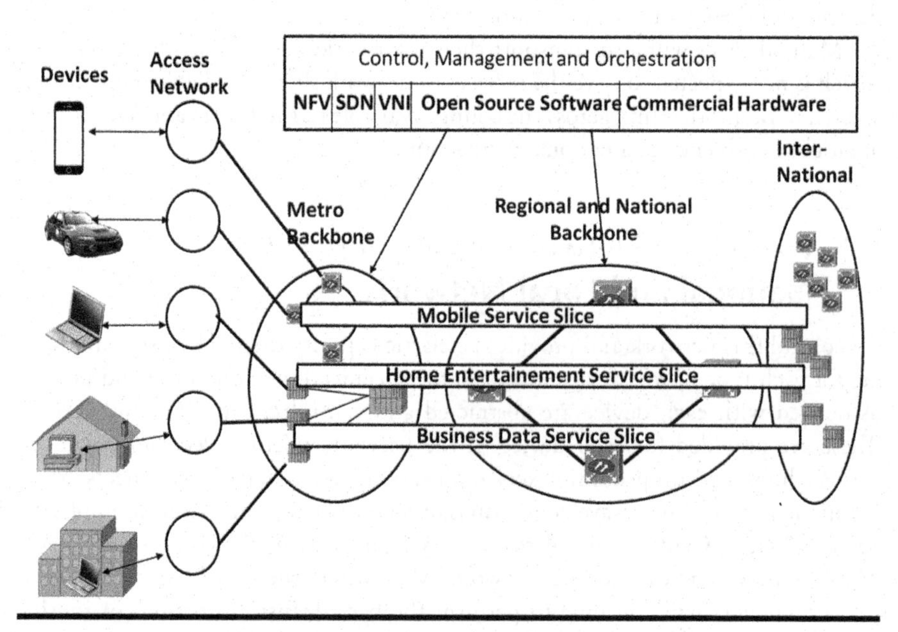

Figure 6.11 Mobile service slices, home entertainment, and business data service slices across the network.

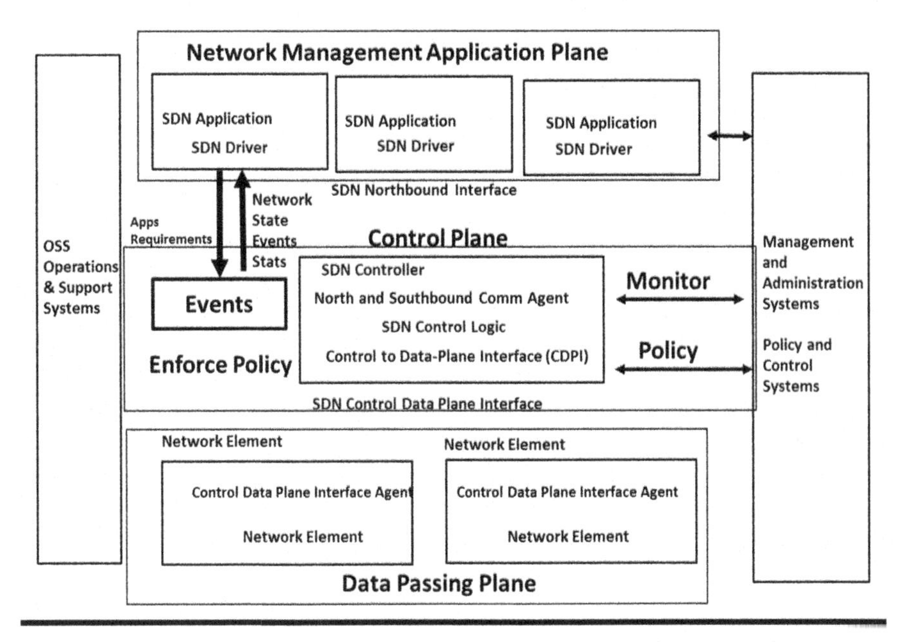

Figure 6.12 The layered architecture of the overall control center software.

There is a vast array of OSSs which provide information concerning all aspects of the telephone network. This information is required by the Orchestrator in performing its overall management and control of the Carrier network. Figure 6.13 depicts the broad range of such systems required to operate a national network (Groom, 2001b).

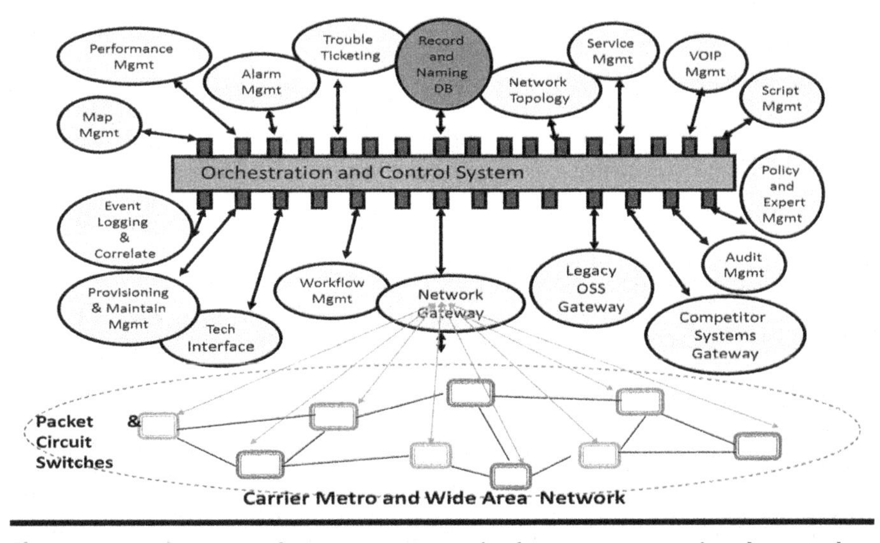

Figure 6.13 The array of OSS systems required to operate a national network.

6.8 Summary—The Anticipated Cloud Network of the Future

Similar to the process where customers can register as users of a Cloud-Computing Vendor's services, specify what services they require and dynamically change those services on demand through an online portal as their needs change, the network of the future is envisioned as a Cloud Network. The carrier will be able to elastically expand and contract the facilities available to each customer based upon a process of virtualizing all of the component specifications and capabilities and defining of all services as "Software Defined." Services will be made available and controlled by a central facility which offers them, activates them, and manages them. After placing a sizable investment in components and software to support the expected vast numbers of IoT devices, the carriers anticipate billions of connected IoT devices will be demanding services in the near future. The carriers envision a profitable future as a consequence of the introduction of 5G and the resulting and the massive number of anticipated IoT devices requiring network connectivity.

Bibliography

5G PPP Architecture Working Group, View on 5G Architecture, 5G-Architecture-Contributions@5g-ppp.eu, https://5g-ppp.eu/white-papers, July 1, 2016.

Black, Uyless, *MPLS and Label Switching Networks*, Upper Saddle River, NJ: Prentice Hall, 2001.

Donovan, John and Krish Prabhu, *Building the Network of the Future*, Boca Raton, FL: CRC Press, 2017.

Griffin, Timothy, An Introduction to MPLS, griffin@research.att.com, http://www.research.att.com/~griffin, November 21, 2002.

Groom, Kevin, AT&T, Carrier Metropolitan Connection to Customers, Optical networking in the MAN Conference, June 14, 2001a.

Groom, Kevin, AT&T, Network Management and OSSs for High Bandwidth Metro and Wide Area Services, And OSSs for High Bandwidth Metro Services, *Optical Ethernet Conference*, San Francisco, CA, August 23, 2001b.

GSMA, An Introduction to Network Slicing, https://www.gsma.com/futurenetworks/wp-content/uploads/2017/11/GSMA-An-Introduction-to-Network-Slicing.pdf, 2017

Ijaz, Ayesha et al, Enabling Massive IoT in 5G and Beyond Systems: Physical Radio Frame Design Considerations, *IEEE Access*, 2016.

Jover, Roger Piqueras and Ilona Murynets, Connection-less Communication of IoT Devices over LTE Mobile Networks, *12th Annual IEEE International Conference on Sensing, Communication, and Networking*, 2015.

Li, Guangzhi and Jennifer Yates, A GMPLS Based Control Plane Testbed for End-to-end Services, 0-7803-9277-9/2005/IEEE, 2005.

Marques, Paulo and Cladio Cicconetti, 5G Radio Network Architecture, http://www.ict-ras.eu, 2017.

NGMN, 5G White Paper, Next Generation Mobile Network Alliance, https://www.ngmn.org/5g-white-paper.html, 2017.

Ohlen, Peter et al, Orchestration and Control Solutions in 5G: Challenges and Opportunities from a Transport Perspective, *IEEE Software Defined Networks*, 2017.

ONF, *TR-526 Applying SDN Architectures to 5G Slicing*, Open Network Foundation, April 2016.

Ordoney-Lucena, Jose et al, *Network Slicing for 5G with SDN/NFV: Concepts, Architectures, and Challenges*, IEEE, 2017.

Rost, Peter et al, Network Slicing to Enable Scalability, and Flexibility in 5G Mobile Networks, H2020-ICT 2014-2 Project 5G Norma, 2014.

Poudelm, Swaroup, Internet of Things: Underlying Technologies, Interoperability, and Treats to Privacy and Security, http://scholarship.law.berkeley.edu/2016/btlj/vol31/iss2/24, 2017.

Rappaport, Theodore, et al, Millimeter Wave Mobile Communications for 5G Cellular: It Will Work. *IEEE Access*, May 10, 2013.

RAS, 5G Radio Network Architecture, Radio Access and Spectrum, FP7- Future Networks Cluster, http://www,jct-ras.eu/, December, 2017.

Sundhar, Kalyan and Lawrence Miller, 5G for Dummies, www.ixia.com, 2017.

Talwar, Shilpa et al, Intel, Enabling Technologies and Architectures for 5G Wireless, IEEE, 2014.

Teichtahl, Marc, Deploying MPLS Traffic Engineering, 2019, mteichta@cisco.com.

Chapter 7

Ethics and Policy of IoT

Gerald DeHondt
Ball State University

Contents

7.1 Introduction

The Internet of Things (IoT) can be broadly defined as a global network infrastructure, linking uniquely identified physical and virtual objects, things, and devices through the exploitation of data capture (sensing), communication, and actuation capabilities (Miorandi, Sicari, De Pellegrini, and Chlamta, 2012) with these objects around us being connected to provide seamless communication and contextual services (Lee, 2012) as an integration of the physical world with the virtual world of the Internet (Haller, 2011). In this space, almost anything can be interconnected, widening the scope and depth of the Internet into our daily lives. These connections between the physical and the virtual world have the potential to become exponentially more complex and dynamic than the Internet. Considering IoT as a radically distributed technology, ethical concepts should not be viewed in isolation but in a contextualized form that incorporates the dynamics and complexity of these "things" with asynchronous, distributed connections of subjects and objects (van den Hoven, Weber, Pereira, and Dechesne, 2012). The IoT does not concern objects only; it is about the relations between the everyday objects surrounding humans and humans themselves. With this comes the imperative to consider the ethics of this depth of connectedness.

7.2 IoT Opportunities and Problems

The development towards an IoT is likely to give rise to a number of ethical issues and debates such as loss of trust, violations of privacy, misuse of data, ambiguity of copyright, digital divide, identify theft, problems of control and access to information, and freedom of speech and expression (van den Hoven et al., 2012). At its best, the IoT has the potential to create an integrated ecosystem that can respond to a spectrum of needs, increasing efficiency and opportunity, and empowering people through technology, and technology through intelligence. At its worst, the IoT can open a Pandora's Box of inappropriate and unsafe behavior, unintended consequences, and intrusiveness (Berman and Cerf, 2017).

Placing this into perspective, critical characteristics of the IoT are specified below:

Ubiquity and pervasiveness	The user will be immersed in the IoT with no clear way of opting out.
Miniaturization and invisibility	Computing technology will become translucent and disappear from human sight, although embedded in objects all around us. Special design measures are recommended to make technology visible and open to inspection, audit, quality control, and accountability measures.
Ambiguity and ontology	Distinctions between objects, artifacts, and humans will blur, and we will have to deal with ambiguous criteria of identity and system boundaries.
Identification	Electronic identity will be achieved by tagging of objects. Who will assign, administrate, and manage these identities?
Connectivity	Unprecedented degree of connectivity and data transfer between objects and persons.
Mediation and autonomous agency	Humans will interact in concert with artifacts, devices, and systems with potential spontaneous intervention by these systems as agents.

Embedded intelligence and extended mind	Intelligence may become an extension of the human body and mind.
Seamless transfer	Information flow within IoT will become effortless, with low transaction and information cost.
Distributed control	Due to the vast amounts of nodes and hubs, governance will be decentralized without a centralized authority.
Big Data	IoT will generate a tremendous amount of data.
Unpredictability and uncertainty	Evolution of IoT will lead to new behaviors with relevant knowledge unknown to the user.

Defining Features of Internet of Things (*Adapted from*: van den Hoven, J., Weber, R., Pereira, A., and Dechesne, F. (2012). Fact Sheet—Ethics Subgroup IoT—Version 4.01.)

Considered in this context, IoT combines several existing and emerging technologies such as Big Data, miniaturization, semantics and ontology, connectivity, and artificial intelligence. It is this broad-based perspective, cutting across and combining multiple areas of technology while integrating with everyday objects, that allows IoT to act as a synergistic force that provides greater impact than any of these technologies would enable individually.

Given the significant current and potential impact of this technology, it becomes imperative to consider the digital divide that may occur, in effect a chasm created between the technological "haves and have nots." This was identified in a survey conducted by the Pew Research Center as one of the major concerns of IoT (Ravindranath, 2014); and that it may eventually divide society between the tech-savvy and others. Ravindranath (2014) continues that the digital divide of the 1990s was solely about access, whereas IoT will be more about those who understand what is going on and the broader impact in terms of privacy and security. Those who understand this will be better enabled to protect themselves while in turn benefitting from the advantages IoT can provide in terms of data analytics. Additionally, those in developed countries may be able to capitalize on the benefits of IoT in regard to public utilities and energy usage, while those in developing countries—with the most critical need for this technology and knowledge—will be left behind without the knowledge or access of developed nations (Ravindranath, 2014).

The IoT is an expanding and developing area of discovery, research, and definition. Even those defining the space have little more guidance to provide than simply "things." As a new area that is emerging, evolving, and even being redefined, it becomes critical to develop a framework of what is right and what is wrong. A couple decades ago, nobody would have voluntarily agreed to carry a device that would track their movements and provide up-to-the-second information on their whereabouts. The right to be anonymous was considered fundamental. Today, nobody leaves home without their cellphone. Various map apps provide information not only about where people are but also where they are going. The ability to share this information with others enables friends to know when you will be there. Other apps allow you to check-in or even poll you for your experiences at a restaurant, store, or other destination. What is considered commonplace today would have been viewed as an unnecessary intrusion just a few years ago.

7.3 Private vs. Public Space

As we move through our day, we view space through different lenses. In the privacy of our homes, we may act and do things we would never consider doing in a public space. Some will sing only in the shower even though their voice may rival a choir. We may dance like nobody's watching. We may even play with our children and take on personas that would be an embarrassment in front of even our closest friends.

With the IoT, we are connecting additional devices within our homes and in our garages to provide up-to-the-second information on areas that were once considered personal, private space. Consider this, the Google Echo is constantly listening to our conversations and will store sounds for up to 1 min (Moynihan, 2016). When it hears specific terms, it knows to jump into action; in the absence of these keywords, it simply remains vigilant. Could we unknowingly be allowing third parties access to intimate details of our lives? Are we allowing these third parties access to private, personal conversations?

In the real estate market, it is not uncommon for sellers to place video or recording devices throughout the house to gather information or bargaining points on potential buyers. When browsing another's house for the purpose of potential purchase, is this considered *your* private space, while in *someone else's* home? Is private vs. public space a question of use or purpose instead of location of the space?

In the public space, we realize that our actions are being observed. Banks, convenience stores, and government buildings are a few of the places that will monitor their patrons and visitors from entry to exit. In this public space, there is recognition that actions are being observed, and this monitoring is accepted. This, of course differs from the same level of monitoring that may be unwittingly occurring within the privacy of our own homes.

"Nanny Cams," toddler video monitors, or alarm system video cameras are installed in our homes to serve a need, yet these devices are easily misused or can be co-opted for malevolent intent. Any networked device is vulnerable to attack and may unknowingly be used as a window into our private space.

Considering this prior discussion, is it unethical for third parties to have access into our private space, or is it up to those to be aware of their surroundings?

We bring these devices into our homes—our private space—yet this functionality has been carried with us throughout our daily routines on our phones since the introduction of Google Assistant.

Beyond video cameras and listening devices, it is the perception of how these IoT devices may be used to impinge upon our privacy. Consider the case of smart metering technology proposed for electrical efficiency. (Al-Abdulkarim, Lukszo, and Fens, 2012). The Netherlands proposed to install smart electricity meters nationwide to improve the efficiency of their electrical grids and meet European Union (EU) CO_2 reduction targets by 2020. These smart meters would provide detailed information regarding household electrical consumption that could be used to predict peak periods and learn how to better balance usage. Over time, these smart meters came to be seen as a spying device that would take snapshots of individual household electrical consumption every 7s and provide detailed information about the activity in the homes of its citizens. With some effort, it became possible to tell which movie was being watched on a given night using these smart meters. By the time this proposal was brought to Dutch Parliament for a vote, public uproar regarding the privacy implications of this device was so prominent, the proposal was rejected on data protection grounds.

With the wired and connected refrigerator, our eating habits have come on display. While it is helpful to know whether we need eggs or milk when making an unscheduled stop at the store, can this information be tapped by health care providers or insurance companies to track our healthy lifestyle?

It is important to make the distinction between public and private space when considering IoT (van den Hoven, et al. 2012) and to recognize what IoT devices are already in our private space. It has long been considered that our homes are considered our private space, shared with others who we select. Another private space would be our car, also shared with others who we select. As IoT develops, these private spaces are being eroded, and they are becoming more a source of unintended data and information on our activities and habits. What are our rights to privacy in the IoT? Have the rules and expectations of privacy changed with the evolution of technology, or are these concepts absolute? Is there a way to provide the benefits of these new technologies without experiencing the invasiveness? Consider the newly enacted General Data Protection Regulation (GDPR) in the EU. One of the basic tenets is the right to be forgotten. Macauley (2017) discusses this as an idea more than a policy and notes that it is almost impossible to enforce. The legislation only provides for data erasure when "no compelling reason" exists for its processing.

Since the advent of OnStar, our regular travels have become a matter of corporate record, one provided by default to a company that answers only to shareholders rather than elected officials. Ethical considerations and decisions in this instance are made for the betterment of the shareholders of the company, not society in general. Should private entities—even though public companies—maintain this type of information on their customers? How can this data be used, or misused, for the betterment of the organization or their potential business partners? More importantly, is it ethical for organizations to use ancillary data, gathered in the course of providing a service to their customers, for additional uses never intended when the data was gathered?

Consider the use of health-monitoring equipment in any of its various forms, be it pacemakers, fitness trackers, or any devices that provide ongoing monitoring of physical attributes and activities. A Spanish company has invented a "smart mattress" that could help the user optimize sleep patterns using sensors that detect motion and restlessness during the night (Colclasure, 2016). The intended benefit of this technology is to help users optimize sleep for the most restful recuperation possible by modifying factors that could contribute to nocturnal restlessness.

An unintended consequence of this smart mattress is that it could also be used to detect motion patterns tied to sexual activity and send a warning to the partner's mobile phone. Beyond the monitoring of intimacy or sleep patterns, who "owns" the data that is provided through these health-monitoring applications?

GingerIO markets an app that runs on a user's smartphone (Weir, 2012). This app gathers data on a customer's daily patterns and routines. It has long been known that those entering into a period of depression will tend to forgo medication, and these patterns of depression are recognizable. For instance, persons will tighten their pattern of travels, staying closer to home, and partake in fewer normal daily activities. Having established a recognizable pattern, deviation will be noticeable. The company may then be able to notify pre-identified contacts or the particular individual just to check-in and make sure everything is okay. Persons forgoing critical medication experience more significant issues than depression.

7.4 Consideration of Trust

IoT will be an amalgamation of both presently defined and yet-to-be-defined "things." This ambiguity demonstrates that we are applying present ethical frameworks to devices and situations that cannot yet be predicted or imagined. As a conglomeration of disjointed and disparate systems and devices all sharing information, a breakdown in one part of the system can potentially corrupt or disrupt the rest of the system. Consider your automobile as a system. All parts must be functioning in unison or the car will not perform its role. A breakdown in one part of the system can cripple the entire system, regardless of how well the other parts perform their job. A flat tire will cripple a car even if the engine has been finely tuned.

Since IoT systems will all be interlinked and sharing information, the user must—by default—*trust everything* in the chain where the data will be shared. It is not enough to be able to trust only the device that is being interacted with, all subsystems must be trusted. As few as one device or "thing" that is not trustworthy will corrupt the entire system. This single point of failure will provide a ripple effect throughout the entire system. Presently, when one retailer is breached, there would not be an effect on the remaining retailers with whom one does business. In the

IoT environment where devices exchange and share information, corruption of one connection—through error or design—can propagate throughout the system and serve as a data leak for the remainder of the system.

Beyond the virtual world, the interconnected nature of all things in IoT can manifest into problems in the physical worlds. A virus or hack on a computer, Internet Application, or corporation would only affect that entity in the virtual realm, without consequences in the physical world. A hack or virus implanted into a smart car can jeopardize the safety of the passengers or others around that vehicle.

7.5 Summary

As a technology that is emerging, developing, evolving, and being defined, the IoT will impact society in ways yet to be imagined. Touching all aspects of our lives in the future, IoT will revolutionize life as we know it for intended good but also for unintended negative. Similar to the microwave, we will look back on the pre-IoT period and wonder how we lived as we did. This is a revolutionary time that will date a pre and current period. As the Internet ushered in a blurring of our households and the outside world, so too will IoT continue to blur this divide between public and private space.

These devices, and their interconnected nature, will affect our lives with a level of incursion unimaginable earlier. Will privacy become a thing of the past with only public and semiprivate space available? Today, we are able to turn off our devices and disconnect. Will this even be possible in an IoT future? Or will we even want to disconnect? Numerous studies have been performed about the evolutionary impact of our current level of connectedness. How will IoT change our evolutionary path?

So too, consideration must be given to those who may fall further behind—the digital "haves and have nots"—who through circumstance may be unable to keep up with this latest wave of technology.

How will the data be handled? Who owns this data—user data, your data? That posted freely into the public domain becomes fair use. What about data provided to companies through everyday actions? Will we even have access to this or be able to correct wrong information?

This chapter has discussed possible ethical implications of this revolutionary technology. As with any new discovery or technology, its effect on our lives and the relevant ethical considerations evolve along with the technology.

7.6 Case Study: Their First House

Pat and Jennifer are a young couple who have decided to purchase their first house. They have been investigating the real estate market in the areas they are interested in living and have been determining what kind of a house would best meet their needs. They have attended several open houses are ready to begin working with a real estate agent.

Their agent has reviewed multiple listings with them, and Pat and Jennifer are ready to begin seeing available properties. One practice that has become relatively common with home sellers is to install video surveillance systems either as a home security measure or as a means to gather information on potential home buyers that may be useful in negotiations. The sellers list the property as having a Home Security System, a feature often seen as beneficial to home buyers. Pat and Jennifer's experienced real estate agent is familiar with this practice of "disclosing" that people viewing a property may be monitored by the owners by citing a Home Surveillance System in the property listing. Potential home buyers would view this as a positive feature of the listing, without realizing alternative uses. Disclosure of the Home Surveillance System in the listing is accepted practice as notification to potential buyers that the property is under video and audio surveillance as they tour the house.

As inexperienced home buyers, the young couple is not aware of this practice and view the house making several comments as to the suitability of the house to their needs, financial capabilities, negotiating strategies, and the like. All of this information is retrieved by the home sellers upon their return and used to strengthen their position during negotiations, including learning of the buyers receiving large gifts from both parents and Jennifer's favorite Aunt Helen to help them with the down payment as they purchase their first home together.

The sellers became familiar with the resources available to this couple through covert surveillance of their discussions while viewing the property in an intended private space. This information was used to receive a higher price for their property during negotiations.

7.7 Discussion Questions

This case study reveals several items regarding the ethical nature of IoT devices that may be intended for one purpose but can be used for alternate purposes.

1. Is it unethical for the sellers to monitor potential home buyers while viewing their property?
    a. Answer-During a showing, it is universally recognized that the property is owned by another party, and potential home buyers are guests in another's residence. They are viewing the property as a means to make better informed decisions as to the suitability of the property to their needs. Free discussions would occur between home buyers regarding pros and cons of the property, suitability for use, and affordability. These discussions would occur in what is expected to be a private space, free from watching eyes.

2. Is it incumbent upon the buyers' real estate agent to inform home buyers of a common tactic used by sellers to gather information on potential buyers?
   a. Answer-Whether or not surveillance systems are readily apparent, the disclosure in the house listing should tip off the experienced realtor that their conversations and actions may be monitored while viewing the property.
3. Should Pat and Jennifer have performed research on common current home seller practices?
   a. Answer-Brief investigation may have revealed this common practice. What do you think?
4. Is it ethical to monitor employee activities using organizational assets?
   a. Answer-Employee Monitoring. It is commonplace for employees to use corporate assets (laptops, smartphones, devices) when working offsite at a client, at a coffee shop, at home, while traveling, or any other number of places. As example, an employee sitting using time in an airport terminal catching up on e-mails or reports. For convenience, they may also use the corporate laptop to pay bills, send personal e-mails, or run a side business. Outside employee is specifically prohibited by the employer.
5. While businesses may consider it helpful to customers to let them know about sales or special events and promotions at their locations, is this an unnecessary intrusion into the privacy of customers?
   a. Answer-Explicit Consent. Consider the location information that is given to our smartphone service provider or third-party application vendors as we go about our daily routines. This information can be provided—for a fee—to local businesses as a means to push advertisements to potential customers throughout the day (i.e. restaurants at mealtime) or to customers who may regular shop at particular stores. What information can be shared, or does it require our explicit consent?
6. Data ownership.
   a. Answer-Implanted medical devices, by their very nature, gather data at regular intervals to help doctors diagnose and treat their patients and enhance their health. Should these devices uncover potential issues unrelated to the malady being monitored, what duty does the device manufacturer have to its patients or doctor to notify of a potential additional malady discovered or at least suggest further investigation? Should the manufacturer be liable if they withhold information about their patient that may have prevented a heart attack, stroke, or other severe medical condition? Should the patient and/or their doctor be able to receive a copy of all the information gathered by the medical device? Note that having access to the information gathered by the device might be used to determine particular issues with the medical device and could serve as a liability for the manufacturer.

References

Al-Abdulkarim, L., Lukszo, Z., and Fens, T. (2012). Acceptance of Privacy-Sensitive Technologies: Smart Metering Case in The Netherlands. In *Third International Engineering Systems Symposium CESUN 2012* (June 2012).

Berman, F. and Cerf, V. (2017). Social and Ethical Behavior in the Internet of Things, *Communications of the ACM*, Volume 60, No. 2, pp. 6–7.

Colclasure, S. (2016). On the Ethical Use of Data vs. the Internet of Things, *Forbes*, December 21, 2016.

Friedewald, M. and Raabe, O. (2011). Ubiquitous Computing: An Overview of Technology Impacts, *Telematics and Informatics*, Volume 28, No. 2, pp. 55–65.

Haller, S. (2011). The Things in the Internet of Things. *Poster Paper Presented at Internet of Things Conference 2010*, Tokyo, Japan. www.iot2010.org/

Lee, G. (2012). The Internet of Things—Concept and Problem Statement. Internet Draft, Internet Research Task Force. http://tools.ietf.org/html/draft-lee-iot-problem-statement-05.

Macauley, T. (2017). What is the Right to be Forgotten and Where Did It Come From? *Techworld*, September 14, 2017.

Miorandi D., Sicari, S., De Pellegrini, F., and Chlamta, I. (2012). Internet of Things: Vision, Applications and Research Challenges, *Ad Hoc Networks*, Volume 10, No. 7, pp. 1497–1516.

Moynihan, T. (2016). Alexa and Google Home Record What You Say. But What Happens to that Data? *Wired*, December 5, 2016.

Ravindranath, M. (2014). Some See Possible Drawbacks in "Internet of Things", *Washington Post*, May 25, 2014.

van den Hoven, J., Weber, R., Pereira, A., and Dechesne, F. (2012). Fact Sheet—Ethics Subgroup IoT—Version 4.01.

Weber, R.H. (2010). Internet of Things—New Security and Privacy Challenges, *Computer Law and Security Review*, Volume 26, No. 1, p. 23–30.

Weir, B. (2012). How 'Big Data' Can Predict Your Divorce, *ABC News Nightline*, December 5, 2012.

Chapter 8

Empowering Older Adults with IoT

Rebecca Lee Hammons
Ball State University

Contents

8.1 Introduction

This chapter provides a context for Internet of Things (IoT) devices and their use by an aging adult population. Gerontechnology is a term coined in 1988 to describe the study of technology and aging, and its scope is far and wide (Kwon, 2017). IoT for elders falls into this area of research, and the opportunities to study the ways in which IoT can be used to support aging are limitless. Each day we are introduced to new technological solutions that aid in healthcare, aging in place, and other lifestyle adaptations. Many of these solutions are designed by Millennials or Gen Xers within the technology industry, and the solutions may lack the user experience (UX) attributes helpful to those who are not technologically astute or have cognitive or physical challenges. In this chapter, we discuss the demographics of aging in America, issues faced by an aging population and IoT options, and several key aspects of the IoT technology UX for our elders, as identified through a recent IoT research activity.

8.2 Aging in America

Growing old in America is not an easy task for a variety of reasons. We live in a youth-obsessed culture that emphasizes consumerism and the achievement of wealth and status over human connectedness, learning, and social and family values. It is tricky to age gracefully in our culture, and the constant transformation of networking and communications capabilities can isolate those who cannot get on board with new technology. Those elders who have an opportunity to learn about and experience new tools and techniques have a lifestyle advantage.

Our aging population continues to grow, and this demographic, comprised of Baby Boomers and the Greatest Generation, creates an IoT market and revenue opportunity. Almost 16% of our population is 65 years or older with about 20% more women in this cohort than men (CIA Factbook, 2017). The elderly dependency ratio in the United States is 22.1, meaning that there are 22.1 elderly for every 100 people of working age (15–64); this is important because there are increases in pension and healthcare costs for government as this ratio rises. Eighty-two percent of our population lives in urban areas, and urban growth is projected at about 1% per year (2015–2020). Our population's life expectancy is averaged at 80 years.

Fewer people in our country attend church or synagogue than prior generations, and this impacts the sense of community that may be available to many aging adults. In 2007, 39% of Americans reported attending weekly church services and that number has declined to 36% by 2014 (Pew Forum, 2018). That trend is important because, historically, church membership provided a form of community and means of connectedness in the United States. Now, however, 76% of the population use the internet—that's nearly 250 million people, and we're fourth in the world for internet usage. According to the Pew Research Center (2018), 82% of 65- to 69- year-olds are internet users. This implies considerable opportunity for the application of IoT solutions to address issues of connectedness, communication, safety, security, entertainment, and health and well-being. While the opportunities for innovation are great, there is likewise a need to assure that IoT solutions for the aging are ethical, responsible, and address the UX needs of this population.

8.3 Issues Faced in the Aging Process

As we age, we face a variety of challenges such as hearing, memory, and vision loss in addition to mobility and health issues. Diabetes and obesity are commonplace in our culture and even more so with the elderly, as noted in the table below. Aging adults are often isolated due to these challenges and may then suffer from a loss of communication with others. While these issues pose significant lifestyle challenges, technology offers some opportunities to improve overall quality of life for this very large, and growing, cohort of elders (Table 8.1).

According to the information from the Kaiser Family Foundation, a quarter of noninstitutionalized adults over 65 have a disability and approximately 22% have diabetes. A whopping 64% of adults are overweight or obese. Each of these health challenges can result in the need for lifestyle adaptations. The IoT industry

Table 8.1 Adult Health Considerations (Kaiser Family Foundation, 2018)

Year	*Kaiser Family Foundation Health Data*	*Age*	*Percentage*
2014	Percentage of U.S. adult population with diagnosed diabetes	45–64	13.2
		65–74	22.2
		75+	21.2
2016	Percentage of U.S. adult population with cardiovascular disease	18+	6.6
2016	Percentage of U.S. adult population who are overweight or obese	18+	64.8
2015	Percentage of U.S. adult population, noninstitutionalized, who reported a disability	18–64	10.4
		65–74	25.4
		75+	49.8

has created numerous devices, and continues to create more, that can aid in aging gracefully and supporting lifestyle and healthcare needs. In addition to the features offered by Amazon and Google (Echo, Home, etc.), there are robots to keep one company and provide services, tools that sense motion to provide alerts if patterns change, sensors and monitors for blood pressure and blood sugar, and devices to manage the household. Such IoT devices can include smart garage door sensors, systems to lock and unlock doors, smart thermostats, window treatment controls, smart refrigerators and air conditioners, and the list goes on and on. Many IoT devices are controlled via a smartphone application, and data including usage trends reside in the Cloud as depicted in Figure 8.1.

My students and I conducted several UX research activities during the spring semester of 2018. "The main goals of a good user experience are to provide some combination of efficiency, empowerment, enrichment, and enjoyment to the people who use the product" (Camara & Zhao, 2015). We identified that some very basic needs to be addressed for this cohort before integration of IoT devices into homes and other living environments can be helpful. The student teams also identified some additional solutions to support IoT for aging in place. These projects are described later in the chapter.

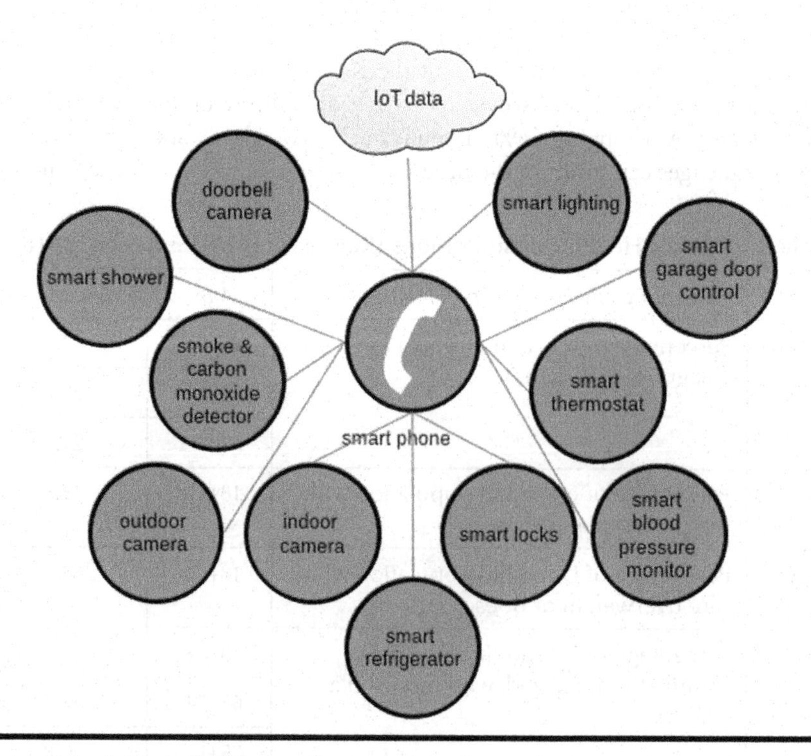

Figure 8.1 IoT possibilities.

8.4 Impediments to Technology Adoption

Our research provided us with insight into IoT usability issues faced by elders. First and foremost, some elders were opposed to the use of any technology that provides a monitoring capability within their home. Not only are many elders distrustful of technology for privacy reasons, they are also concerned that no one be able to hack into a device from a security standpoint to perform illicit monitoring. Wikileaks exposed the reality that the government can surreptitiously monitor email communications, and many new technology "advances" are viewed as having the potential for harm by those with extensive life experience. These devices include services such as GPS/location identification, social media, web cameras, smartphones, and hosted applications in general. Some elders reported receiving gifts of IoT devices that remain in their packaging in a closet, unused. Caregivers have their own concerns about active monitoring equipment in their presence and may be opposed to the use of IoT, and adult users are often rightfully concerned about the security of devices made in China (of which there are many).

While IoT devices can indeed solve lifestyle problems, we found that the elderly struggled to grasp the techniques to implement, master, and maintain the devices. The concept of the Family Technologist was described (Huber et al., 2017) by researchers as a critical support role for those elders who are willing to step into the world of technology. While the authors defined this role for the elderly who wish to embrace technology including computers, smartphones, and the internet, this role is even more critical for those wishing to integrate IoT into their home wireless network infrastructure due to the complexity of the UX. "As technology moves into elder care, there is a need for a family technologist to help aging loved ones and the caregiving team adopt and use technologies that support health, well-being, and decrease caregiver burden" (Huber et al., 2017, p. 58). Despite the fact that the master's students conducting the UX evaluations on IoT for Aging were tech-savvy, they reported their own challenges in designing, implementing, and maintaining solutions for their study. Two teams took an approach, based on elder inputs, of creating hands-on workshops for elders addressing the basic functionality of the Amazon IoT devices that were the focus of the research. Figure 8.2 describes the continuum upon which most individuals, including elders, fall in their adoption of IoT devices.

There will always be a percentage of the population, in all age groups, who choose to live without the influence of IoT products or services in their lives. They might be Luddites, have religious reasons, live in remote areas without internet service, have disabilities which inhibit use, or other sound reasons. Some elders, as we found in our research, may be unaware of the availability of IoT devices that could support their lifestyle needs, while others may have awareness but choose not to embrace

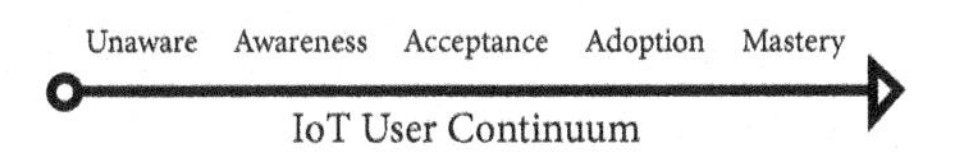

Figure 8.2 IoT user continuum.

the technology. Once awareness is created, some elders may move into a phase of acceptance of IoT devices to age in place, as long as there is support for learning, configuration, maintenance, and usage of the devices. We also learned about this segment of the user continuum in our research; numerous students reported they held the role of family technologist for their parents and grandparents. Several subjects were eager to learn about the core functionality of their IoT devices and how to use them more effectively. Some elders are tech-savvy and eagerly adopt devices that improve their lives, such as IP-based thermostats for warming up their second home while en route or fitness trackers to monitor and improve their exercise and fitness outcomes. This is likely a small percentage of elder users. Finally, some elders may demonstrate mastery of the technology and create a robust smart home environment to optimize the fulfillment of their needs based on the value they see in the cost/benefit analysis. This value could be in terms of time, energy or cost savings, living a green lifestyle, managing remote properties from afar, convenience, or social attributes (such as being "the first"). It should be pointed out that a robust smart home environment requires an investment in and understanding of several key infrastructure elements such as networking equipment, a laptop, a smartphone, and a good Wi-Fi service. Figure 8.1, IoT possibilities, provides a visual of many of the devices that can be used in a smart home environment.

Other concerns about the use of technology in the home include the loss of human contact, the cost versus perceived years of use, and ethical issues related to the use of scare tactics (such as in advertising or by family members) to get elders to adopt technology products or services (Huber et al., 2017). We heard stories from some elders about devices that they would resist no matter what the benefits might be due to the offensive or patronizing nature of the marketing pitch (such as *Help! I've fallen, and I can't get up!*).

According to Huber et al. (2017), technology is not meant to replace human relationships yet the automation of communications, for example, makes some elders fear that this will occur. In actuality, the authors state that, "Older adults can use tech as a reason for spending time with loved ones" (Huber et al., 2017, p. 66) as the family members implement and support their technology use. Maintaining interpersonal relationships is important to those aging in place, and yet many prefer a face-to-face or telephone-based form of communication over new tech methods. Texting is now a common form of interpersonal communication, in part because it is easy and maximizes the use of one's time. For retired elders, time is not a critical variable in the lives of many and thus texting is not seen as an optimal communications vehicle, even if done through the assistance of Alexa (at home) or Siri (even while driving). Texting can also be problematic for those with manual dexterity issues, and it lacks the rapport of interpersonal voice communications.

Huber et al. (2017, p. 66–67) found that "Most aging adults are unwilling to pay for any type of technology." When IoT devices make it into the home of an elder, it is likely that the device was a gift. Elders do not necessarily perceive that they will get much use over time from technology devices, and they do not want to invest the time, money, and possible frustration in learning the technology. It is very important to identify and communicate the value proposition for IoT for aging in place if one wants to market to this segment of the population.

8.5 The IoT UX Evaluation for Elders

There are some fascinating products available in the IoT space to support aging in place. Many older adults are unfamiliar with these devices and creating awareness is a good first step toward future adoption.

Three of our class UX projects focused on learning how the Amazon Echo, Show and Dot solutions could be integrated into the living environments for those over 65 to serve some basic lifestyle needs. Sample functions included being able to check the weather, get recipes, get medication reminders, listen to music or the news, and call for assistance. Each team selected one of the Amazon devices and had the opportunity to include other IoT devices in the configuration, as noted below. Two additional UX project teams sought to implement the Amazon Skillset Application Programming Interface (API) to automate functions using the Amazon Echo, Show or Dot.

The objective was to assess the UX for a variety of elders in using these devices and identify usability and design feedback that could create opportunities for IoT improvement. Many technology devices are designed with Millennials and Gen-Xers in mind, and their usability skills and needs are very different from the elder cohort. One very basic example is the printing on IoT packaging and devices; if too small or too light, this inhibits the elder from successful configuration and setup. Another example is when a product lacks written instructions and the user must seek information online.

8.5.1 Scope Statement for Our IoT and Aging Adults Projects

Select one of the following projects. You will analyze the automation/integration opportunities for these products in combination with smart phone devices and analyze and improve the usability to meet the needs of aging adults.

Project 1	Amazon Dot™	Nest Thermostat™ and Nest™ Smoke and CO_2 Alarm
Project 2	Amazon Show™	Nest Cam™ Security Camera and Belkin™ WeMo Insight Switch
Project 3	Amazon Echo™	Foobot™ Home Air Quality Monitor and iRobot Roomba™
Project 4	Amazon API	Automate new functions using the Skillset API
Project 5	Amazon API	Automate new functions using the Skillset API

The first three projects included the following activities throughout the semester:

Critical inquiry—brainstorm the possible lifestyle enhancements that the technology can support; interview at least five adults aged 65 or over; ask about their lifestyle and household needs and if or how the targeted IoT device would assist them due to hearing, memory, visual, or mobility issues; document user stories including automation opportunities

Prototyping—write test scripts for the adults to execute the target functionality via the IoT device; write any desired automation using Amazon Skillset API; implement the initial designs

Evaluation—evaluate the UX, identify recommendations for improvements to the UX

Redesign/implement—refresh/redesign the scripts and rerun the usability review to assess value of improvements.

8.5.2 Lifecycle Process and Findings

Figure 8.3 depicts the lifecycle process used in UX design. Our projects were conducted over a 16-week semester, and each team had the opportunity to work with elders to conduct the activities including critical inquiry (needs assessments) and analysis, design concepts, prototyping/implementing, and design evaluation. What did we learn?

- Some elders have IoT devices but are not using them or are not using them to their full potential, due to challenges in learning about the devices. Online user documentation is limited. Elder learners are less likely to explore and experiment with the technology on their own. Some of the students who already owned Amazon Echo products learned about capabilities of which they were previously unaware.
- For IoT workshops provided to the elderly, it's important that the activities are hands-on, that each learner has their own device to work with, and that

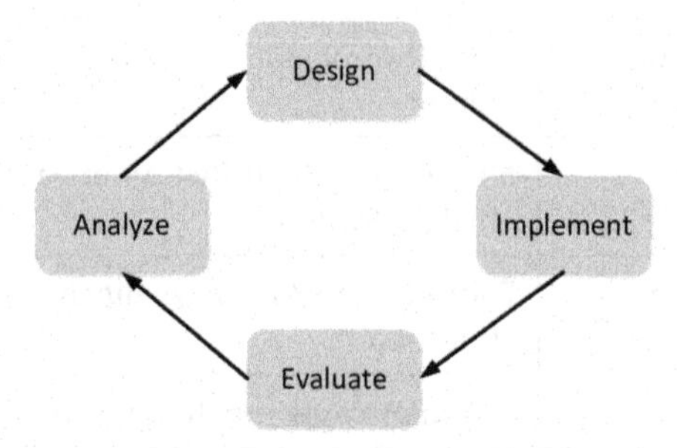

Figure 8.3 UX life cycle.

the pace be tailored to the individual. If some learners are quick to pick up new information, others may withdraw from the activities and just watch rather than participate. Allocate time to review the material several times over. The elderly lack a technology language that supports ease of learning. For example, routers were a new concept to several of our participants. It is important to teach the terminology as much as the actions.

- There are limited affordances to support the needs of the elders in learning the products. An affordance is embedded in the product design to provide a clue to the next thing to do. As an example, the Ring™ video doorbell has a very "clean" design with no labeling of buttons on the device itself. One must review a setup and installation guide which is packaged with the device to learn how to install and configure the Ring™. It includes a five-step process and seems simple at first glance, until one realizes the need to also download and configure a smartphone application to manage the device and get the device on the wireless network so that it is functional with the smartphone application. This becomes a multipronged configuration activity to achieve setup success, and one must be cognizant of the need to perform the additional setup steps outside of the Ring™ installation alone. Affordances, such as labeling on buttons, are a key element to the ease of use for non-tech-savvy users.
- Issues arise in the wireless configurations of IoT devices if the network goes down; reconfiguration of devices is often required. Network stability is of the utmost importance. Students reported their frustration with the reconfiguration due to outages.
- One team investigated the possibility of implementing and testing the Amazon Dot™ and Nest™ solutions in a local nursing home facility and the team met with the staff for input on the design of the solution. The team was inhibited by the lack of a robust public wireless system within the nursing home facility that could support the addition of the devices, and there was some concern by the administrators about the ongoing support needs for residents. Nevertheless, the teams found several key ways in which the Dot™ and Nest™ could enhance the lives of the residents:
  - Voice recognition to help control the Nest™ thermostat and other IoT devices
  - Designate contacts to (voice) alert if resident is in need of help
  - Capable of placing phone calls through voice command
  - Ask Alexa to read books aloud
  - Ask common questions
  - Find your cell phone if it has been misplaced
  - Read messages aloud and send messages to family and friends
  - Play any type of music just by asking to hear it
  - Family members can check up on the resident if they have access to the Nest™ camera application.

- Some of the identified constraints include the need for precision and clarity (a strong voice) in the commands used with Alexa, the inability to call 911 in an emergency through voice command, the need for a wireless network capable of supporting the additional network load, and someone within the nursing home facility with the skills to install, administer, and maintain the IoT technology as well as teach residents how to use it.
- A team of students, working with elderly stakeholders, designed a prototype of a wearable IoT voice-enabled device to cater to the personal needs of an elder by acting as a communications device between loved ones and the IoT device. The prototype is intended to be paired with existing IoT devices, such as the Amazon Echo™ and Nest™, to support monitoring of surroundings while also having the ability to use auditory commands. The prototype concept came from an analysis of the functions provided from the Amazon Echo Show™, Nest™ Cam Security Camera, and Belkin WeMo™ Insight Switch. Stakeholder feedback indicated that the IoT devices are not necessarily tailored for the elderly, thus the team sought ways to simplify the interface through a voice-enabled, wearable device. They assured, in the design, that the wearable had a simple design and was easy to use. Additional features in the prototype included a button to be pressed to activate listening mode of the Echo™, a light surrounding the button and a chime to indicate when the listening feature is activated, and a design that does not capture user data.
- Two teams focused on automation of IoT devices through the use of the Amazon Skillset API. One team developed scripts for automation of a smart home environment. Task controls that were automated included shower, sound system, security cameras, door locks, temperature control, smart toilet, washer/dryer start, smart TV, and garage door. This team created the scripts to control automation of these IoT devices via a smart phone. They found the scripting process to be fairly easy to follow, and it did not require coding skills. Testing of the scripts was constrained by a lack of access to all of these IoT devices. One important consideration in creating a smart home environment for aging in place is the compatibility of devices; product compatibility is not always transparent from the vendor information made available.
- In the context of Don Norman's emotional design considerations (Norman, 2005) of visceral, behavioral, and reflective design, each team learned more about the emotional responses of the elderly to IoT design.
  - At a visceral level of design, IoT devices can be perceived as complex and off-putting by elderly users and it needs to be clear to them if the functionality is worthwhile to learn and implement. Some of the participants saw no value in augmenting their lifestyle with IoT and some were repelled by the thought of using technology to age in place.
  - At a behavioral level of design, elderly users want ease of use and intuitive functionality. They have concerns about the intrusiveness of devices such as "listening in" to their conversations or other privacy concerns.

Additionally, they need IoT devices that are perceived to have value to them and are easy to maintain. This implies, for example, the need for robust quality attributes in the design such that users do not have to reconfigure a device each time there is a wireless network outage. Think of Apple's wireless technology—the ease of use and plug and play features which make it appear more intuitive and seamless to users.

- Finally, at a reflective level of design, an elderly user will benefit from IoT solutions that are not physically intrusive, exploitative in any way, measurably enhance their ability to age in place, and do not reduce or replace human connectedness.

8.6 Opportunities with IoT

The IoT industry can consider a more diverse user base in the design of IoT devices. Consideration of the usability needs of aging users would provide an expanded marketing opportunity for IoT vendors, rather than the current ad hoc integration of such devices into homes, often with the assistance of a family technologist. An example of such design is the Greatcall Jitterbug Smart™ smartphone (greatcall.com), which has a simple home screen; large fonts and icons; and the ability to make calls, text, and play games to stimulate the mind. It can also be used as a lifeline, with optional health and safety services, connecting users to a live operator and/or medical professionals. Another application reminds seniors to take their medications.

An application called Oscar Senior™ is available for Android and iOS phones at a low cost, and it enables the family technologist to remotely access and manage an elder's tablet from their own smartphone, for application and contact management purposes. This type of solution makes it much easier to provide assistance to an elder from any location and could be useful in managing IoT devices that are configured via the elder's tablet.

Aside from the need to assure a high level of product and service quality for IoT solutions that target the elderly, there is an opportunity to emphasize Poka Yoke in the design process. Poka Yoke is a term originating in the Japanese auto industry of the 1970s and 1980s, and it means to error-proof a process such that one cannot perform the task incorrectly. As an example, Apple's iOS makes it simple to correctly set up a wireless network connection, while Microsoft Windows OS requires considerably more information from a user to configure a network without error. The same situation exists for adding printers to the configuration. IoT solutions for users of all ages, and especially for elders, will benefit from designs that include Poka Yoke considerations to prevent frustration and error conditions. This will also improve the UX and emotional satisfaction with the setup process.

The technology industry has, in recent years, reduced its investment in user documentation, including that which ships in the product packaging and instead relies on user "help" text embedded in a product and online support tools like wikis or videos. There is an opportunity to craft educational tools, even local workshops, which meet the needs of elders in being guided in their learning about IoT devices. Graafmans (2017) referred to the issue of elders learning new technology with insights that are not commonly known:

> The position of aging people in changing, innovative environments has been characterized by Powell Lawton (1998) as individual and sociocultural lag, the origin of which is that natural adaptation to technological and other environments stops at about age 30 and is then replaced by explicit learning directed at daily needs (e.g., at work). Here, it is worthwhile to note that retraining efforts at work in general stop at age 45 or even younger and that the average age of retirement is still around 60 or earlier. This implies that formal training or work experiences are not a resource for older people to adapt or adjust to the rapid cascade of technological innovations (Graafmans, 2017, p. 6–7)

What are the implications for elders and the adoption of IoT? This text reinforces the thought that it becomes more difficult to learn new technologies as one ages, and that voluntary adoption of IoT technologies might be rare for elders. Graafman's research confirms that the young learn new technologies more easily. Despite the format or availability of effective user documentation and other learning tools, most elders may require the support of a family technologist. Given that such a person may not be available in the lives of many, there may be a future business opportunity for a service that pays home visits, similar to an electrician or plumber, who installs, maintains, supports, and instructs elders on IoT infrastructure.

Several new services are being marketed to the elderly that meet some of these provisions. For example, Intel's Care Innovations® QuietCare® smart sensor technology led to the development of a new service for use in the home, "Health Harmony Home Sensing monitors changes and trends in patients' and residents' activities of daily living (ADLs) that may indicate changes in health"[1]. Sofihub (2018) "uses Artificial Intelligence to identify when something unusual happens in the home and will raise an alert automatically." Sofihub is interesting because it does not require the user to speak to the device, only to be within the room, in order to receive messages, reminders, alerts, etc. Using artificial intelligence (AI), Sofihub learns the patterns and habits of the user and will send an alert to family if an anomaly occurs.

Airstream provides a new means of "tiny home" living adjacent to caregivers with the new "Care-A-Van," which contains the technology required to monitor all vital signs, maintain personal connections with family and medical staff, and support assisted living needs. Called a "pod," the Care-A-Van has the patient's medical records and personal history. This is a remarkable IoT breakthrough concept to enable those who need assistance to reside adjacent to the homes of loved ones while benefitting from the latest in IoT technology support.

The OhmniLabs Telepresence Robot[2] is also targeted to support the elderly and those who need assistance. It provides virtual accessibility to the user for family and home health care personnel. The company is creating a developer's kit that adds controllable arms to the telepresence robot in addition to the base functions, along with an enhanced developer's API. The addition of arms to the robot can increase its effectiveness for assisting the elderly and disabled. The case study provided at the end of the chapter includes information about the UX reported in an evaluation of the OhmniLabs Telepresence Robot during a class project.

Some very basic needs for elders with mobility issues, who wish to age in place, can include home and room monitoring through commercially available IoT devices with alerting capabilities, smart thermostats, smart garage door openers with cameras, smart doorbells with cameras, and smart door locks.

8.7 Summary

One should not underestimate the value of a good, if not great, UX and the application of UX with elders in the IoT design process. Areas of emphasis can include affordances; design simplicity; ease of implementation and configuration; and accommodations to the visual, auditory, and memory issues that affect many elders. Consideration of design implications for caregivers and family technologists is also important when developing IoT concepts for aging in place. Another significant area for research and design enhancement includes senior living environments and the benefits of providing voice-controlled AI devices, smart watches, and in-room monitoring and alert systems. While most senior living environments likely lack the technical support experience and knowledge to design and implement such equipment, third-party turnkey solutions might provide significant enhancement to elders' quality of life.

Introducing IoT devices into the lives of elders should not be disruptive or overwhelming. Great planning will alleviate many issues; clearly define the problem to be solved with IoT technology, consider how you will handle end-user training and support, determine if the solution is affordable and will enhance quality of life, and consider taking small steps toward a larger objective. For example, one can start by learning to use a smartphone and a wireless network; subsequent steps can address the implementation of smart locks or video doorbells. Have a good plan and get buy-in from your target audience. Create awareness, develop acceptance through familiarity, and enable adoption by providing training and support. Also consider in your planning the variety of infrastructure elements that are needed to configure the desired implementation (smartphone, laptop, network, Wi-Fi, etc.) as well as the security precautions for the equipment. Remember that not all IoT devices are interoperable and seek solutions that assure compatibility. Finally, keep abreast of new research and tools through publications from groups like the SeniorPlanet.org and the International Society for Gerontechnology. Technology moves fast, and new IoT innovations may become available that address the very challenge that stumps you and your loved ones.

8.8 Case Study

8.8.1 Introduction

Telepresence is a tool that has been in existence in the world of business for some time. With new age of technology advancing faster and farther, there are more opportunities. Ohmni Labs, a company out of Silicon Valley, California has created the Ohmni Robot. Not your typical telepresence device, *The Ohmni Robot* is a fully functional telepresence device movable through user control via laptop computers. Figure 8.4 displays an image of what the device looks like. Transcending typical telepresence devices in the fact that it can be used in other situations like homecare, healthcare, and educational settings, the Ohmni Robot is remarkable.

8.8.2 Operating Capacities

AI and robots are becoming more sophisticated in their abilities. The goal is to make such technologies adaptable to the human environment becoming more normal regarding human interaction and daily functions. As these machines are becoming more prominent, especially as we continually progress into the IoT era—they have expanded their operating capacities to healthcare, business, academia, and in-home care domains. A Ball State University graduate team named their telepresence robot

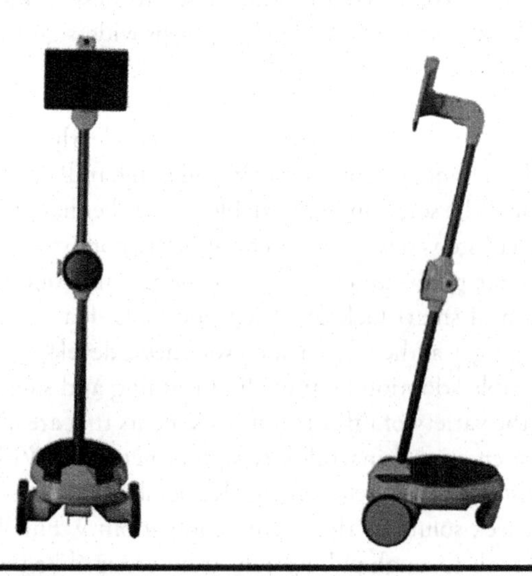

Figure 8.4 Rosie's structure.

"Rosie" (a Jetsons reference), and the team evaluated her abilities in some of these areas focusing on the healthcare and business realms.

8.8.3 Healthcare

The possibilities are endless when it comes to the potential value provided by robots and AI. In learning the functions of Rosie, including factory applications preloaded on her interface, comparing "reality" to "lab research," Rosie proved to be beneficial for many involved concerning healthcare. Due to her wireless connectivity ability she allows instant access. Syncing her to the multitude of online doctor platforms, patients or users may use her telepresence function to video call the doctor. Additionally, the telepresence aspect is advantageous for caregivers. Caring for a loved one is accompanied by a list of tasks and responsibilities and at times you are not able to be there 24/7. With Rosie, caregivers, especially distant ones, are able to dial into their robot companion to "check-in" on their family members. Caregivers can also use this robot to program reminders and alerts for taking medications, appointments, physical therapy, visitors, and more. Overall, such functions are becoming more crucial as there is a rise with in-home healthcare services. Furthermore, Rosie may act as a companion for those receiving in-home care, especially for the elder generation who tend to be less mobile.

Beyond her value to home healthcare, Rosie's maneuverability and data capacity affords the opportunity for usage within healthcare institutions as well. Operating across distances, this robot can be used to conduct certain tasks at the front end. Such tasks may include updating pre-check information, relaying messages from doctor to patients upon their wait, or even room checks by nurses. For example, hospitals are known for being nonstop action, receiving requests left and right from patients and those visiting patients. If Rosie were to be implemented for operable use, she could commit room checks, transfer information from one nursing station to the next, or evaluate the premises.

Note that Rosie must be operated by a laptop or desktop computer. This could be perceived as a hindrance considering society's push for mobility; it would be ideal to have the ability to operate Rosie from a smartphone or mobile device. Second, like most wireless devices, she is dependent on the network connection. Therefore, if the connection is poor or supports an overwhelming amount of traffic, she may be susceptible to interference or "noise."

8.8.4 Operability for the Disabled

When one is either mentally or physically disabled, there is a piece of their life that has been altered. Beyond the healthcare benefits that are provided, for those battling with disablement, Rosie presents a companion aspect. If purchased with the developer kit that contains modular arms, once assembled to the main body of Rosie, the developer kit can be used like that of physical human arms. With this addition, physically disabled individuals can maneuver these parts in a comparable fashion of human arms. Giving individuals such level of functionality that were previously denied offers them a sense of independence.

8.9 Business

8.9.1 Meeting Attendance

Rosie provides a much-needed service to meeting attendance that you cannot get with traditional phone or web-based chat technology. The robot allows employees to be present at meetings, forcing them to stay engaged in conversation. It allows you to be in the room even if you are remote. There are many people who can agree that they have been distracted during a Skype call or put a conference call on mute while working on other projects or are generally not paying attention to the meeting. According to a study done in 2018, 34 billion dollars were wasted due to distractions during conference calls in the United States and United Kingdom (Loopup, 2016). Rosie makes it easy to not only connect but to stay engaged in a conversation.

8.9.2 Out-of-State Interviewing/Training

In a study done by 1,000 organizations in the United States and United Kingdom, 88% of organizations felt that video conferencing was effective for training and interviewing.[16] Rosie provides a way for people to "meet" without the cost of travel. Having an initial interview with someone via video conference allows you to see and interact with the person; something you are unable to do with a phone call. For training purposes, this allows companies to train people in a one-on-one session, without the need of travel expenses. This is especially helpful for companies that have a training manager in a few locations, for a worldwide company, or a company that has remote workers. It allows employees more opportunities to better themselves within their career. Furthermore, it gives your training staff more opportunities to help colleagues grow.

8.9.3 Movable Workspaces

The idea for movable workspaces is becoming more popular in today's working industry. With workspaces constantly changing to prevent employees from becoming stagnant, it is important to have the equipment that can be easily moved as well. Quality of space projects are being implemented in organizations to help maximize and use the most space, without employees sitting elbow to elbow. Rosie can provide an extremely mobile device that is needed to meet the needs of a moveable workspace and a functional video conferencing tool. Rosie allows for the workspace to change with the needs of an organization, while providing an easy way for business to have a functional technology for video calls, interviews, and meetings.

8.9.4 Functional Capabilities

Rosie provides more than just a video call. Her other functions allow for a better experience for users within a work space. Starting with the bottom of the robot and working up, the movement system is robust and allows for it to change locations with ease. The ability to change speeds makes it easier to keep up with someone you may be having a conversation with who is moving. The movement allows you to quickly adjust your direction to get through small spaces like a meeting room. The speaker is another great feature for businesses. Depending on your work environment, the space you are working in may be loud and need the volume to be increased. It may also be necessary to have louder volume in larger rooms. The opposite is true as well. If you are in a location with many people on the phone, Rosie can have its volume adjusted to meet the needs of the area. The tablet on the top is small enough to not be a distraction but large enough to display a person's image clearly. It is also easy to navigate and has a quick response time when using the touch display. The picture quality is clear on the tablet for both the main screen and smaller screen. The front- and bottom-facing cameras provide an excellent display for the user. It is easy to see the people in the room with the front-facing camera and the bottom camera makes it much easier to navigate and maneuver through small spaces. Finally, the ability to adjust Rosie's tablet and camera up or down is helpful when needing to adjust to what is being shown to the robot. Rosie can easily look down at documents and back up to speak with the person directly.

8.9.5 Agile Workspaces

Agile workspaces are becoming much more common in the workplace. Many small teams are working together to achieve goals quickly and more effectively. The need for easy and mobile video communication is necessary for remote workers or managers with agile practices. Many of these teams will work in different areas and Rosie will be able to take the video call to these areas. By having a mobile video technology, teams can show remote workers what they are working on to help them get a better understanding of processes. It allows workers the flexibility to be in daily agile meetings without being on premise and being able to go to meetings that may be held in different locations within a building. It also allows for remote workers to meet with colleagues individually and in their own workspaces.

8.10 Discussion Questions

1. If your parents or grandparents plan to age-in-place in their own home, will they consider using IoT devices to make their lifestyle safer and more convenient? If not, what are the inhibitors? If so, what can you do as a "family technologist" to support their needs?

2. Review the website Seniorplanet.org, which supports elders in New York City. Would your community benefit from such a resource for elders? Why or why not? If so, identify some important elements of such a site.
3. What types of IoT devices do you think will be helpful to you or your loved ones through the aging process? Why? What support resources will you need?
4. What type of privacy concerns do you think telepresence robots will present in the home? In schools? In medical settings? Are there ways to alleviate any privacy concerns?
5. What are some of the important characteristics of a "family technologist"? How can we better prepare students for this role in the future?

End Notes

1. www.careinnovations.com/home-sensing/
2. https://ohmnilabs.com/#section-ohmni-forever-transforms-stay-connected-family-miles-away

References

Camara, C. and Zhao, Y. (2015). *The UX Learner's Guidebook*. Skokie, IL: Deuxtopia, Inc., p. 9.

CIA Factbook. (2017). Retrieved from www.cia.gov/library/publications/resources/the-world-factbook/geos/us.html

Graafmans, J. (2017). The History and Incubation of Gerontechnology. *Gerontechnology*. S. Kwon, Ed. New York: Springer Publishing Company.

Huber, L.L., Watson, C., Roberto, K. and Walker, B. (2017). Aging in Intra-and Intergenerational Contexts: The Family Technologist. *Gerontechnology*. S. Kwon, Ed. New York: Springer Publishing Company.

Kaiser Family Foundation. (2018). Retrieved from www.kff.org/state-category/health-status/

Kwon, S. (Ed.) (2017). *Gerontechnology: Research, Practice, and Principles in the Field of Technology and Aging*. New York: Springer, p. 1.

LoopUp. (2016). Enterprise Conferencing: User Behavior & Impact Report.

Norman, D.A. (2005). *Emotional Design*. New York: Basic Books.

Pew Forum. (2018). Retrieved from www.pewforum.org/religious-landscape-study/attendance-at-religious-services/

Pew Research Center. (2018). Retrieved from www.pewinternet.org/2017/05/17/technology-use-among-seniors/

Sofihub. (2018). Retrieved from https://sofihub.com on 6/12/18.

Chapter 9

Data Analytics with IoT

Frank Groom and Angelia Yount
Ball State University

Contents

9.1 Introduction

This chapter will deal with the analysis of data that comes from IoT devices. Sometimes this is called *Big Data Analysis* as the data that IoT derives is big in all three of the Vs (velocity, volume, and variety). The old methods of data analysis will just not work on this type of data. Accordingly, new methods of analysis have been derived and are being developed.

Because this is such a vast topic and one that is steeped in complexity, fluidity, and depth, we will have two sections to this chapter, each written by a different author. The first section outlines the scope, direction, and speed of the Big Data field, and the second section defines the technology, moving parts, and mathematics of the field. We felt this appropriate given the complexity of Big Data.

9.2 Section 1—Scope and Direction of Data Analysis

The Internet has been a valuable asset for individuals to acquire rapid information. However, the use of the personal computer (PC) to access this information has decreased due to the increase of individuals using non-PC machines.[1] The generation of large amounts of data is limited by the computer power; therefore, other devices are being used to establish large volumes of data.[2] These other machines generate large volumes of Internet traffic of data to servers.[1] Other machines include devices such as smartphones, tablets, Wi-Fi sensors, wearable devices, and household appliances.[3] This entire system of non-PC traffic signals between machine to machine (M2M) and machine to users is referred to as the Internet of Things (IoT).[1] Kevin Ashton is credited with creating the term "IoT" in 1999.[4]

It was at a 1999 presentation by Ashton for Proctor and Gamble on "Internet of Things" that the IoT trend began.[4] This type of Internet traffic of data to other machines has continued to grow. In 2014 experts claimed that there were 27 billion connected devices in the world and that 3 trillion dollars will be spent on IoT in 2020 to produce 2.5 quintillion bytes of data per day.[9] In addition, by 2020, approximately 25–50 billion devices will be connected to the Internet.[10] According to a Cisco report, 46% of the Internet traffic is generated by PC. The access of information by the PC will decrease to 21% by 2021.[5] Smartphones are predicted to account for 33% of the total IP traffic by the year 2021[5] while, televisions, tablets, and M2M will account for the remainder.[5] The number of devices connected to the Internet now exceeds the number of individuals in the world.[3] By the year 2020, the estimated network of interconnected devices will grow to approximately 26 billion units.[2] This growth will also generate approximately $300 billion in revenue.[2]

9.2.1 Big Data

The analysis of the data that is transmitted and remotely stored is possible using a number of techniques and algorithms developed specifically for dealing with enormous, disparate, and frequently unstructured data, generally termed "Big Data." This downstream analysis of collected and stored IoT data is a powerful benefit to developing smart IoT devices and new applications for integrating that data with other sources of information.[10]

Big Data has deep roots in new software paradigms developed by Internet and social media enterprises such as Google, Facebook, or Yahoo.[9] Big Data refers to datasets that are massive, complex, and high in variety and velocity.[6] Therefore, "Big Data" is very difficult to analyze using traditional tools and techniques.[6] As a consequence, many new tools have been developed to store, search, integrate, and analyze unstructured "Social Media" data and relate it to other more traditional structured, record-oriented information. Furthermore, traditional database (DB) management systems (Relational and Object-Oriented) are hardly capable of storing and mining the data generated today.[6] According to Ahsan and Basis, IoTs have increased the amount of volume of data traffic from various devices.[1] The challenges in IoT are the functioning of the devices due to the high transmission rate and the proper analysis of Big Data.[1] This issue can lead to far greater problems if the Big Data does not have algorithm to properly analyze it.[1]

9.2.2 Steps Involved in Analyzing Big Data

The primary steps involved in addressing Big Data are collection, transmission, storage, mining, processing, integrating with other data through pattern recognition, and the generation of accurate results.[1] Big Data storage of raw sensor data is typically done using an Extract, Load, and Transform (ELT) technique.[1] ELT is not flexible and does not work well with new data sources, which is not well suited for existing networks. However, such data can be stored in many of the Non-Relational DB systems that have been developed in recent years. Particularly those which are well adapted to the distributed and massive parallel processing (MPP) employed in Cloud Data Centers.[18,20] This range of storage technologies provides a range of flexible alternatives for storing data from new stationary and mobile devices, some with extensive processing power.[1] In addition to its need for storage, Big Data has to be processed. Magnetic, Agile, Deep (MAD) analysis separates and analyzes the storage and management of data processing.[1] In addition, MAD is highly desirable due to the high data processing speed and because it also allows for a huge IoT network.[1] There are multiple parallel processing techniques that can be used for the enormity of Big Data.[1] Parallel processing techniques like Message Passing Interface (MPI), MapReduce, and General Purpose Graphic Processing Units (GPU) are suitable for a centralized Big Data system, and many additional features and modifications have been created to make them suitable for the IoT structure.[1]

In some situation, the goal in storing and analyzing sensor and agent data is to store and process that data locally on a local host or a remote enterprise server.[1] One of the possible algorithms for localized processing is Localized Cooperative Access Stabilization (LCA).[1] Using the LCA algorithm, the nodes are not only aware of the main server state but are also aware of neighbor nodes.[1] There are four different schemes in which the LCA can be implemented in IoT, including exact, ad hoc, hierarchy, and hierarchy+ad hoc.[1] Below is an outline of how the schemes are used:

Exact-based scheme

- Devices send and receive data locally using LCA.
- Acts as a bridging node for communication between other devices.

Ad hoc-based scheme

- Sensors can act as a node for Big Data transmission of data from original to destination node.
- In order to calculate the shortest path, all the nodes frequently broadcast a short message, and then from the response time of each node, it calculates the shortest available active path.
- Not only utilizes all the available resources but is also more efficient in terms of data transmission rate than exact scheme.

Hierarchy-based scheme

- Devices are subdivided into upper and lower layers.
- Upper layer devices act as nodes to transmit data for lower layer devices in addition to the transmission of their own data, but lower layer devices only act as a sender or receiver of their own data.
- Has higher data transmission rate as compared to ad hoc scheme.

Hierarchy + ad hoc

- Similar to hierarchy scheme, but the devices in lower layer can also communicate with their immediate neighbors locally which makes it the most efficient scheme among all other in terms of data rate and energy consumption.[1]

All schemes are able to receive data continuously. However, there are cases where IoT devices do not need continuous data transmission. In manufacturing environments, modified TEEP (Trusted Execution Environment Protocol) and ISEP (Industrial System Engineering and Products) protocols are the best option for the IoT for interacting with local wireless communication technology and searching stored data using LCA algorithm (Lowest Common Ancestor search is a key work search algorithm which is efficient for local searching of simple web-style data storage). Also, since the amount of data associated with IoT is immense, there are several advantages in using the Cognitive Radio Technology (CRT).[1] One advantage is that limited bandwidth and spectrum issues may be resolved as needed as 5G implementation expands the employment of the higher mm wavelengths which research may also prove useful to manufacturing in building communication.[19]

The second advantage is that the IoT network can use multiple spectrum frequencies for different devices, which reduce interference between signals.[1] Next, the data can be transmitted through CRT for heterogeneous devices.[1] Finally, CRT can change frequency to allow data to transmit at a higher speed.

According to Zhang, Sun, and Cheng, in their discussion of radio frequency identification (RDIF) in manufacturing and similar settings, the appropriate architectural structure needed for IoT is made up of six layers.[7] The six layers used include the coding layer, perception layer, network layer, middleware layer, application layer, and business layer.[7] The architectural layers have different functions: the first layer, the coding layer, provides identification to IoT and also communicates algorithm to maximize data rate.[1] The perception layer collects Big Data and is known as the middleware layer.[1] Big Data analysis happens in the middleware layer. Data is received in the application layer.[1] Then, the network layer serves as a bridge between the perception and generates more responses of the processed data.[1] So, the layer architecture is critical in reducing data transmission to the Internet as it is processed in different layers.[1] Once the protocols are in place, testing is necessary. Therefore, in order to implement an IoT in the real world, it is important to make sure the environment is safe and secure.[1] Figure 9.1 shows one example of an architectural structure.

9.2.3 Four Data Mining Models for Processing IoT Data

Working with common data mining applications and research data that are similar to IoT data,[10] researchers have derived interesting and useful patterns and relationships. They used data from e-commerce, industry, healthcare, and city governance applications and employed commonly available data mining software to study this data.[10] These researchers have proposed four data mining models for processing IoT data that are listed below in Figure 9.2 with a description of the model.

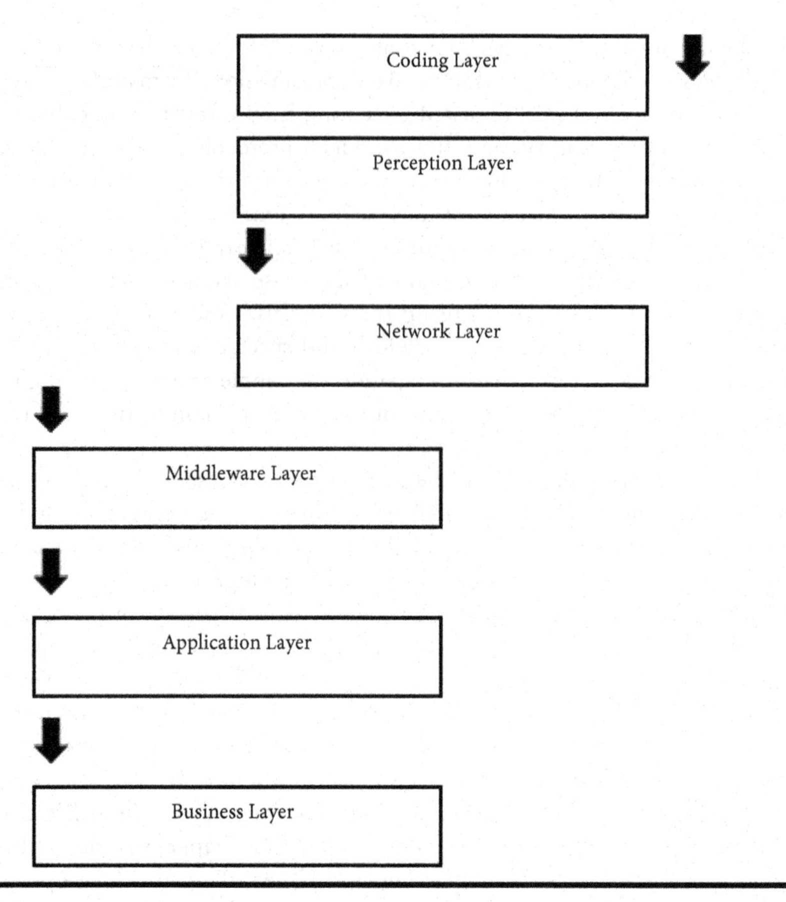

Figure 9.1 Layers of IoT.

9.3 Section 2—Categorization, Storage, and Analysis of IoT Device Information

9.3.1 Introduction

Agents, sensors, and devices in the time of the IoT decade each have a special purpose. Some respond to a stimulus by taking a direct action. Others respond by transmitting a unit of information, such as a code. Sensors are built into units and respond to the environment or another component such as a conveyor belt. More complicated devices may process an array of information, store a result, and transmit either wirelessly or wired to a central site. From there, either further action is taken or information is stored to be later merged with other data to conclude a more complete process. Therefore, to address analysis on data emanating from the variety of thousands and possibly millions of types of such components, we need

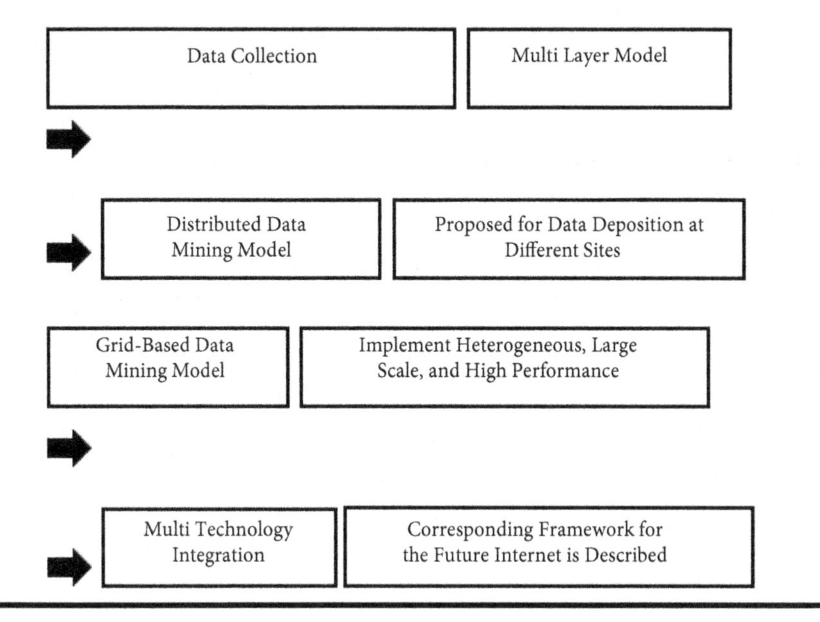

Figure 9.2 Four data mining models.

to first arrange them in a set of categories so that we can describe what analytics tend to work well with an array of devices that are generally similar in nature and purpose.[17,21,24] Furthermore, since such devices are now categorized into one large category of the "Internet of Things," transmission modes must be considered.[22] In this case, "Internet" stands for all means of transmission from local wireless, distant wireless, wired, and fiber transmission and may transmit across a room or assembly-line floor or internationally to a distant website or Cloud Data Center. This section intends to display such variation and to indicate a few of the storage and analytic techniques that might be employed to glean useful information or conclusions that may then be employed in directing further processes whether they be marketing, engineering, social, political, or personal. We will address this by moving from a broad IoT device categories, to the kind of data they produce, to transmission technologies they may employ, the data storage means that are appropriate, to finally settling on a set of traditional and more recently deployed analytic techniques. We concentrate on those useful in culling conclusions from this vast array of deployed diverse devices.

9.4 Some General Classifications of IoT Devices

IoT device classifications are initially organized as to whether the device is stationary or mobile. If the devices are stationary, we might divide them into home use, business office use, manufacturing, warehouse, retail, or data center use. If the devices are mobile we might divide them into ground-based, close to the ground, aerial unmanned drone, manned aircraft, or water-born.

9.4.1 Stationary Devices

9.4.1.1 Home Use

We first consider IoT devices located within the home. These devices may be used for home automation including lighting, heating and air conditioning, media, and security systems. Security may employ a security device at the doors and windows which wirelessly connects to a central node, which further transmits a signal over telephone or the Internet to local authorities if necessary. Or the device may be located in kitchen appliances, entertainment equipment, shades, windows, garage doors, or other similar home apparatuses as part of a home automation.

9.4.1.2 Business Use

1. The office

   Stationary devices for businesses have been in existence since the early 1960s with the introduction of terminals to connect to distant IBM Mainframe computers. In the early 1980s, this changed with the introduction of PCs and local area networks connected to telephone networks and the Internet in the 1990s. However, with the flattening of the organization structures in the late 1990s and the work-at-home and on-the-road as well as work by smartphone, the stationary business traffic has moved sharply to mobile communication.
2. Manufacturing, warehousing, and retail operation

   With the automation of factories, warehouses, and distribution centers, fixed sensors and other algorithm-driven agents are increasingly being deployed. An extreme example is the Tesla automotive assembly line that is almost completely automated. The stationary sensors, agents, and robotic arms must sense, communicate to a central processing location, and operate on a continuous basis. Digital control systems are implemented to automate robotic operations, process controls, operator interaction, and the service control information systems to insure safety.[8] Such operations are pervasive in factories, warehouses, fulfilment centers, trucking, and packaging locations, and retail operations are systematically replacing human labor. These agents and sensors transmit a coded radio-frequency identification (RFID) using electromagnetic fields to automatically identify and track tags attached to agents, sensors, and other objects. These tags contain electronically stored information. Passive tags collect signals from strategic locations in the facility while mobile supervisors communicate within the facility and with the local office by means of smartphones.[11]

3. Hospitals and medical care

   Many of the patient-attached and machine-controlled sensors are stationary, but the personnel are continuously mobile working with smartphones and tablets as they go from room to room.

   Health monitoring devices ranging from blood pressure and heart rate monitors to devices that communicate with implanted pacemaker abound. Some hospitals have even begun investing in intelligent beds which can detect a range of patient activity and modify the bed components to accommodate the patient, while communicating results to nurses' stations.

9.4.2 *Mobile Use*

With the decade of deployment of "smartphones" of the fourth generation (4G), 4G LTE (long term evolution), and, in 2020 the introduction of the fifth-generation (5G) technology (which some call the IoT generation), there has been and will be an explosion in the uses of mobile devices. The land line to home has largely been replaced by use of the ubiquitous iPhone or Galaxy smartphone for telephone, text, e-mail, search, news, picture taking and transmission, social media, and chatting. This has required the upgrade of the towers and support access networks. The anticipated move to the fifth generation of devices and services has required a complete upgrade of the cellular access network, from the device, to the radio transmission technology, to the towers, and the forwarding facilities.[23,27]

Mobile users with smartphones and mobile automobiles will share these towers with cellular and Wi-Fi short-range wireless communication. As unmanned automobiles begin to invade our highways, they need to augment their moment to moment decision-making by transmitting their location information and receiving information about upcoming traffic conditions which are outside their scanning technology capabilities. Figure 9.3 portrays the Access Tower and Relay facilities to transport mobile traffic to the city or metropolitan facilities.

9.5 Transmission Alternatives for the Various Classes of IoT Devices

9.5.1 *Local Wireless and Distance Wireless*

Over the past few decades, a number of transmission technologies have been created. Table 9.1 portrays 28 varieties of local in-room or in-building wireless technologies as well as the generations of more distant transmission through cellular wireless transmission and even point to point distant wireless transmission technology. For each IoT device deployed, there are a set of possible transmission technologies available to support its usage and a variety of vendors providing them.

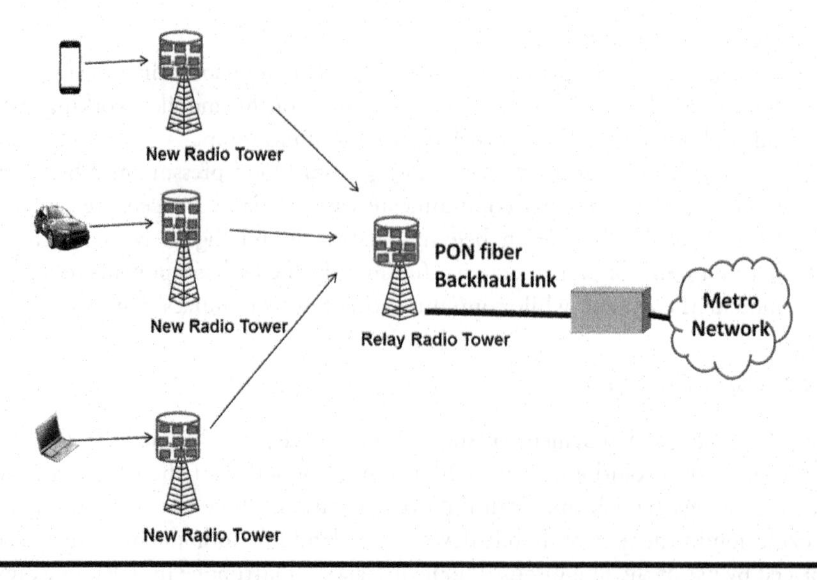

Figure 9.3 The new radio and relay towers to forward cellular traffic to the networks.

Table 9.1 Some Wireless IoT Network Protocols

Bluetooth	BLE	Zigbee	Z-Wave
6LoWPAN	Thread	Wi-Fi	Wi-Fi___33-ah
2G Cellular	3G and $G Cellular	LTE-M1 Cellular	5G Cellular
NFC	RFID	SigFox	Ingenu
Weightless-N	Weightless-P	Weightless-W	ANT, ANT+
DigiMesh	MiWiIngenu	Dash7	NB-IoT
EnOcean	WirelessHart	WiMax	LoRaWAN

Furthermore, when data is transmitted across a city, a state, a nation, or internationally, there are a number of first-mile to 15-mile access approaches. These are existing cross-city telephone and Internet facilities and services from private line to carrier ethernet and cross country by MPLS, carrier ethernet, or the public Internet. These options are portrayed in Figure 9.4.

The varieties of more distant transmission technologies available for IoT transmission are further discussed in Chapter 6. That chapter also discusses the Cellular 4G, LTE, and 5G distant access technologies as well as the standard private line, fiber-based, and point-to-point wireless technologies that provide local access to the metropolitan networks, the public Internet, and the national telephone company backbone facilities.

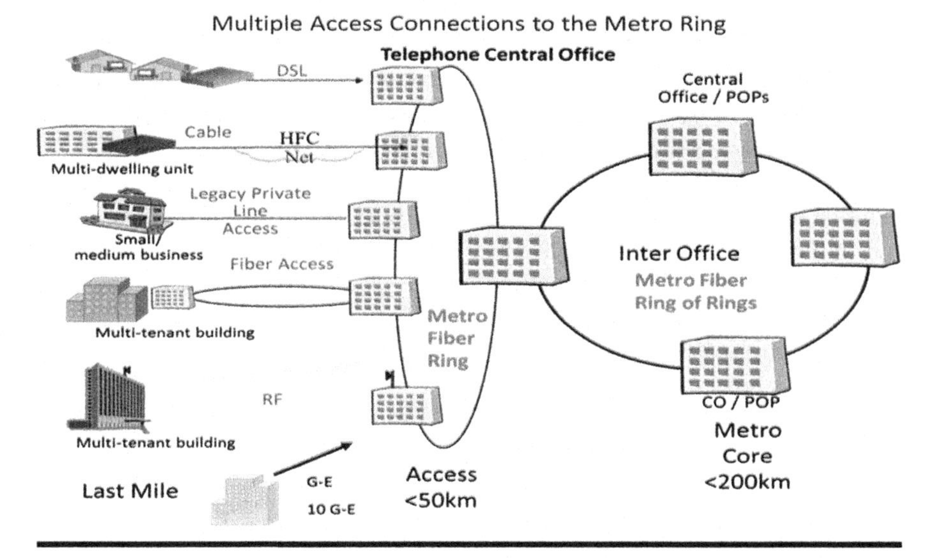

Figure 9.4 Connection alternatives for businesses and residences cross city and cross country.

9.6 Local and Distant Actions Initiated by Device Types

In manufacturing, warehousing, and retail enterprises, each IoT sensor or agent tends to recognize a situation and transmit a message locally to an intelligent processor which triggers a local action by another agent and frequently uploads all resulting information to some local or distant storage and analytic center. Although this accumulation of information tends to be continuous, its content has limited utilization beyond its immediate purpose.

However, in other situations, the collection of information from such locally triggered actions might be useful to the enterprise itself on employing, extending, or modernizing the automation and accompanying operation. It may further be valuable to industry specialists and researchers and have implications far beyond the factory floor, store, or warehouse. Thus, distant storage and relations with data in other DBs might prove valuable.

Other devices might exist in mobile situations including people on the go as well as vehicles such as cars, trains, buses, aircraft, drones, and watercraft. Agents in such mobile vehicles need to continuously react with the vehicle as well as the outside environment. And they must continuously connect to a stream of antennas to transmit and receive information from external sources. This vehicular information can be transmitted to public transportation agencies, corporate employers, public entities, or personal sites.[29] Each of these may collect some subset of the received information and store it for later analysis and merging with other DBs of information.

9.7 Distant Storage of Transmitted IoT of Device Data

Once data from any of the devices we have discussed arrives over some form of network, it may trigger an action at the destination site. Furthermore, it likely will be accompanied by metadata (data such as time of day of event, who triggered it, what action was triggered, who is responsible for dealing with the action, and any billing resulting from the triggered action). Then this transported information is stored in usually some form of database. If the data is structured in a common repeatable record structure it likely will be placed in a Relational DB. If it is unstructured but extensive, it may be placed in an object database such as Amazon AWS Simple Storage System (S3). If it is only to be retained temporarily, it might be stored at the processing device in solid-state storage (SSD). Since much of IoT data is unstructured with no formal record structure, it is commonly stored in flat, unstructured, NoSQL DBs.

9.7.1 Storing IoT Big Data

One of the traditional storage methods, dating back to E. F. Codd's work at IBM, has been the storage of structured records in a standard column-based Relational database. Payroll records, parts records, and student records, all in a fixed formatted and repeatable structure with a key to find and access the records were efficiently addressed. These fixed structures were searchable with a standard Structured Query Language (SQL). An example of an SQL request might be:

(Select File A and B, from Table C, where Field 1 > 12 and Field 2 < 35, having characteristics c, d, and e)

Among these Relational Databases (DBs) searchable with SQL are IBM's DB2, Amazon's Relational Data System (RDS), and Oracle's Relational Database Management System (RDBMS). However, as the Internet progressed, simple, extendable, unstructured, and "flat" files were employed. These were frequently termed NoSQL Databases which flourished where the data could be extended out end to end and queried by a stored set of key word indexes.

Among the NoSQL type databases suitable for unstructured data are those that are document-oriented, those that support column-oriented, and those that support key value-oriented information.[12] Examples of each of these types are presented in Table 9.2.

Furthermore, as images, videos, sound, and other varieties of material are to be stored, each has its own characteristics and size variations. Soon object-based storage systems emerged and then came the mixed Object and Relational Database systems. Among these are (Table 9.3).

Each of these Database systems supports a specific set of languages but frequently not the same languages. Among these programming languages are usually found C++, Java, Java Script, Cache, and Perl. However, Python and Ruby have become increasingly popular and some DB systems also support ObjectScript, Smalltalk, Telos, Python, and JDBC to name a few of those in evidence.

Table 9.2 Some Examples of NoSQL Database Systems

Type	*NoSQL DB Example*
Document DBs	MongoDB, CouchDB, Terrastore
Column-oriented DBs	Google's BigTable, Hadoop's HBase, HyperTable, Cassandra, Amazon's Dynamo
Key-value stores	Scalaris, Redis, Voldemori

Table 9.3 Object-Based and Object Relational DBs

Cache	ODABA
ConceptBase	OpenAccess
Db4	OpenLink Virtuoso
Gemstone	Perst
ObjectDB	Picolisp
ObjectDatabase	Versant Object Database
Objectivity/DB	WakandaDB
ObjectStore	Zope Object Database

The difficulty is not that we have a shortage of storage systems appropriate for individual types of data but that we have collection problems of vast amounts of individual recordings. We then have storage issues due to the volume of such data at locations where we can process and analyze that data, and then the data must be edited, "cleansed," and extracted in units suitable for analysis.[25,28] These issues are dealt with in the following fashion:

1. The vast number of moment-to-moment collection of each type of data emanating from the variety of IoT devices quickly overwhelms available storage media. With factory sensors, the frequent recording of the repeated indicator quickly delivers massive amounts of the same data. Usually, a sampling approach is used taking every 10th or 30th actual recording and uploading it. Furthermore, we then generally store counts of individual data types (rather than the actual data itself) along with metadata about each type, dramatically reducing the amount of storage required. This is a lesson learned from the telephone companies which would find it impossible to store all the telephone messages even in DBs arrayed across the Milky Way Galaxy.

2. Rather than process the data on one computer and data storage unit, we first distribute the data across hundreds and even thousands of data storage units in a procedure termed "Sharding" the data. We then similarly execute the copies of the same program across hundreds or more rack-mounted commodity-type computers in a standard distributed processing methodology as performed in most Cloud Data Centers.
3. The power of Data Mining and Relational Analysis is in examining across a wide variety of the stores for different data types. Frequently, the important discovery comes from relationships between seemingly unrelated kinds of data. Thus, the analyst must have data from the various kinds of media types.
4. The answer seems to be to extract a "sample" (probably not a random sample) from a variety of data stores and then determine if some preliminary conclusions can be made, particularly whether some underlying patterns of groupings and relationships exist among certain data types.

   If such patterns are discovered, then larger datasets and new datasets can be studied to see if these newly discovered patterns of relationships and groupings continue to apply. A system of sequential data extraction and analysis can deal with the difficulty of stranded available Data Analysis tools having low levels of data size tolerances.
5. NoSQL Data Management Systems, with their flexibility to handle unstructured and heterogeneous data, appear to be the most popular data modeling and storage system when we store the IoT data at a distance, usually Cloud site. NoSQL is likely to be also appropriate for intermediate storage sites at a company or manufacturing factory site. The local, temporary storage on the factory floor, or in an adjacent room, is likely to employ simple web-based flat file systems with some standard RFID codes or similar content classifications.[13]
6. We must be aware that analytic approaches depend on "clean" data. After data is extracted from a data storage facility and prior to analysis, a set of programs must first make sure the files of data contain content that is expected for that type of source. A standard range of tolerable codes must be stored. Then the absence of any code or count among a set of such data must be either eliminated or standardized in some acceptable fashion. Blanks, out-of-range numbers, and nonstandard text can make the analysis unworkable. Only after an acceptable set of "cleansed" data emerges, can such data be entered into an analytic program.

9.8 Mining Stored IoT Big Data

Data mining has become a popular venture for corporations who store large amounts of data. That data and customer behavior when combined with more general population data and more recently acquired social media data may provide

Table 9.4 Some Commonly Used Data Mining Tools

Oracle Data Mining	Weka	Qlik Sense
Data Melt	Knime	Birst
IBM Cognos	Sisense	ELKI
IBM SPSS Modeler	SSDT (SQL Server Data Tools)	SPMF
SAS Data Miner	Board	GraphLab
Mahout	Dundas BI	Mallet
Teradata	Intetsoft	Alteryx
Rapid Miner	KEEL	Mlpy
Rattle	R Data mining	
Orange	H_2O	

lucrative opportunities that were not previously perceived. Table 9.4 presents just a few of the mining software which companies currently employ.

9.8.1 Pattern Recognition When Scanning IoT Big Data

Most of early pattern recognition software was designed to be used with structured records since those were what all organizations processed and stored such data. Table 9.5 shows some analytical tools for structured data that unstructured data has now become dominant. And IoT-originated data will soon dwarf both structured and our current version of unstructured data.

Table 9.5 Analytic Tools for Structured Data

Grouping K-means
Probability Bayesian
Prediction and cause
Component and factor analysis
Forecasting
Trend analysis
Periodic seasonal and cyclical decomposition
Triple exponential smoothing
Multiple ARIMA

9.8.1.1 Dealing with Structured Data

The nature of "Big Data" has rapidly changed from the structured, record-oriented data of companies, local, regional, and national organizations and of government entities. Structured data contains a defined data type, in a repeatable structured format. This is the traditional record-based data which is employed in online application processing and is stored in fixed record tables of a Relational DB. Such structured data continues to grow at a modest but continuous fashion as populations grow and the organizations that support them expand. However, with the dawn of social media in the later part of the 2000s, unstructured data has exploded in an exponential fashion. As the 5G implementation of a set of network protocols and technologies progresses, connectivity for a whole new array of new devices will become feasible.[26] Now, mobile use of previously stationary devices may prove feasible, and the introduction of virtual reality components may enhance everyday living in surprising ways. The resulting vast ocean of IoT device information threatens to make the exponential growth of unstructured data appear rather unsupportable as the oceans of IoT information flood through our networks. This will be a challenge. Figure 9.5 presents the relative expansion of these three waves of information growth.[14]

The opportunities created by storing the content and metadata about all IoT device information collected and eventually stored in a central repository result from the ability to relate and merge with other data. Sometimes profound results occur when data is analyzed for relationships with wholly unlikely other

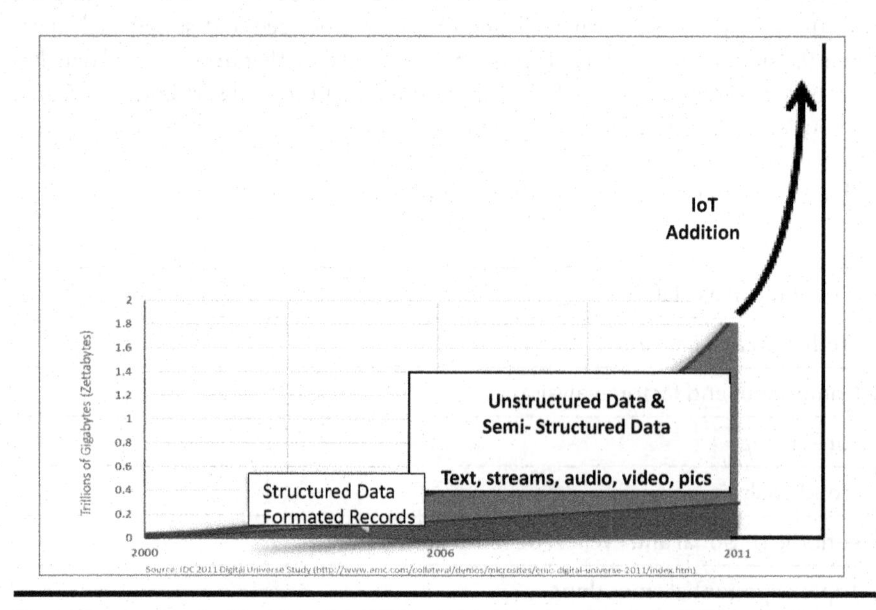

Figure 9.5 Massive IoT data in addition to structured and evolving unstructured data.

data. A number of preplanned analytic routines can be performed on such sets of data looking for the strength of such relationships and the possibilities for action on the resulting information. However, the more surprising opportunities come from the possible pattern discerned. When we have available a trend series over time of individual generally unrelated data, we occasionally find hidden patterns which provides profound new understanding of how people and industrial operations perform over time. We may discover patterns more correlated than previously understood. This relationship may lead to new actions that may benefit society and clusters of individuals which might normally not be thought of.

9.8.1.2 Dealing with Unstructured or Semi-Structured Data

Semi-structured data consists of textual data files with a discernable pattern which allows the researcher to parse the text for grouping, clustering, and analysis. Web-based XML data files follow a semi-structured scheme. Other such data is also composed of textual data but which has erratic data formats and may contain some inconsistencies in data values. This may require the researcher to expend time and effort in reformatting the text prior to analysis.

Unstructured data has no inherent structure such as text documents, PDFs, images (JPEGs, GIFs, PNGs), video clips (MPEGs), and even full movie videos. Such data types are usually stored as objects (with an Object ID, metadata about the data, and the actual data itself). These might be stored in an Object DB which is a continuous "Flat" file system such as Amazon ASW's S3 Object DB which can contain a variety of different types of unstructured files.

Although some new tools are employed in dealing with unstructured and semi-structured data, such as Hadoop, many of the Clustering, Grouping, and Times Series Analysis routines work equally well with this data. However, some preliminary reformatting and indexing of the data may be required as a preliminary step. Table 9.6 lists a few of the analytic tools that might be employed with such unformatted material.

9.8.2 Analyzing Unstructured Text Material and Pattern Analysis with Hadoop

Hadoop is an overall software framework containing a set of applications to process large datasets in parallel and allowing users to create some of their own applications for specialized purposes (Table 9.7). This processing frequently occurs in parallel in Cloud Data Centers on hundreds and sometimes thousands of rack-mounted, commodity computers against data stored on thousands of Raid Disks using Hadoop's HDFS (Hadoop Distributed File System).[15]

Table 9.6 Analytic Tools for Semi-structured and Unstructured Data

Grouping K-means
Hadoop-MapReduce, HDFS, Big table
Probability-Bayesian analysis
Prediction and causation
Regression
Component and factor analysis
Trend analysis and forecasting
Periodic seasonal and cyclical decomposition
Triple exponential smoothing
Multiple ARMA

Table 9.7 Hadoop Standard Family of Application Programs–Apache Version

<table>
<tr><td colspan="3">Hadoop Family Components</td></tr>
<tr><td>PIG Scripting</td><td>MANHOUT Machine Learning</td><td>HIVE Query</td></tr>
<tr><td colspan="2">MAPREDUCE</td><td>Distributed processing</td></tr>
<tr><td colspan="2">HDFS</td><td>Distributed storage</td></tr>
<tr><td colspan="2">HBase NoSQL</td><td>Unstructured data base</td></tr>
<tr><td colspan="2">REST/ODBC/SQOOP</td><td>Data integration</td></tr>
<tr><td colspan="2">AMBARI</td><td>Management & monitoring</td></tr>
<tr><td colspan="2">ZooKeeper</td><td>Coordination</td></tr>
<tr><td colspan="2">Ooozie</td><td>Workflow & scheduling</td></tr>
</table>

Using the MapReduce component of the overall Hadoop package, a central MapReduce process separates the input dataset into independent pieces, which are then sent (mapped in Hadoop terms) to different computers in the array, processed separately, and in parallel on different computers in the distributed array, sorted, and merged (reduced in Hadoop terms) back into an overall conclusion and then stored back in the original distributed file system. On the nodes where data is already present, results in high aggregate transfer bandwidth across the cluster.

Furthermore, we can then apply our new information about interrelationships to real-time operations and action, we can have immediate benefit and results rather than delayed results.

9.9 Further Statistical Analysis of IoT Big Data

Since the number of IoT devices and the 5G networks that are being enacted to transmit their information, the ideas associated with Big Data come into consideration. In most cases, the amount of such data is enormous, and a decent portion of it may need to be transmitted to a distant storage center. However, in many cases that data need not be stored for its own purposes. We then need to consider if we can discern new information be combining such Big Data with other behavior data and personal information data such as search sites selected, social media sites and content selected, purchased, or sites browsed but items not selected.

From such combined information, individual behavior patterns may be deduced for a benefit to concerned parties if the component data is stored and then made available internally or externally by entities such as marketing, research, educational, or political concerns.

When approaching analysis, the researcher first must decide which problems she wishes to address, then the appropriate category of technique to be used, and then a specific technique is chosen from a large number of alternatives. Technique has proved fruitful in specific areas such as psychology but may have proved more difficult to utilize in areas such as medicine and neuroscience. Each specific technique has a set of program algorithms and useful file structures appropriate for accomplishing the analytic tasks. These are presented in Table 9.8.

The process undertaken in dealing with text-based information begins with a process of parsing the sentences and phrases and code of the unstructured information whether writings, phrases, or streams such as Twitter communications with its specialized coding and indicator structures, which are unique to that media. Then, a file of common Word and common symbols files "Bag of Words" are employed and used to look up synonyms, unique social media words, or phrases. Buzz tracking of popular blogs may also be searched for matching and understanding. A search and retrieve process of words and phrases are then undertaken to create a set of search arguments.[30] These are then employed in the process of mining the text files for content, meaning, and similarities. This may be done in an orderly fashion or in an intuitive approach particular to the researcher. Finally, the material is grouped into a set of categories and classifications. Then the specific data subjects are sorted into the categories that best fit each subject. Some of these techniques appropriate for dealing with certain general classes of problems are portrayed in Table 9.8.[16]

After beginning by determining the specific type of problem the researcher wishes to address, the researcher will narrow down the approach to a series of tasks to accomplish. These may be attempting to separate subjects in groups, determine

Table 9.8 Solution Techniques for Individual Problem Types

Type of Problem to Solve	*Category of Solution*	*Solution Techniques*
Desire to group items by similarity. Desire to find commonality structures in the data	Clustering	K-means clustering
Desire to discover relationships between actions and items	Association rules	A priori
Desire to determine relationship between the input (independent) variables and the outcome (dependent) variables	Regression	Linear regression Logistic regression Ridged regression
Desire to assign (known) labels to object	Classification	Naïve Bayes decision trees
Desire to find the structure in a temporal series of data (daily, monthly, quarterly, annually)	Time series analysis	ARIMA ACF, PACF, Exponential smoothing
Desire to analyze unstructured text data	Text analysis	Regular expressions, document representation (bag of words), TF-IDF, Hadoop program family
Find the "Term" occurrence frequency, document frequency of a word series of documents	Document analysis	TF-IDF

Table 9.9 A Range of Analytic Methods Appropriate for Analytic Problem Areas

Key Analytic Approaches	*Key Analytic Methods*
Algorithm and technical foundations	Categorization
	K-means clustering
	Association rules
Key cause and effect and projection use cases	Regression
	Linear
	Multivariate
	Logistic
	Ridged
Grouping, classification, diagnostic, and model validation approaches	Classification
	Naïve Bayesian classifiers
	Decision trees
Algorithms dealing with data spanning time	Time series analysis
	Decomposition and curve fitting
	Smoothing-triple exponential
	Multiple ARIMA
Fitting scoring, validating, relating, and storing, querying, and analyzing unstructured text	Text analysis Hadoop family MapReduce, Manhout, Hbase, HDFS, Yarn, Pig, Hive

cause and effect, or forecast a series of available data out into the future where the problem needs to be addressed. Table 9.9 separates such analytic approaches down to specific activities and appropriate algorithms.

9.9.1 Groupings and Clustering

Although a marketing firm may wish to address a particular customer with action, this is too expensive to do individually for specific "customers." One of the most prominent techniques employed when dealing with the voluminous amount of what is termed "Big Data" is to attempt to classify the information into similar groupings. This is further required by the data analyzed being unstructured and even having coded content. The content may be something that appears as ABCD or 12 CD or !@#$, or some combination of these along with numbers and text. This coded content is not particularly useful on its own but may be used to help classify associate people of actions into similar groupings. If we can classify a large number of people into a set of overall groupings and subgroupings, further actions can be taken towards those members of particular groups as a unit. This makes resulting actions affordable since expenditures can only be expended towards highly likely and appropriate parties.

9.9.2 Cause and Effect

One of the principle conclusions that statistical research desires to extract from selected material is whether there are any instances where a certain collection of variables tends to relate to a collection of other variables of interest in such a consistent fashion that we can reasonably conclude that these "independent" variables actually cause the behavior of the principle "problem" variable of interest. This observed cause and effect must be consistently stronger than merely indicating that there is a relationship between those variables, but that one set causes the behaviors of another. This is a difficult threshold to exceed. The family of statistical tools employed to attempt to reach such conclusions is listed in Table 9.10.

Table 9.10 Causation, Prediction, and Projection Statistical Tools

Linear regression	Probit
Simple regression	Multinomial probit
Ordinary least squares	Ordered probit
Polynomial regression	Nonlinear regression
Ridged regression	Nonparametric
Factor analysis regression	Semiparametric
General linear model	Robust
Generalized linear model	Quantile
Discrete choice	Isotonic
Logistic regression	Principal components
Multinomial logit	Least angle
Mixed logit	Local
Ordered logit	Segmented
Poisson	

9.9.3 Projection

In some cases, we may desire to use these regression tools not merely to determine cause and effect but whether we can project either into the future similar results by extending the fitted formulation. In other cases, we might wish to interpolate within a time frame of the data in blank area values that fit the formulation achieved by the regression analysis. Simple projection is the beginning technique to forecasting future patterns from existing ones.

9.10 Summary

IoT, with its vast array of devices, unique content for each type, and possibility of each set of data to be related to so many other traditional and emerging unstructured social data, provides a blossoming opportunity for discovering new insights into human behavior. Organizations continue to optimize their operations by collecting, mining, and analyzing this treasure of information. Fortunately, with current statistical clustering and grouping techniques, we are able to analyze how individuals and device placements logically separate into large common groupings. This allows decision makers to design strategies to address a smaller set of groups rather than each individual. General strategies can thus be devised which can be applied to a group of people or devices placements who share common traits making the action conclusion from analysis more efficient and action combinations more manageable.

Furthermore, as artificial intelligence allows for more sophisticated analysis and machine learning allows for the deployment of devices and software to progressively learn from their daily operation, the results will enable companies to automate at levels well beyond our expectations.

We must now enter into discussions concerning what we want to happen, how we control what happens, and how we regulate the employment of IoT, Big Data, cross database mining, and implementation of artificial intelligence to enhance our lives while simultaneously enabling the efficiencies that make us prosper.

9.11 Case Study

Medical personnel in year 2020 wish to employ a mobile device while driving an ambulance in a city. That device connects to a 5G Cellular Network while driving and delivers and receives time-sensitive information while in motion. This mobile device connects to a specialty hospital within the city and to a national database at a Federal site in Atlanta, Georgia.

The mobile medical device delivers precise information concerning the patient being transported to the local specialty hospital, provide notification and precise diagnostic information to that local hospital while in transit.

The medical personnel in the ambulance supports the patient in transport, interact with the local specialty hospital, while communicating with the national disease control personnel to perform notification, specification transmission, diagnosis, and communication with other hospitals dealing with similar infectious cases.

Describe what statistical problems the hospital and the ambulance personnel might address while in transport and describe the statistical tools that each would employ to determine the cause and the effect of various drugs applied and handling of the patients during transport to the hospital.

9.12 Discussion Questions

1. What are some new types of the home uses of stationary IoT devices that you can think of?
2. What are some of the situations where businesses might transmit data from stationary IoT devices?
3. What are some new, unthought of, mobile uses of IoT devices?
4. What are some of the Wireless Transmission alternatives for transmitting data from stationary devices in a local room or area?
5. What category of data storage system is appropriate for storing unstructured IoT data?
6. What are some of the tools available for mining stored IoT Big Data?
7. What software is used to analyze unstructured text material?
8. What are some of the problems to statistically address for IoT Big Data?
9. What are techniques that are appropriate for addressing these problems?
10. What techniques are appropriate for dealing with cause and effect?

References

1. Ahmed, Ejaz, Ibrar Yaqoob, Ibrahim Abaker Targio Hashem, Imran Khan, Abdelmuttlib Ibrahim Abdalla Ahmed, Muhammad Imran, and Athanasios V. Vasilakos. "The role of big data analytics in Internet of Things." *Computer Networks* 129 (2017): 459–471.
2. Ahsan, Umar, and Abdul Bais. 2016. "A review on big data analysis and Internet of Things." In *Mobile Ad Hoc and Sensor Systems (MASS), 2016 IEEE 13th International Conference on*, pp. 325–330. IEEE.
3. Ashton, Kevin. "That 'Internet of Things' thing." *RFID Journal* 22, no. 7 (2009): 97–114.
4. "Cisco visual networking index: Forecast and methodology, 2016–2021," www.cisco.com/, September 2017, [Online; accessed 11-July-2018].
5. Elgendy, Nada, and Ahmed Elragal. "Big data analytics: a literature review paper." In *Industrial Conference on Data Mining*, pp. 214–227. Springer, Cham, Switzerland, August 2014.

6. Jesse, Norbert. "Internet of Things and big data–the disruption of the value chain and the rise of new software ecosystems." *IFAC-PapersOnLine* 49, no. 29 (2016): 275–282.
7. Zhang, Minghui, Fuqun Sun, and Xu Chen. 2012. "Architecture of internet of things and its key technology integration based-on rfid." In *Computational Intelligence and Design (ISCID), 2012 Fifth International Symposium on*, pp. 294–297, vol. 1. IEEE.
8. Mahdavinejad, Mohammad Saeid, Mohammadreza Rezvan, Mohammadamin Barekatain, Peyman Adibi, Payam Barnaghi, and Amit P. Sheth. "Machine learning for Internet of Things data analysis: A survey." *Digital Communications and Networks* (2017).
9. Nobre, Gustavo Cattelan, and Elaine Tavares. "Scientific literature analysis on big data and internet of things applications on circular economy: A bibliometric study." *Scientometrics* 111, no. 1 (2017): 463–492.
10. Shah, Mohak. "Big data and the internet of things." In *Big Data Analysis: New Algorithms for a New Society*, N. Japkowicz, and J. Stefanowski (Eds.), pp. 207–237. Springer, Cham, The Netherlands, 2016.
11. Wu, Xingdong, Xingquan Zhu, Gong-Qing Wu, and Wei Ding. "Data mining with big data." *IEEE Transactions on Knowledge and Data Engineering* 26, no. 1 (2014).
12. Li, Tangli, Yang Liu, Ye Tian, Shuo Shen, and Wei Mao. 2012 "A Storage Solution for Massive IoT Data Based on NoSQL." In *2012 IEEE International Conference on Green Computing and Communications.*
13. Kang, Yong-Shin, Il-Ha Park, Jongtae Rhee, and Yong-Han Lee. "MongoDB- based repository design for IoT- Generated RDID/sensor big data". *IEEE Sensors Journal* 16, no. 2 (2016).
14. EMC. *Data Science and Big Data Analytics Student Guide*, EMC, Hopkinton, MA, 2013.
15. DeZyre. Hadoop Ecosystem Components and Its Architecture, June 4, 2015. www.dezyre.com/article/hadoop-ecosystem-components-and-its-architecture/114, [Online; accessed 15-February-2019].
16. EMC. *Data Science and Big Data Analytics*, Wiley, Indianapolis, IN, 2015.
17. Bin, Shen, Liu Yuang, Wang Xiaoi. 2010. Research on Data Mining Models for the Internet of Things, IEEE, 978-1-4244-5555-3.
18. Cai, Hongming, Boyi Xu, Lihong Jiang, Athanasios V. Vasilakos. "IoT-Based big data storage systems in cloud computing: Perspectives and challenges." *IEEE Internet of Things Journal* 4, no. 1 (2017).
19. Chang, Fay, Jeffery Dean, et al. "BigTable: A distributed storage system for structured data." *ACM Transactions on Computer Systems* 26, (2008).
20. Fazio, M., A. Pulifito, M. Villari. 2015. "Big Data Storage in the Cloud for Smart Environment Monitoring". In *6th International Conference on Ambient Systems, Networks, and technologies*, ANT.
21. Halpere, Fern. 2015. Operationalizing and Embedding Analytics for Action, Best Practices Report Q1 2016, TDWI and SAS.
22. IEEE. 2014. A Cognitive Oriented Framework for IoT, Big Data Management Prospective, IEEE.
23. Jiang, Lihong, Li Da Xu. "An IoT-oriented data storage framework in a cloud computing platform." *IEEE Transactions on Industrial Informatics* 10 (2014).
24. LaPlante, Alice. *Analyzing Data in the Internet of Things*, O'Reilly, Sebastopol, CA, 2016.

25. Marjani, Mohsen, Fariza Nasaruddin, Abdullah Gani, Ahmad Karim, Ibrahim Abaker Targio Hashem, Aisha Siddiqa, Ibrar Yaqoob. "Big data analytics: Architecture, opportunities, and open research challenges." *IEEE Access* (2017). DOI: 10.1109/Access.2017.2689040.
26. O'Leary, Paul, Matthew Harker, Roland Ritt, Michael Habacher, Katharina Landl, and Michael Brandner. 2016. "Mining Sensor Data in Larger Physical Systems." In *ACM/IEEE International Conference on Cyber-Physical Systems.*
27. Rizvi, Sanam Shahla and Tare-Sun Chung. 2010. Data Storage Framework on Flash Memory Based SSD RAID 0 for Performance Oriented Applications, 978-1-4244–5586-7/10, IEEE.
28. Russom, Phillip. 2011. TDW and SAS, Big Data Analytics—The Best Practices Report, Fourth Quarter.
29. Xu, Li Da, Wu He, and Shancang Li. *Internet of Things in Industries: A Survey*, IEEE Transaction on Industrial Informatics, IEEE: Online, USA, 2014.
30. Zaslavsky, Arkady, Charith Perera, and Dimitrios Georgakopoulos. *Sensing as a Service and Big Data*, Australian National University, Canberra, AU, 2018.

Chapter 10

Redefining Artificial Intelligence through IoT

Aaron Khoury

Contents

10.1 Introduction

A recent advances and research reports suggest that artificial intelligence (AI) has reached an inflexion point. Over the past 50 years, there have been several points where AI appeared poised to take off in sophistication and impact. The cost of enabling technologies and energy, and the state of supporting processes and algorithms, held back this projected advance. Today, computation power is widely and cost-effectively available. Parallel graphics processors, first developed for games, have proven easy to adapt for widespread AI use. Decades of steady efforts developing algorithms and self-learning code have brought the field to a true take off point.

As author and futurist Arthur Clark once noted of virtual reality technology, the first is already here, it is just not evenly distributed—and one might add fully visible. Today, machine learning algorithms and other subsets of AI are rapidly altering many forms of Internet of Things (IoT) devices, from home thermostats to autonomous vehicles. If one scans the current technical horizon, one can see that nearly every trajectory of IoT is being shaped by the power and potential of AI.

Angel investor and CEO of Bootstrap Labs, Nicolai Wardstom, says, "In my personal experience, I haven't really come across any technology for industry that could have a greater impact than AI."[1] More famously, Elon Musk, founder of Tesla and SpaceX, has made speculations of the risks in not regulating AI; the billionaire even went as far as saying that, "We need to be super careful with A.I. Potentially being more dangerous than nukes."[2] These comments are debatable but what is not is the rapid development of AI and its implications and applications across all segments of technology. It is vital to ask: "So, what is AI"? Where is it headed, and what are the implications for technical ecosystems, especially IoT?

10.2 AI Basics

AI is often commonly defined as "the capability of a machine to imitate intelligent human behavior."[3] This idea of creating a machine with human intelligence is not novel, rather it can date back to when Alan Turing and John Von Neuman proposed the idea of building a machine capable of mimicking the human brain. This idea was based on a medical knowledge base that the brain was functionally just an electrical network. In the field of AI and machine learning, this human-focused definition and approach represents just one domain. A broader definition of AI focused on the creation of "any device that perceives its environment and takes actions that maximize its chance of successfully achieving its goals."[4] The pursuit of human-level intelligence remains an active and fruitful area of research. Over the coming decade, however, there will be an almost weekly advance in the development of intelligent agents and machine learning algorithms connected to the network (IoT). To understand and contribute to the innovation occurring at the nexus of AI and IoT, it is helpful to have a baseline understanding of key enabling tools.

10.2.1 Machine Learning

Machine learning is the practice of using statistical techniques to allow an information system the ability to "learn" with data without being explicitly programmed. It is the process by which a system is given a set of examples that set parameters for responses; from these parameters, the machine may then interact with novel data and be rewarded or otherwise based on the goals of the program designers. This process is most commonly performed in what is called, somewhat misleadingly, a supervised or unsupervised environment. In supervised learning, a machine is given a set of data (or boundary conditions) which it is meant to sort and categorize—each piece of data receives a score or output based on determined relevancy. The goal of the machine is to categorize these data—pictures, variables, files, etc.—and achieve the lowest rate of error. The machine will continue to adjust its parameters of evaluation progressively trying to reach an ideal means of evaluation. A simple analogy for understanding this would be giving a student a series of paired number: 1:1, 2:4, 3:9, etc. The child (or machine) would make logical inferences about the relationships of those numbers. In unsupervised learning, the scores or outputs are not given. The machine must recognize patterns in data on its own. An example of inputs might be pictures of a bench, a sofa, a desk, and a chair. The machine would recognize patterns in the pictures' structures and begin to recognize items that are used for sitting.[5]

10.2.2 Neural Networks

If you say the word "elephant," what other thoughts might appear in your head associated with that term? Maybe you think of the terms *tusk*, *circus*, or *peanuts*—you may associate the word with ice cream because of the elephant statue outside your local ice cream store. Now take the words you have associated with elephant and repeat the exercise—what words are associated with those words? At its simplest form, you just practiced creating a neural network. Unconsciously, you filtered millions of experiences in your life and ended with a handful of results—all from the simple input of the word elephant.

A neural network is the result of machine learning. It is represented by a network of nodes. Each node in the neural network is connected to many other nodes in an intricate web of interconnections. When an input is received, the machine analyzes the data and activates nodes which it has been taught to relate to the input. If the next node or association has an output with a high enough score, it will propagate another connection with its related items. If it does not, the connections stops and further associations are not made. From these inputs, the network can adjust based on frequency of node interaction. Returning to the elephant example; you likely did not associate the animal with ice cream, but if you had the treat every time you saw a massive pink elephant statue, you may have a stronger association between the two words. In these ways, neural networks are often described as mirroring the function of the human brain. Because of their capacity for self-refinement, neural networks are a key component of current IoT and AI technical developments.[6]

10.2.3 Deep Learning

The term "deep" is often used by companies when naming AI systems or technologies. For example, Google's DeepMind Technologies which specializes in the development of AI. These brand terms are rooted in the computational approach of the same name. When using the term deep learning, we are quite literally referring to the depth at which data is analyzed by a system. Deep learning is essentially the process of stacking neural networks together to develop more complex analyses for given inputs. If you were designing a system to recognize pictures with stop signs you would not base your criteria on the image solely having the right shade of red in it? Rather, you would have multiple networks trained to recognize letters, cars, traffic, etc. These networks can quite literally be stacked to create a three-dimensional network which has more elaborate and hopefully more accurate results. The more variables—or deeper—a system can observe, the more it can potentially learn.[7]

A more recent breakthrough in AI research occurred in the 1990s when graphics processing units or GPUs gained traction in computer technology. These chips were previously used for playing graphically intensive video games. Since then, graphics processors are now also used for complex engineering software and testing AI neural networks. The GPU is ideal for neural networks because of the high density of microprocessors they can support. In a GPU, there may be thousands of cores, whereas the highest quality Intel CPU has 28 cores. Individual nodes of the neural network can thus act on each core of the GPU and maximize the rate at which processing occurs. The importance of the neural networks and graphics processors points to two of the driving forces in AI innovation—the accelerated development of massively parallel processing and the advance of self-learning networks or algorithms.

10.3 Applying AI in IoT

The terms machine learning and AI may have struck you as complete novelties; however, it is more likely that you interact with or see the products of such every day. Websites like Pandora Radio and Amazon are great examples of basic, but common, uses of machine learning. Pandora Radio curates the songs that are played based on musical attributes and user interactions with the songs—if you have ever listened to one of the 7.6 billion played tracks on Pandora or a similar service, you have helped with the training of an AI algorithm.[8] Amazon works similarly with books, movies, and shopping recommendations. However, it is not to say that AI only has its place in cyberspace. AI and IoT have made huge strides in logistics, education, healthcare, and customer service.

10.3.1 Healthcare

When it comes to complexity, the healthcare field exposes some of the major challenges that AI integration will face, such as ethics and government regulations.

This is not to say that AI does not have its place in the industry—healthcare providers are some of the largest data generators after all. One of the most prominent innovations in healthcare has been the use of wearable IoT devices. While many may think that wearables are limited to being pedometers and current biometric devices, they have strong potential when paired with the use of AI analysis. Wearable devices offer the potential to monitor any range of physiological attributes: sugar levels, heart rate, physical activity, etc. Global Industry Analysts, Inc. has predicted that wearable health devices alone will have a market valuation of $4.5 billion by 2020. By using AI to monitor and analyze the pools of data created by IoT devices, doctors will have a larger deeper view of health interactions. Further, the manipulation of these data allows for quicker and more automated responses in terms of healthcare decisions.[9]

10.3.2 Logistics

Another area of application for AI and IoT integration is the ability to automate tasks which would otherwise require human moderation. A common practice in supply chain management is the use of radio-frequency identification (RFID) sensors which allow for quick scanning and identification of goods; these IoT sensors are cheap to produce and implement. In conjunction with AI and robotics, the production, routing, and delivery of goods will become a more streamlined, hands-off process. Additionally, AI is especially important when predicting market trends, optimal and least cost routes, and the most efficient means of transport. IBM claims that the use of AI is now being used to remove the guesswork in determining the fastest delivery routes in both daily transits and cross-continental deliveries. By creating systems that choose the most efficient air freight forwarders or freeways, more time is created for companies to focus labor on tasks that require direct human interaction.[10]

10.3.3 Education

We have discussed “what we can teach AI.” In coming years, we will more often ask, what can we be taught by AI. Jennifer Rexford, the head of Princeton’s computer science program, believes that AI can help with human analysis and lead to discoveries regarding how humans best learn. Using magnetic resonance imaging (MRI) machines and video-guided lectures, she hopes that machine learning algorithms will be able to identify facets of education that best engage students.[11]

Bill Gates believes AI may help enable an education system where students have a constantly available and open resource. Gates claims that by creating virtual tutors, students will worry less about asking questions or being confused by complex subjects. Additionally, the use of AI in education can help personalize education; this task is important especially considering people all learn at different paces and in different styles.[12]

10.3.4 Interactive Assistants

Many of the most visible IoT devices that leverage AI are voice assistants and other customer interaction/service agents. The use of machine learning with voice recognition has far separated us from the days of frustrating phone calls with nonresponsive customer service answering machines. IBM reports that 52% of observed customers hang up rather than waiting for live agents to answer their calls and 50% of customer calls go unresolved. Using AI for customer interaction is now in the forefront for call center responses—the use of AI can help diffuse wait time for customers and benefit call center agents. A device such as Alexa creates real-time support interactions with users. By facilitating conversations and interactions with other help support systems, agents are put under less stress to fill call quotas; rather, more attention can be directed to interactions addressing complex issues. Further, 70% of customers report preferring customer chat services which are heavily leveraging AI to resolve common customer complaints. The disruption which AI has created in customer care is not only creating better customer experiences but will hopefully also reduce the 40% employee turnover rate that many call centers face.[13]

One of the greatest challenges in making AI a mainstream utility is identifying creative and effective areas where the technology will be effective. It is projected that, for the next several years at least, thousands of new AI-related products and services will be created each year in the IoT field alone. AI has the potential to reduce jobs that focus on tedious work. But we are creating more time and opportunities for the workforce to develop creative, personalized solutions to issues that previously received less attention than needed. AI is and will continue to be a driving force in delivering innovative, quality products of the human mind.

10.4 Case Study: AI and Applicant Screening

One major concern posed by AI is how the technology will impact the global job market and whether there will be enough work for humans to fulfill. In 2013, it was predicted that nearly 47% of U.S. employment may be at risk to computerization.[14] However, here we will discuss how AI performs in a way to both find and train individuals in jobs.

Recruiting and job placement are key elements in matching individuals with jobs where they will succeed and enjoy the work being done. Finding, hiring, and training individuals can be costly; it is therefore in the best interest of both parties that these processes are performed efficiently and accurately. Human resource personnel are now being equipped with third-party tools like X.ai, ClearFit, and Filtered. These tools use AI algorithms to assess technical skills with tasks like coding challenges and brain teasers. This process moves away from the traditional college campus visit or resume review and rather assesses abilities through direct interaction. IoT voice agents, robots, and similar systems are transforming the HR screening process. The data of the individuals who partake in these recruiting activities can then be evaluated for the machine to better recognize individuals who will succeed in any range of positions.[15]

Unilever—the conglomerate known for products like Axe, Dove, and Lipton—adopted the use of AI-recruiting efforts and saw amazing results. The process Unilever has adopted allows for individuals to use recruiting service provider software that analyzes them as they partake in neuroscience-based games. The service's algorithms are calibrated by having standout employees take the tests prior to the recruiting effort. During the test, individuals are assessed based on keywords, fluctuation of tone, and body language. From the benchmark scores, which current employees produce, candidates can be assessed and considered. If the candidate is found to be comparable to exceptional employees, they then move on to parts of the in-person portion of the recruiting process—this includes visits to the employer, in-person interviews, and interviews over video chat. From using these technologies, Unilever saw the amount of time spent on a candidate recruiting decrease from 4 months to only 4 weeks—the rate at which offer acceptance also increased from 64% to 82%.[16]

Following the hiring process, AI can then be used to help personalize the process of onboarding new employees. Systems like Axonify are able to offer personalized training programs based on identified personal/company goals, existing skills/knowledge, and job-specific skills which need further development. This system benefits both the company and employer as it allows for the reduced costs and time spent in personalized/specific training which would have otherwise needed countless hours of labor and direct contact with the individual.[17]

Using AI in human resource services is not meant to replace the human aspect of the job. Rather, this technology allows for a broader view of candidates. AI allows for companies to peer inward to the aspects of their most valued employees and seek out like-minded individuals. When hiring, it is also a key aspect to analyze for individuals who show the ability to change their roles, skill-sets, and knowledge base—a pivotal employee feature as more company tasks are automated every year. These systems not only look for individuals who fit the role but also help transform individuals to best perform upon entering the workforce. As algorithms are better engineered to identify key factors for job specifics, recruiters have more time and opportunities to analyze the cognitive and social features of applicants—components of hiring which are best performed by human interaction.[18]

10.5 Case Study: AI and OpenAI

AI is capable of many great things, more now than ever before. One of AI's greatest potentials lies in the ability to learn and solve complex problems. A perfect example of the capabilities of AI learning can be found in the efforts of the team at OpenAI. They are a nonprofit AI research company that is focused on creating AGI or artificial general intelligence.[19] Through the company's efforts they have managed to apply their AI learning techniques to the world of professional gaming.

They wanted to create an AI that could stand its own against some of the best players of Dota 2, one of the most complicated strategy games. Dota 2 is a multiplayer online battle arena (MOBA) that pits two teams of five players each against each other to see who can destroy the others ancient, a structure located on either side of the map, first. The game boasts a character roster of over 100 heroes to play all with their own set of abilities and statistics. These heroes can hold up to six different items from a pool of over 100 as well. This all builds up to an exhaustingly complex game with a constantly changing dynamic as the games progress according to the players' abilities. Looking at the setup, it would seem almost impossible for an AI to understand the complexities of this game and be able to best a human player, let alone some of the best players the community has to offer. It may seem impossible, but it is not.

The team over at OpenAI started small at first. Instead of creating a whole team of AIs to fight in a regular game of Dota 2, they created one that could hold its own in a 1v1 situation. They employed the tactic of self-play to get the AI to learn how to play the game. They would simulate a fresh game of Dota 2 and pit the AI against itself (i.e., have the AI tool play against itself). They gave it basic rules to not die and try to win the game. They would simulate hundreds of games at a time, equating to about 300 years of experience a day, to let the AI explore its environment and learn how to survive and thrive. As time passed, the team built a reward system that helped refine the AI's decision-making process.

The reward system is easily comparable to the same reward system we have in our own brains. When we do something good, we are rewarded with a chemical feeling in our brain as feedback. The AI works in a similar fashion except with a simple point system instead of brain chemicals. The AI's goal is to increase its points at every moment it can. It would actively weigh the outcomes of its decisions and predict its next moves to maximize its total points by the end of the game.[20] This reward system laid the foundation for the AI's growth.

At first, it would just run around the map until it was killed, but, after a few hours of playing, it began to pick up the basics of the game that would equate its skill to that of a novice player. After a few days, the AI figured out how to use and defend against basic human strategy. After its thousands of games against itself, they finally unleashed their AI onto the community of Dota 2 at the 2017 International tournament.

The Dota 2 International tournament, or TI for short, is an annual eSports event where the greatest players in the Dota 2 community come together to battle at the highest level of play. OpenAI brought their creation to the tournament to pit it against the professionals and show off just how advanced it had become. The AI stomped almost all the professionals that challenged it. Its timing was almost perfect, its strategy was well crafted, and it adapted to the other players strategy almost instantly. While it may have bested many in a 1v1 match, the goal of a completely AI team still loomed in the back of OpenAI's collective head, so they got to work on their next project.

The project would come to be known as OpenAI Five. The project would pit five AI-controlled characters that, in theory, should be able to hold their own against human players. To accomplish this, they needed to deepen the learning portion of the AI. Not only would each AI have to play effectively to garner its own success, but it would also have to work in tandem with its teammates to ensure their success as well. To accomplish this, they implemented a system called Team Spirit that played on their initial reward system.

Team Spirit coordinates with the ultimate goal of the game, which is destroying the opponent's base. Each AI makes regular checks into the Team Spirit value to determine whether it should continue its natural selfish behavior or if it should start sacrificing its own score to benefit the team as a whole and get closer to a victory.[21] This all culminates in a set of five AIs that all work alone and together to achieve victory over the other team, but now there is one final question: How did they do?

On August 5th, 2018, during a benchmark session, OpenAI Five won a best of three against five top players in Dota 2, four of which were active players in the professional scene.[22] OpenAI Five was able to strategically choose from a pool of 100 unique video characters. The program defeated the human opponents in the first two rounds and was defeated in the last round when forced to play with randomly chosen characters. OpenAI Five managed to adapt in varying situations. It would later go on to an international tournament and hold its own against a professional team of Dota 2 players. While it did not win, it did manage to adequately compete with the professionals in an impressive display of adaptable skill. Open AI Five lasted roughly 51 min against one team and 45 min against another team an average game at the professional level lasts around 45 min.[23] This shows OpenAI Five's ability to express high-level play on the fly almost like a human player.

OpenAI's work is an example of niche use of AI in the technology industry. Though the case above pertains to a video game, the potential for AI is limitless. If anything, OpenAI Five shows that an AI can be implemented in an entirely unfamiliar situation and learn from naught. The capabilities of AI are only limited by industry leaders' willingness to embrace and explore its application.

10.6 Discussion Questions

1. What are some areas that a generalized AI model, like the one described in this case, may serve a purpose? How would it change those areas with its implementation?

2. What are some concerns that could be raised in relation to the implementation of a generalized AI?
3. What are some emerging technologies that would benefit from something like a generalized AI?
4. How would something like the generalized AI described above change the technology landscape in the next 5–10 years?

References

1. Raghav, Bharadwaj. "The Analogy of AI and the Fourth Industrial Revolution." TechEmergence. June 19, 2018. Accessed October 13, 2018. www.techemergence.com/analogy-ai-fourth-industrial-revolution.
2. Strange, Adario. "Elon Musk Says Artificial Intelligence Could Be 'More Dangerous Than Nukes'." Mashable. August 04, 2014. Accessed October 13, 2018. https://mashable.com/2014/08/03/elon-musk-artificial-intelligence/#bBCNXZ0INgqi.
3. "Artificial Intelligence." Merriam-Webster. Accessed October 13, 2018. www.merriam-webster.com/dictionary/artificialintelligence.
4. "Artificial Intelligence." Wikipedia. October 12, 2018. Accessed October 13, 2018. https://en.wikipedia.org/wiki/Artificial_intelligence#CITEREFRussellNorvig2003.
5. Watson Health. Artificial Intelligence, Machine Learning and Deep Learning: What Do They Mean? March 20, 2018. Accessed October 13, 2018. www.ibm.com/blogs/watson-health/what-do-they-mean/.
6. Hardesty, Larry, and MIT News Office. "Explained: Neural Networks." MIT News. April 14, 2017. Accessed October 13, 2018. http://news.mit.edu/2017/explained-neural-networks-deep-learning-0414.
7. Copeland, Michael. "Deep Learning Explained." Ebook. Nvidia. 2018. Accessed October 13, 2018. www.nvidia.com/content/dam/en-zz/Solutions/deep-learning/home/DeepLearning_eBook_FINAL.pdf.
8. AItrends. How Machine Learning Helps Pandora Find the Music of the Moment. April 6, 2017. Accessed October 13, 2018. https://aitrends.com/machine-learning/machine-learning-helps-pandora-find-music-moment/.
9. Singh, Sangita. "Artificial Intelligence and the Internet of Things in Healthcare." *Journal of Mhealth*. April 6, 2018. Accessed October 13, 2018. http://thejournalofmhealth.com/artificial-intelligence-and-the-internet-of-things-in-healthcare/.
10. Gesing, Ben, Steve Peterson, and Dirk Michelsen. 2018. "*Artificial Intelligence in Logistics*." Troisdorf, Germany: DHL Customer Solutions & Innovation.
11. Rexford, Jennifer, and Rik Kirkland. "*The Role of Education in AI (and Vice Versa)*." McKinsey & Company. April, 2018. Accessed October 13, 2018. www.mckinsey.com/featured-insights/artificial-intelligence/the-role-of-education-in-ai-and-vice-versa.
12. Newton, Casey. "Can AI Fix Education? We Asked Bill Gates." The Verge. April 25, 2016. Accessed October 13, 2018. www.theverge.com/2016/4/25/11492102/bill-gates-interview-education-software-artificial-intelligence.
13. Vennard, Chris. "The Future of Call Centers and Customer Service Is Being Shaped by AI." IBM. October 20, 2017. Accessed October 13, 2018. www.ibm.com/blogs/watson/2017/10/the-future-of-call-centers-is-shaped-by-ai/.

14. Frey, Carl Benedikt, and Michael A. Osborne. "The Future of Employment: How Susceptible Are Jobs to Computerisation?" *Technological Forecasting and Social Change* 114: 254–280. January, 2017. Accessed October 13, 2018. www.sciencedirect.com/science/article/pii/S0040162516302244.
15. Seseri, Rudina. How AI Is Changing The Game For Recruiting. January 29, 2018. Accessed October 13, 2018. www.forbes.com/sites/valleyvoices/2018/01/29/how-ai-is-changing-the-game-for-recruiting/#7389f43a1aa2.
16. Feloni, Richard. "Consumer-goods Giant Unilever Has Been Hiring Employees Using Brain Games and Artificial Intelligence—and It's a Huge Success." Business Insider. June 28, 2017. Accessed October 13, 2018. www.businessinsider.com/unilever-artificial-intelligence-hiring-process-2017-6.
17. Sathe, Sanjay, and Rise Smart. "How AI Is Radically Streamlining the Onboarding Process." VentureBeat. February 27, 2017. Accessed October 13, 2018. https://venturebeat.com/2017/02/27/how-ai-is-radically-streamlining-the-onboarding-process/.
18. Lake, McCree. Artificial Intelligence Gives HR an Opportunity to Transform the Enterprise. March 20, 2018. Accessed October 13, 2018. www.cio.com/article/3263450/artificial-intelligence/artificial-intelligence-gives-hr-an-opportunity-to-transform-the-enterprise.html.
19. "About OpenAI." OpenAI. Accessed October 13, 2018. https://openai.com/about/.
20. Dfarhi. "Dota_reward.md." GitHubGist. Accessed October 13, 2018. https://gist.github.com/dfarhi/66ec9d760ae0c49a5c492c9fae93984a.
21. Tang, Jie. "OpenAI Five." OpenAI Blog. June 25, 2018. Accessed October 13, 2018. https://blog.openai.com/openai-five/#exploration.
22. 11 Dota, Openai's Team. "OpenAI Five Benchmark: Results." OpenAI Blog. 2018. Accessed October 13, 2018. https://blog.openai.com/openai-five-benchmark-results/.
23. OpenAI. "The International 2018: Results." OpenAI Blog. August 31, 2018. Accessed October 13, 2018. https://blog.openai.com/the-international-2018-results/.

Chapter 11

Launching an IoT Startup

Dennis Trinkle
Ball State University

Contents

The best way to predict the future is to create it.

Abraham Lincoln

11.1 Introduction

According to the Small Business Administration, as of 2017, small businesses account for 99.7% of employer firms in the United States and for 67% of new job creation. Entrepreneurial startups account for the great majority of innovation and early product commercialization in the United States. This innovation trend is projected to increase over the next two decades. Workers of all ages are increasingly seeking the opportunity to develop new products and services, define new workplace cultures, and achieve more flexibility and self-determination in their work-life choices. It is an easy prediction, therefore, that the majority of IoT innovation and new company creation will be driven by entrepreneurs.

Individuals become entrepreneurs for a wide range of reasons. Some develop a new product or service and seek passionately to bring it to the market. Others have identified a problem or challenge that they are confident can be solved with the right organization of resources. Increasingly, many individuals are also starting small businesses for the ability to set organizational values and work patterns. Nearly all entrepreneurs seek to make a positive difference in the lives of others or for themselves.

If you have a novel IoT product or service idea, see great opportunity in the emerging IoT landscape, or/and would like to lead your own organization, starting a small business or nonprofit organization could be the right path for you. But, how do you get started? If you think founding an IoT organization sounds intriguing, how do you determine if it is the right fit at the right time for you? How does one consider the prospects of becoming an entrepreneur? And, if one decides to embark on the process, how does one successfully go about becoming a successful entrepreneur?

These are important questions. Starting and growing a business requires time, commitment, and detailed analysis and planning. The stakes are high. Eighty-five percent of new businesses will fail or be abandoned in their first 5 years. Fortunately for would-be entrepreneurs passionate about the IoT space, both the process for successfully launching a new organization and the most common reasons for failure are well known and understood. This chapter will help you understand the key steps to becoming an entrepreneur and the most common reasons why a new business will fail. Consider it a basic recipe for success. Like the basic recipe for any dish, it will help you get safely started. As you better understand the key ingredients and potential obstacles, you will make the recipe your own.

11.1.1 Step 1: Self-Evaluation

One of the earliest mistakes new entrepreneurs make is just plunging ahead without self-reflection and planning. Even if you are ideally suited to becoming a successful entrepreneur, the Nike approach is not the wisest strategy. Just doing it is a mistake. Before beginning, you need to assess whether you are ready to make the commitment required to become an entrepreneur. Are you comfortable with the obligations? With the risks? Do you have the necessary resources and talents? Determine what roles you are well suited and qualified to undertake in the startup process and where you will need help. Becoming a successful entrepreneur is a team sport—and each player must understand their talents and their role.

According to Sangeeta Bharadwaj Badal and Joseph H. Streur (2014)[1], self-awareness is the conscious knowledge of one's own character, feelings, motives, and desires—and it directly contributes to the success of an entrepreneurial business. Badal and Streur have developed the Builder Profile 10 (BP10) for Gallop to assess an individual's entrepreneurial talent. Leaders with a good grasp of their strengths, weaknesses, blind spots, and biases are better equipped to improve upon themselves and their organizations. According to a Green Peak Partners study of 72 senior executives conducted in collaboration with the Cornell School of Industrial and Labor Relations, "a high self-awareness score was the strongest predictor of overall success." ("What Predicts Executive Success, Green Peak Consulting Study, www.greenpeakpartners.com) This is because executives who are aware of their weaknesses are better able to attract and hire team members who perform well in areas in which the leader lacks.

In a related Gallup study, more than 1,000 entrepreneurs were analyzed to create a list of the ten major qualities of highly successful entrepreneurs. These traits can be found below:

1. **Business Focus:** Do you base your decisions on the potential to turn a profit or improve your business?
2. **Confidence:** Do you know yourself and your abilities well, and can you read others?
3. **Creative Thinker:** Do you know how to turn an existing product or idea into something even better?
4. **Delegator:** Do you know to not try to do it all yourself and build the best team to help?
5. **Determination:** Are you able to battle your way through difficult obstacles?
6. **Independent:** Will you do whatever it takes to succeed in the business?
7. **Knowledge-Seeker:** Do you constantly hunt down information that will help you keep the business growing?
8. **Promoter:** Do you do the best job as spokesperson for the business?
9. **Relationship-Builder:** Do you have a high social intelligence and an ability to build relationships that aid their firm's growth?
10. **Risk-Taker:** Do you have good instincts when it comes to managing high-risk situations?

If you are high in these attributes, you are likely well suited to start a business. Indeed, you may feel restless and less satisfied working in a large organization or for another. But what if you are not already a highly self-aware and reflective person? Have no fear, much of self-awareness is a process that you can master. Like anything else, it just takes practice. Having a framework for self-awareness helps to better understand your strengths and weaknesses.

To build your self-awareness, you may want to spend time contemplating your actions and reactions to them and learning from your mistakes and successes. In the Harvard Business Review article, "Managing Oneself," Peter Drucker wrote, "Whenever you make a decision or take a key decision, write down what you expect will happen. Nine or twelve months later, compare the results with what you expected." Drucker described this process as the "only way to discover your strengths".[2]

Becoming aware of your strengths and where you will need a team to cover your gaps is also critical. To be truly self-aware, you must focus on others as well. Building the perfect team requires you to work around your own strengths and weaknesses. Knowing your natural strengths and weaknesses also makes you a better recruiter and allocator of talent. But you also must be an acute observer of others' strengths and weaknesses. Effective teams are made up of people who both understand and complement each other. By definition, it is impossible for any one individual to be above average across all the business-building traits.

The best teams are rarely made up of similar types. If one is open-minded and objective, different types of people on a team will also help over time to further self-awareness and deepen an appreciation for the variety inherent in patterns of success. When you see people progressing towards a common goal by following different paths from yours, there is an implicit peer-based feedback loop and systemic learning in that observation itself. Having the right complement of people and a supportive learning organization allows you to see more clearly what you do well and what others do well.

Finally, if you would like a gut-level tool to help you judge whether being an entrepreneur is right for you, you can take the Isenberg, 2-min entrepreneur test. Developed by Daniel Isenberg, professor of entrepreneurship at Babson College, this is a useful tool for quickly gauging whether you are well suited to starting and growing a business. Answer each of the questions below with a simple yes or no:

1. I don't like being told what to do by people who are less capable than I am.
2. I like challenging myself.
3. I like to win.
4. I like being my own boss.
5. I always look for new and better ways to do things.
6. I like to question conventional wisdom.
7. I like to get people together in order to get things done.
8. People get excited by my ideas.
9. I am rarely satisfied or complacent.
10. I can't sit still.
11. I can usually work my way out of a difficult situation.
12. I would rather fail at my own thing than succeed at someone else's.
13. Whenever there is a problem, I am ready to jump right in.
14. I think old dogs can learn—even invent—new tricks.
15. Members of my family run their own businesses.
16. I have friends who run their own businesses.
17. I worked after school and during vacations when I was growing up.
18. I get an adrenaline rush from selling things.

19. I am exhilarated by achieving results.
20. I could have written a better test than Isenberg (and here is what I would change ….)

If you answered "yes" on 17 or more of these questions, then you are a solid, natural fit for being an entrepreneur. If you answered yes on 12–16, being an entrepreneur might be a good fit for you with planning and the right team. If you answered yes on 11 or fewer, carefully consider your motivations for being an entrepreneur and the resources you would bring to the team. It might not be the most natural fit for your personality and talents. But, remember, no test is conclusive. Being self-aware can mean having the ability to see areas in your skillset that need improvement and taking the time to improve them. According to Gallup, the key personality traits that make for a good entrepreneur are only between 37% and 48% genetic. More than half of what it takes for success can be learned and mastered.

11.2 Developing and Validating the Idea

Successful entrepreneurs identify a solution that solves a problem. Some IoT entrepreneurs may begin with a novel technology and seek to identify an ideal market. Other aspirational entrepreneurs recognize a problem and seek to solve it with an innovative product or service. Successful entrepreneurs must do both—identify a real problem perceived by customers for which they are willing to pay for a solution and create a solution that resolves the problem at a price deemed reasonable by the marketplace. It is okay to start with either a solution or an idea, but both require research with true potential customers and a willingness to be responsive and flexible.

Many potential IoT entrepreneurs today will be familiar with agile development, design thinking, and other customer-focused, iterative approaches to design. Such iterative approaches are ideally suited to launching a new business. As Tim Brown, Founder of IDEO, the Stanford d.School, and the design thinking approach stated, "Design thinking is a human-centered approach to innovation that draws from the designer's toolkit to integrate the needs of people, the possibilities of technology, and the requirements for business success."[3] Put more simply, the iterative design approach aims to achieve precisely what a successful technology entrepreneur must do: bring together what is desirable from a human point of view with what is technologically feasible and economically viable, as depicted in Figure 11.1.

The IDEO approach to design innovation relies on a four-phase approach to idea development and validation. It starts with identifying a real problem and interviewing and/or observing potential clients on their needs, challenges, and desires. This feedback is considered to develop (or refine for entrepreneurs who had an initial product idea) the product or service to assure it addresses a real problem and meets real needs. This analysis is distilled to the most promising concept(s) which

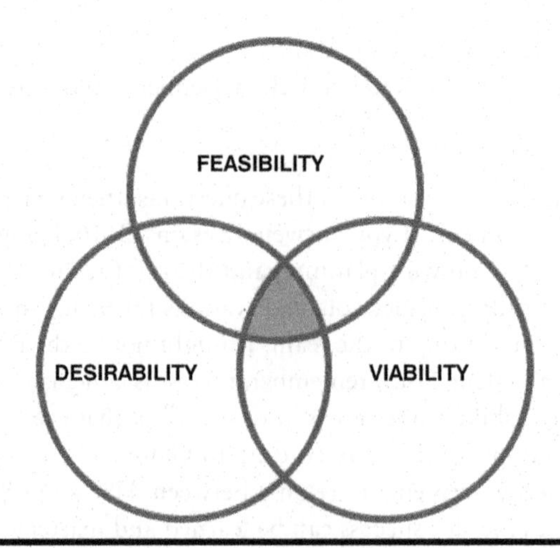

Figure 11.1 The successful IoT entrepreneur finds a solution to a problem that is desirable, feasible, and viable.

are translated into a prototype for further research, feedback, and testing. This process repeats interactively to develop a product or service that can be launched into the market. The development process also provides the powerful human stories that will support and sell the new idea. In brief, the design process is depicted in Figure 11.2.

Inspire new thinking by discovering what potential customers really need

Push past obvious solutions to get to breakthrough new ideas and solutions

Build rough prototypes to test how to make ideas better and iteratively refine your product

Craft a human story to inspire others toward adopting your solution

Figure 11.2 The design process.

By relying on potential customer input in the beginning, testing low-cost product ideas early and often, and iteratively refining solutions to meet real needs and solve real problems, the design approach to innovation helps assure entrepreneurs to avoid many pitfalls in the fire, ready, aim approach to entrepreneurship.

11.3 Building the Business Model

With a solid initial product and target market developed through the ideation process, the entrepreneur can shift to determining the business model for the product or service. Much like developing and refining an idea, crafting a successful business model can be accomplished by following a clear process. A class approach developed by Michael Porter focuses on conducting research to answer fundamental product, resource, sales, and strategic questions.[4] These questions build on the idea phase and address the following considerations:

11.3.1 Size Your Product's Value in the Market

You should match your prices with those of the competitor products. If your product is priced too high, the sales will dip and, if too low, the margins will not be high enough for long-term sustainability.

11.3.2 Acquire High-Value Customers

High-value customers are those customers from whom you gain maximum value while keeping your costs as low as possible. They help you achieve your business targets. You can reach these customers with minimal marketing expenses.

11.3.3 Ensure Sufficiently High Margins

Keeping manufacturing costs low by either outsourcing the manufacturing or having an improved process helps in keeping higher margins. Your product can have features that provide enhanced value to the customer and thus allow you to charge a premium for it. You can even opt for a low-touch model to reduce manpower costs. There are various ways to make sure that the margins are high enough *to sustain the business in the long run.*

11.3.4 See If Your Product Is the Best Solution Available

The prototype phase reveals how most of the stakeholders feel about your product. You have to see that the pain points you are trying to address are solved by your solution and it is the best at what it does. A lackluster product cannot be saved even by the best business model.

11.3.5 Decide on the Channel and Distribution Strategy

A great product will not be great if the distribution or marketing is poor. Test all the elements of your channel strategy in detail. An efficient distribution channel keeps margins high.

11.3.6 Maintain Market Position

A good business model will include plans to make the business sustainable and improve its market position. So, it will have to consider all the growth opportunities and external threats and incorporate a long-term product roadmap. You do not want to rely on just a few customers for most of your product. You also would not want almost the entire distribution to be controlled by your competition or your competitors to be better funded than you. Technology is rapidly evolving making new product development risky, but you would not want to be left behind too. You have to consider all such factors and decide how you plan to maintain and then continuously improve your business model.

11.3.7 Formulate Funding Strategy

Your long-term sustainability depends on funding your business. By funding, we are not referring to only funding by investors. Your company can be bootstrapped, yet it will require a constant flow of funds to sustain itself. Personnel costs, operating capital, and various other overheads are required to be borne by any business, not just a startup. Startups require even more funding as most entrepreneurs have to spend significantly more on customer acquisition and retention than established businesses. Ideally, your initial funding rounds should allow you to multiply your sales so that external funding is no longer required. Your funding options are discussed more fully below.

For many IoT entrepreneurs, a more recent approach to developing and clarifying your startup approach called the Business Model Canvas may be preferable. This clear and highly visual tool makes it very easy to see how all the pieces of a business model fit together. Founded by Alexander Osterwalder, a simple business model canvas can be developed online at low cost at the website Strategyzer.com, which is Osterwalder's business consulting firm.

11.3.8 Business Model Canvas

Many small startups use the Business Model Canvas to search for the right business model. It has nine elements:

1. Customer Segments

   For whom are you creating value? What products and services are you offering to each customer segment?

2. Value Propositions
   What value are you going to deliver to the customer? Which customer pain points are you addressing?
3. Channels
   Which channels are to be focused on to reach the desired customer segments? How are those channels integrated? Which ones are the most cost-effective?
4. Customer Relationships
   What type of relationship do you maintain with each customer segment? What are the expectations of your customers? How to establish them? What would be the associated costs?
5. Revenue Streams
   What are the customers willing to pay and for what value? How would they prefer to pay? How are they currently paying? How does each stream add up to the total revenue?
6. Key Resources
   What key resources do your value propositions require? Your distribution channels? Customer relationships? Revenue streams?
7. Key Activities
   What key activities do your value propositions require? Your distribution channels? Customer relationships? Revenue streams?
8. Key Partners
   Who are your key partners? Your key suppliers? Which key resources are you acquiring from them? Which key activities do your partners perform?
9. Cost Structure
   What are the most important cost drivers in your business model? Which key resources and activities are most expensive?

11.4 Developing Your Team and Operating Model

Once you've defined your product and business model, you need to define your team and operating model. Building your team is a crucial part to being a successful entrepreneur. As much as you might have been solo on this journey, you will need a team with you, especially while you grow. Your team should be there to support you along your way and grow with you. Along with building a team, you will need to make an operating plan. This operating plan will help you focus on your goals. It is important to know what you need to include in your operating plan and how to write the plan itself. Overall, this section will help you focus on building your startup for growth.

11.4.1 Your Team

One crucial step of any startup is your team. You will need to build a team with talents and abilities that fit with your entrepreneurial goals. You will need to know how to grow your team, manage your team, and how to recognize the challenges that come with it. Once you build a strong team, your startup will have a strong foundation for success as well as people who are invested in its growth.

11.4.2 Growing the Team

The first step into having a team behind you is growing it. While it may seem easy to pick a close friend, there are a lot of aspects of growing your team that need to be considered. You want to know who to pick and who not to pick to best fit your company. Along with that, you will need to understand how to grow your team, what positions you need, and how to hire along the way. Your team is crucial and should not be an afterthought. They are the ones who are celebrating your wins with you and picking each other up after your losses. Pick those who have your goals in mind and want to best your company, the same as you do.

11.4.3 What Roles Do You Really Need?

The positions you need for your startup are not ones that can simply be laid out in a book. To realize what positions you need, you need to look into your company. What do you need help with? Where is the most growth headed to? Is there an aspect of business you want an expert in? This should match the structure of your organization and prioritize the positions that you do need. Also, make sure you have the funds to pay for a new employee. You will also need to identify if you can hire a full-time employee or even a contractor. This may help in the beginning when funds are lower and you cannot afford a full-time position. A contractor or adviser can help guide you in the right direction without breaking the bank. Another option is a part-time employee as well. This can help before moving the employee into a full-time role, if needed.

11.4.4 Hiring Right

When it is time to hire, you will first need to find candidates that fit your culture and your company. After you identify these candidates, interview them. Be thorough and make sure to fact-check on your interviewees. This can help prove their credentials and that they are honest with you. Also, if you believe the candidate is a great fit, hire them and see if they work well with your team, your culture, and if they can handle pressure. However, hiring for the team does not end there. You will need to train and help position them for growth as well.

11.4.5 Who to Hire

Due to the nature of a startup, you want to pick individuals who are willing to ride the startup ride with you. This means that you are willing to work closely with them and understand the pressure of the situation. You want someone who will help you overcome obstacles instead of creating them. A good team can make or break a startup due to the "do or die" situation. Cultural fit is a large part of this, if they do not have the same values and goals as you do, it can make for a tough situation. Also, you do not only want positive and super optimistic people for a startup. A startup is a tough situation, and it is okay to have someone who points out the issues. This can push you to overcome them and identify potentially bad situations.

11.4.6 Who Not to Hire

Do not pick the first person who vaguely fits your criteria. While it is hard to fill some positions, it does not mean you pick up the first person who can do it. If they do not value your startup the same as you do or provide too much resistance, this can cause more of a rift. As mentioned previously, do not only pick optimists. This can be bad for your startup, and it is important to have a healthy mix between optimists and pessimists. Do not sacrifice your companies' value for an individual who is extremely talented. You want them to be happy here and for them to fit in with your organization.

11.4.7 Managing Your Team

You have amazing new employees, now what? Managing a team is ideal for your startup for your long-term growth. Disrupting changes can hurt your startup. In fact, 65% of startups are doomed to failure due to people problems. Also, you need to watch out for the way that you treat your team. If you only focus on the end goal and not the individuals, you will cause your employees to become burned out. As their leader, it is your goal to manage them, not only for their success but for your company as well. You want to make sure everyone is on the same track with the same goals. Overall, managing your team well is crucial for your growth as a company.

11.4.8 Setting Expectations

Starting your new entrepreneurial venture with expectations may be one of the last things on your mind. However, creating expectations and talking about them openly is what is needed for managing a team. This does not only apply to you and your employees, but your co-founders as well. While you are all on the same level, you will need to talk this out. Your co-founders may have slightly varying expectations, and this could cause issues down the road. It also creates a solid goal for your team. You will need to ask expectations with topics such as salary, costs, what is expected from others, vacation time, sick days. This will not be solved in one short meeting but will involve tough conversation. Have follow-up sessions and discussion. Write down your expectations so there is a record. Expectations are the comparison point of your team and will help you decide who is ideal for your future as a company.

11.4.9 Accountability

Now that your team knows its expectations, it is ideal to check up with them daily. This does not have to be an extensive rundown of everything the team member is doing. However, it is ideal to ask them what they are working on, what they finished, and if they have any issues. This is a good way to gauge how they are doing and to see if they see any potential issues. This can prevent possible issues, but it can also highlight some potential strengths. As your team goes, you will have to morph your checks to smaller groups. You can also check in over the internet if you have remote workers. No matter the situation, you should make sure to continually interact and check in with your team to promote the end goal.

11.4.10 Meetings

A more intensive meeting could happen weekly or biweekly and highlights the goals completed and the goals for the next week or 2 weeks. Push forward the most important items to receive a solution. Also, provide an agenda so all members are prepared and know what is expected of them. An agenda also helps the meeting to stay on track and end on time, which helps with team's trust. Respecting their time will cause them to respect yours. No one wants to be held over, and there is no reason to hold a meeting longer than it needs to be. During this meeting you want someone to take notes and to have a moderator for the meeting. A meeting without a moderator can cause chaos to infiltrate your discussion. Having a moderator will create consistency and enable everyone to stay on track. Also, do not be afraid to mention your failures and what else you can learn from them. Not talking about your shortcomings is not being honest with yourself or your team. Focus on something to build the team to end the meeting.

11.4.11 The Operating Plan

An operating plan is one that shared your vision and goals for your business. This will help guide your company and address the daily aspects of the business. This will be your handbook for everything that is going on. It allowed your goals to be written and for you to have something to fall back on. It goes hand in hand with your team building and hiring. Knowing what your plans are for your business will help you pick your best team. This is why creating a succinct operating plan is important because it ties all aspects of a business together. But how does one build an operating plan?

11.4.12 Building Your Plan

Ultimately building the plan and understanding what is included is unique to you and your organization. However, there are a few steps that will make creating your operating plan much easier. These three simple steps are getting information, planning, and writing the plan. Once the building process is broken down into these three steps, it will make creating your operating plan less daunting. Before you start creating your plan, you will need your team behind you to help you along the way. Your team is a part of the startup and should have input on what goes into the plan as well. As previously mentioned, even co-founders have differing visions for the future and this will help make the vision unified.

The first step of the process is getting the information. You should ask yourself four questions before diving deep into the operating plan creation. Where are we now? Where do we want to be? How do we get there? How do we measure our progress? Once you identify and answer these questions, you will have a strong foundation of your operating plan. You want to make sure you are thorough when gathering this information and that it is accurate. Inaccurate or dishonest information could sway you in a negative way in your operating plan because it would not be best suited for you. Instead it would include a plan that is not entirely for your organization and could steer you in another direction.

You will want to decide what you are going to include in your plan. You want to make sure to convey your financial aspects, your risks, and your future goals. Along with this you will want to add what your team will look like. What skills are you in need of, what are their expectations, etc. You also want to have someone designated for creating the operating plan. While information gathering and input is a team effort, one sole person will make writing the plan easier. Plan who is going to write the operating plan and their timeline. After this planning step is finished, you are ready to write the plan.

For your plan, you want to make sure all of your information is readily available for your writer to reference. This will ensure that they do not have to struggle to find the information and that the plan reflects what was created in the initial notes. Also, during the writing process, the writer will still need to consult and include the rest of the team. While they are the sole writer, the focus is on the team and company as a whole. Also, when writing the plan, make sure to include the date when it was created, include if it is a draft or final copy, and make sure to add "confidential" or something of that nature. When writing, make it easy to read and follow. Formatting should be consistent and easy to follow. Overall, make it an easy read for the rest of your organization.

11.5 Structural and Legal Issues in Launching an IoT Business

In this section, we examine the various kinds of business structures, as well as some of the major topics involving entrepreneurial legalities. We will provide you with a high-level overview of how to legally protect yourself, your team, and your business assets. Understanding the legal aspect of your company and how to equip your company with the necessary tools and resources may be the difference between success and the flop of your company.

11.5.1 Business Structures

In this section, we will explore the various types of business structures that an entrepreneur may choose to operate their business. Each structure provides advantages and disadvantages for the startup. Due to this, the entrepreneur should be sure of the needs and future vision of the company. They should also have a business plan detailing the company's operating and financial plan to help provide insight while making this important decision.

11.5.2 Sole Proprietorship

Sole proprietorship is one of the simplest and quickest business structures to select for the operation of your business. Due to its instant and inexpensive structure, it is very popular among entrepreneurs. This type of unincorporated business is owned and run by one individual, in this case, the entrepreneur. The sole proprietor (the entrepreneur) does not need to worry about paying unemployment taxes on themselves and is free to mix both personal and business assets. There is little discrepancy between the business and the entrepreneur. While the positives do lie in the lack of formalities and simplicity of this type of structure, there are important considerations that may negatively affect the owner. In sole proprietorship, the owner is responsible for all company associated debts, losses, and liabilities.

11.5.3 Partnership

A partnership structure is the legal operation of the business between more than one individual. These partners would share administrative responsibilities, profits, debts, and liabilities, respectively, depending on the type of partnership. The two types of partnerships are general partners and limited partners. In a general partnership, all responsibilities and assets would be divided and maintained by both individuals. Each partner has an invested stake in the company. In contrast, a partner in a limited partnership would function only as an investor. This type of partner has no administrative control within the company. The investor partner is also not responsible for any company assets, besides their invested contribution. When deciding on a partner, it is important to consider the potential partner's background and experience. Regardless of the type of partnership, legally protect yourself and your company by creating a nondisclosure agreement. A nondisclosure agreement is a contract between partners that specifies knowledge and intellectual property of the startup that cannot be shared with external parties. This way, even if the partnership is to go sour down the road, your intellectual property and company assets will be protected. It is also important to have the partner specify if they are currently bound in any external party contracts that may affect their partnership with you.

11.5.4 Corporation

The corporate business structure is one of the most complex forms of business operation. A corporation must be filed with the appropriate state Secretary of State and involves more intricate tax responsibilities. The investors of this type of structure will often purchase shares of the company stock. This acts as "ownership" of the company. This ensures that owners have limited liability due to the corporate entity being separate from their personal investments. These company shares can later be used as capital within the company or may even be used to transfer ownership of the company through shares. The shares also help lengthen the life of the company, as ownership can transfer easily among past and future administrative heads. Many of the consequences of choosing a corporate structure lie in its tax complexities. Owners will experience double taxation due to paying income taxes as well as taxes on any dividends received. This will result in excessive tax filing and the possibility of needing to seek external tax-filing services.

11.5.5 Limited Liability Company

Within a limited liability company, more commonly known as an LLC, one or more owners will create the business through a written agreement. This agreement is called an operating agreement and details out the specific administrative operating agreements of the company. It often includes the financial structures, responsibilities of each party involved in the agreement, and any specific information in relation to the new business. LLC agreements are legally required in California, Delaware, Maine, Missouri, Nebraska, and New York. Even if it is not required in your state, it is highly recommended that the owners sit down to create an agreement to act as an outline of accountability for everyone involved in the company down the road. Online legal services, such as LegalZoom and Northwest, are available to assist with the creation of an LLC. Once an agreement has been created provide each party with a copy of the legal document.

11.5.6 Licensing

The type and number of licenses needed for the new startup depend on the type of business it is and the state in which the business is being opened. At the most basic level, the entrepreneur will need to obtain a business license. To do so, it is best to contact your local business licensing department to find specific application information as each state's licensing requirements and applications vary. It may be valuable for the entrepreneur to do ample research in their field and industry to discover what type of licenses may be needed besides the basic business license. The U.S.

Small Business Administration website may be a good place to begin your research if you are unsure of what licenses and permits are required for your new business. The website features many business guide resources for both federal licenses and permits as well as state licenses and permits.

11.5.7 Working with an Attorney

If you are feeling overwhelmed in trying to figure out the multitude of newly discovered legal precautions that exist in your new entrepreneurial project, you're not alone. Most entrepreneurs choose to enlist the assistance of an attorney to guide them through the legal aspects of the startup process. While this may be pricey, this will be a worthy investment. Hiring an attorney will ensure that the company is legally protected and will provide the entrepreneur with a consistent source for any legal questions or situations that may arise. When choosing an attorney, it is beneficial to seek professionals who have past entrepreneurial-based legal experience. This way, the attorney will already be aware of common legal roadblocks, necessities, and procedures.

11.5.8 Employment Laws

As a new entrepreneur, you will want to equip yourself with a basic knowledge base of employment laws. With the assistance of your attorney, you will be responsible for ensuring that employees are working in a fair and legal matter. Two important laws, the Fair Labor Standards Act and the Federal Equal Employment Opportunity Law, are summarized below. While these are only two of many laws, they will give you a good foundation to begin creating the employee section of your operating plan.

11.5.8.1 Fair Labor Standards Act

The Fair Labor Standards Act of 1938 20 U.S.C. § 203, also known as the FLSA, sets the legal standards for labor in the United States. For the company to remain compliant, the entrepreneur must set all wages, overtime, and record-keeping procedures based off this legal document. It should guide the entrepreneur as they begin to make decisions regarding their hired employees. If the business is not compliant with this document, the company may face violation fines of up to $10,000 per employee or criminal prosecution. To find more information regarding the Fair Labor Standards Act visit the U.S. Office of Financial Management website at https://ofm.wa.gov/.

11.5.8.2 Federal Equal Employment Opportunity Laws

The Federal Equal Employment Opportunity Law, also known as the EEOC, enforces a majority of federal laws that are related to discrimination. The EEOC helps to protect employees from any type of discrimination regarding their race, color, religion, sex, national origin, age, disability, or genetic information. This enforcement is active throughout the employee's presence of the company including the processes of hiring, firing, and promotion. If at any time the employee feels as though they are being discriminated against, they may report the employer and cannot be met with any type of negative consequence under the EEOC. An example of laws covered by the EEOC include Title VII of the Civil Rights Act of 1964, The Pregnancy Discrimination Act, The Age Discrimination in Employment Act of 1967 (ADEA), and the Title I of the Americans with Disabilities Act of 1990 (ADA). For a more complete list of federal laws covered under the EEOC, and to learn more about the EEOC, visit the U.S. Equal Employment Opportunity Commission website at www.eeoc.gov/.

11.5.8.3 Tax Laws

Depending on which business structure the entrepreneur selects to operate under, taxes can become overwhelmingly complex. Researching some common new business tax topics, as well as consulting with your attorney, will enable you to be well prepared when tax season approaches. Be proactive when initially creating and later when gathering your company's tax information for processing by the Internal Revenue Service (IRS). This section will cover how to obtain an Employer Identification Number (EIN) and some important tax considerations for new business owners.

11.5.8.4 Employer Identification Number

The Employer Identification Number, also known as the EIN, or the Federal Tax Identification Number, is used to identify your business entity. This nine-digit number is used by the IRS to identify tax accounts of the employees associated with your business. To apply for an EIN head to the U.S. Internal Revenue Service website at www.irs.gov/. Here you can find an online application where you will determine eligibility, find more information on the process, and complete and submit your application.

11.5.8.5 Tax Responsibilities

As your new startup continues to prosper, you will need to understand your current and future tax responsibilities. These tax responsibilities will include social security and state/county payroll taxes for your employees, as well as taxation of your selected business structure. As an employer, the entrepreneur will need to match employee social security tax contributions. If an employer fails to contribute to the employee's social security fund, hefty penalties will be levied on the company. The entrepreneur will have to research what is included on the employee's state payroll taxes. There is often an unemployment tax, and at times, a disability tax. The employer will be responsible for taking that out of the employee's pay and submitting the funds to the state. Your additional business taxes will depend on the type of business structure that was selected by the entrepreneur. Taxes on corporations, partnerships, and sole proprietorships vary greatly. To find more information on the type of taxes applicable to your selected business structure visit www.irs.gov/.

In this section, we enabled you to learn more about the legal aspects of your future business. As an entrepreneur you will need to be proactive in setting up the company's legal foundation. Be mindful and insightful when selecting a business structure for operating as it will affect all current and future company decisions, such as the type of licensing that owners should seek. Equipping your company with a permanent attorney will settle many of the unknowns of the entrepreneurial journey. Although this may be expensive, it is worth the cost as issues with taxation, employees, and general business issues arise.

11.6 Money Matters

To successfully begin your entrepreneurial adventure, you must have a solid understanding of finance. In this chapter, we will begin to explore the basics of finance including different funding acquisition strategies as well as key financial terms and statements. The chapter will help guide you to better understand the financial components of your startup.

11.6.1 Market Analysis

It is wise to conduct a market analysis prior to beginning your own entrepreneurial project. By analyzing the market, you can observe consumer trends related to your product or service. Understanding if there is need or desire from your consumer market is very important for the future financial health of your company. A market analysis will also provide you with a snapshot of the current industry environment. Although markets are incredibly dynamic and ever-changing, this type of analysis can provide important insight. For example, insights such as existing and emerging competitors assist administrators to determine critical financial decisions within the company.

11.6.2 Acquiring Funding: Internal and External Sources

After completing a thorough market analysis, it is now time to take funding into consideration. Where will you receive the money to begin and sustain your new business? This question can cause major stress. It is one of the most important questions to ponder in the early stages of your startup. Luckily there are a few key types of stakeholders and finance methodologies that may be able to assist during the acquisition of necessary funding. Financial sources can be classified as two types: external and internal.

11.6.3 Internal Sources of Finance

Figure 11.3 shows the varied forms of internal sources which include personal investment, sale of fixed assets, sale of stock, and retained earnings.

11.6.3.1 Personal Investment

Personal investment is funding that comes from the entrepreneur's savings accounts or other resources. Personal investment may also come from close friends and family who are affiliated with the founder. Although this method can be quicker and more flexible than external sources, it should be noted that this type of acquisition may cause stress and issues within the founder's personal life.

11.6.3.2 Retained Earnings

Retained earnings refer to the portions of the business's profits that are reinvested back into the business, hence retained. As your startup becomes more successful, earnings may be utilized towards dividends for stakeholders. However, when first starting up, it is wise to simply reinvest any incoming earnings back into your business.

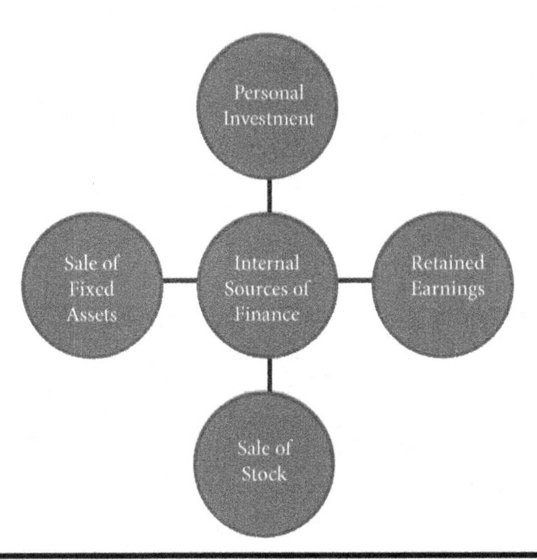

Figure 11.3 Sources of financing.

11.6.3.3 Sale of Fixed Assets

You will need to invest in assets needed for operation when first beginning your startup. Fixed assets are items that are generally used for 5 or more years specifically for business operations. Examples of fixed assets can include machinery, basic technical infrastructure, or even the building that you will be operating out of. As your business begins to progress, you will need to refine your products and service for cost-saving outcomes. As this refining continues, certain assets may no longer be necessary and may be sold to generate additional funding sources. A founder's personal fixed assets, such as property, can also be sold to generate startup funding.

11.6.3.4 Sale of Stock

If your startup begins to experience the signs of growth, it may be time to begin selling stock in your company. When the startup chooses to sell stock, the founders control begins to become diluted. In contrast, dilution means that there is opportunity for a gain in capital, usually translating to quick and successful company growth. In a case study done between two companies, Derek Pilling states "if you raise capital, you are selling stock. Via the sale of stock, the company gets capital and with that capital, the company funds growth initiatives, creating the opportunity to grow and become a more profitable enterprise. By choosing not to raise capital or to raise less capital, the reverse is true; the company has essentially sold future growth potential by not selling (essentially buying) stock" (2010).

11.6.3.5 External Sources of Finance

Figure 11.4 shows the varied forms of internal sources which include loan capital, venture capital, business angels, and share capital.

11.6.3.6 Loan Capital

Loan capital is funding that is acquired but must be repaid at a later date. These funds are almost always long term and are often obtained from banks, finance companies, or from debt–equity investors. Debt–equity investors will provide monetary assistance in exchange for preferred stock within the company. For the investor's protection, this is often accompanied with a fixed or floating charge that is directly tied to the company's assets. While this acquisition of funding may be relatively quick and simple, there is risk of debt if the startup is unable to produce earnings. There is also a possibility of the founder losing partial ownership of the company if debt–equity investment is utilized.

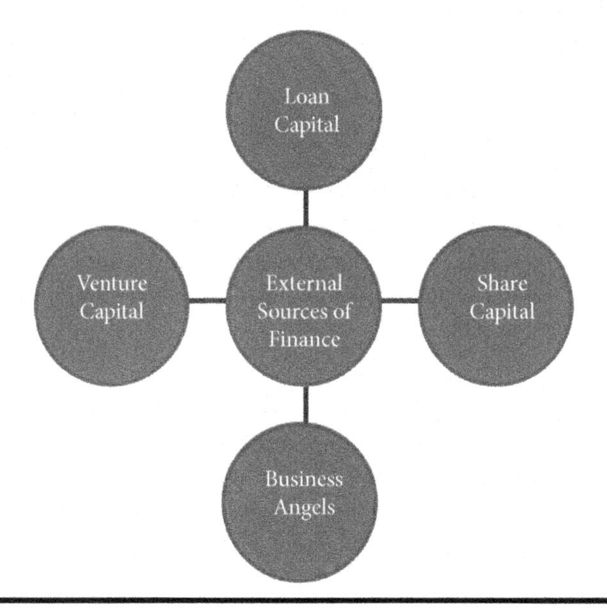

Figure 11.4 External financing sources.

11.6.3.7 Venture Capital

Venture capital funding is one of the most popular sources of financial means for entrepreneurs and their startups. This type of funding comes from individuals or groups of individuals who are interested in growing their current wealth. Venture capital investment groups are drawn to the high-growth and high-risk potential that is found in most startup environments. After an investment prospect research period, a limited partnership is often formed between the venture capitalist and entrepreneur.

11.6.3.8 Angel Investors

Angel investors are like venture capitalists in that they are individuals who are looking to invest to grow their own personal wealth. They are often friends or family. Unlike venture capitalists who are estimated to invest about $7 million per investment, angel investors generally invest lower amounts, with the most common amounts being between $25,000–100,000.Angel investors also do not spend as much time conducting preliminary research prior to investment and often are the sole decision maker in the investment process. This contrasts to the committee decision-making style found in most venture capitalist environments.

11.6.3.9 Equity

Equity is a form of funding that comes from those willing to invest in the company in exchange for shared. Investments might be in cash or in-kind through the contribution of time, expertise or other valuable assets. In exchange for investment, portions of the company in the form of shares are granted to the investor. Although this is an external source of finance, the entrepreneur is often a key shareholder of the company's capital. This enables the founder to have full control of the company. As new shareholders enter the company, control begins to fade, and decisions fall out of the hands of the founder

11.6.3.10 Bootstrapping

Methods such as bootstrapping can also be utilized to alleviate financial pressures. Bootstrapping is exactly as it sounds—building a company from the ground up with as little external investment as possible to maintain control and focus. Bootstrapping requires a heavy focus on resource use and creative thinking and problem solving. Generally, entrepreneurs who use this method rely on little to no external finance support during the initial years, meaning that each financial decision must be carefully deliberated. It may be tempting for a startup to want to pour most of their funding into having the most attractive or best product on the market; however, this is often costly and unnecessary. Remember to stay within your budget. There is always opportunity to innovate and update design as your business grows and develops. Consider the following steps when attempting to bootstrap your business:

1. Select a market, product, or service that you are passionate about.
2. Surround yourself with a team that is more interested in the startup than initial financial gain.
3. Innovate and build within the budget.

11.7 Understanding Revenue and Expenses

11.7.1 What Is Revenue?

Revenue is a financial term that a founder may hear, but what exactly is it? Simply it is the sum of the money that is generated from the company's products and services. Revenue is typically measured per year or per quarter. Total revenue of a company may be calculated using the following formula:

$$\text{Total Revenue} = (\text{Current Price Per Product}) \times (\text{Current Number of Products Sold})$$

Essentially the more product or service that is sold, the greater the total revenue. An example of this formula would be as follows:

If we sold smart phones to a group of consumers at $363 per phone, and 78,000,000 units were sold in 2017, the total revenue in 2017 would be $28,314,000,000.

$$\text{Total Revenue} = (363) \times (78,000,000)$$

$$\text{Total Revenue} = \$28,314,000,000$$

While this number may seem high, it is important to remember all the factors that affect this number once other accounting principles are applied. When total expenses are subtracted from total revenue, which will be discussed in the section below, this will indicate your net income. Net income, also known as net profit if it is positive and net loss if it is negative, is an important indicator of the financial health and success of your startup. Your goal as an entrepreneur should be to always have a positive net profit at the end of each year or quarter. This will provide shareholders confidence in their investment and indicate that the business is performing efficiently and effectively.

11.7.2 What Are Expenses?

Expenses are costs that occur due to the operational activities of the company. Like revenue, expenses are typically measured per year or per quarter. The types of expenses are as follows:

1. Actual cash payments (e.g., employee salaries)
2. Expired portion of an asset (e.g., depreciation of supplies, equipment)
3. Reduction in revenue (e.g., bad debts)

Examples of expenses include costs to produce the product or service, depreciation of fixed assets, supplies, property taxes, insurance costs, and many others. It is important to also note that while all expenses are costs, not all costs are expenses. A cash payment does not always mean that an expense has occurred. For example, if an entrepreneur takes out a long-term bank loan and pays $5,000 towards that loan, the cash within the business is minimized. However, this type of activity does not technically count as an expense.

11.8 Financial Statements

11.8.1 Income Statement

The income statement will become one of the most important financial statements that your startup will produce. This statement provides insight into your financial performance during a given accounting period. It includes revenue, expenses, net profit, and net loss. This type of statement will give stakeholders an understanding of the current operating costs, as well as a comprehensive look at the financial standing of the company. The analysis of these numbers through varying ratios such as return on equity, return on assets, and gross profit help administrators and stakeholders make informed decisions regarding strategy and growth. An example of an income statement can be seen in Figure 11.5.

INCOME STATEMENT,
SAM'S APPLIANCE SHOP

SALES REVENUE		$1,870,841.00
Cost of Goods Sold:		
Beginning Inventory 1/1/8X	$ 805,745.00	
Purchases	939,827.00	
Goods Available for Sale	1,745,572.00	
Less Ending Inventory 12/31/8X	455,455.00	
COST OF GOODS SOLD	1,290,117.00	
GROSS MARGIN		$ 580,724.00
Operating Expenses:		
Advertising	$ 149,670.00	
Insurance	56,125.00	
Depreciation		
Building	18,700.00	
Equipment	9,000.00	
Salaries	224,500.00	
Travel	4,000.00	
Entertainment	2,500.00	
TOTAL OPERATING EXPENSES		$ 464,495.00
General Expenses		
Utilities	$ 5,300.00	
Telephone	2,500.00	
Postage	1,200.00	
Payroll Taxes	25,000.00	
TOTAL GENERAL EXPENSES		$ 34,000.00
Other Expenses		
Interest	$ 19,850.00	
Bad Check Expense	1,750.00	
TOTAL OTHER EXPENSES		$ 21,600.00
TOTAL EXPENSES		$ 520,095.00
NET INCOME		$ 60,629.00

Figure 11.5 Sample income statement. (Source: www.slideshare.net/sanjay_jhaa/financial-reports-and-accounting-ratios on September 2, 2018.)

11.8.2 Balance Sheet

The balance sheet is an important tool for the startup's shareholders to utilize. It includes information on the company's assets, liabilities, and equities. It provides shareholders and other key decision makers a brief overview of what is owned and owed by the company. The main formula that is utilized with the balance sheet is as follows:

$$\text{Assets} = \text{Liabilities} + \text{Shareholders' Equity}$$

Assets are company-owned resources which can be expressed as a cash value. Assets can be either tangible or intangible. Examples of tangible assets include cash, supplies, and property. Examples of intangible assets include patents, brand, and intellectual property. A liability is something that is owed by the company. As discussed earlier, this may be linked to an external finance source such as a bank loans, venture capitalists, and other types of investors. Shareholders' equity is the amount of the company that is owned by the founder and other investors. This may be uniquely segmented depending on the types of funding acquisition that was utilized during the business's startup period. An example of a balance sheet can be seen in Figure 11.6.

Paul's Guitar Shop, Inc. Balance Sheet December 31, 2015				
Assets			**Liabilities**	
Current Assets			Current Liabilities	
Cash		32,800	Accounts Payable	49,000
Accounts Receivable		300	Accrued Expenses	450
Prepaid Rent		1,000	Unearned Revenue	1,000
Inventory		39,800	Total Current Liabilities	50,450
			Long-term Liabilities	99,500
Total Current Assets		73,900	**Total Liabilities**	149,950
			Owner's Equity	
Long-term Assets			Owner's Equity	
Leasehold Improvements	100,000		Retained Earnings	11,950
Accumulated Depreciation	(2,000)	98,000	Common Stock	10,000
Total Long-term Assets		98,000	Total Owner's Equity	21,950
Total Assets:		171,900	**Total Liabilities and Owner's Equity**	171,900

Figure 11.6 Sample balance sheet.

11.8.3 Statement of Cash Flows

As they say in the world of entrepreneurship, cash is king. This statement is an important factor in understanding where and how your company is acquiring its profit. The cash flow can be divided into three main components:

1. Operating activities
2. Investing activities
3. Financing activities

A positive cash flow indicates that your business is running well and that profit is being made. If cash flow is negative, it may indicate issues with your businesses lack of earned profit, which will result in not being able to pay off expenses. At times, a negative cash flow can indicate a mismatch of income and expense, and a recalculation should be completed. An example of a cash flow statement sheet can be seen in Figure 11.7.

Cash Flow Statement For the Year Ended December 31, 2016	
Cash Flow from Operations	
Cash receipts from customers	86,772
Cash paid for inventory	(7,400)
Cash paid for wages	(53,000)
Net Cash Flow from Operations	**26,372**
Cash Flow from Investing	
Cash receipts from sale of property and equipment	13,500
Cash paid for purchase of equipment	(17,500)
Net Cash Flow from Investing	**(4,000)**
Cash Flow from Financing	
Cash paid for loan repayment	(5,000)
Net Cash Flow from Investing	**(5,000)**
Net Increase in Cash	**17,372**

Figure 11.7 Sample cash flow statement.

11.8.4 Understanding Your Numbers

Perform an in-depth analysis by going through all necessary financial statements and ratios. This will give you a holistic understanding of the current state of the startup's finance. This type of analysis will come in handy when completing strategic planning for upcoming years. The trends that are observed will become invaluable for future product and service launches, as well as the way in which operations within the company are completed. For example, if there is a loss on an operating expense that can easily be minimized, this change can create less loss for the upcoming year. The most important tip to remember is to be proactive, not reactive, with your companies' financial planning and decision-making. Understanding your market environment, consumer trends, expenses, revenues, and cash flows can allow for predictions for upcoming years.

11.9 Why Businesses Fail: The Most Common Startup Derailers

It is appropriate to end this chapter with an overview of the most common reasons why new ventures fail. Much of the prior discussion is the mirror image of the following derailers. Proactively doing the right things at the right times will allow an IoT entrepreneur to anticipate and head off most of these obstacles, thereby considerably reducing the chance of failure.

For power of understanding, the primary challenges are grouped into four categories—financial, leadership, marketing, product, and team derailers. At times, these challenges will overlap and be dependent on one another. Therefore, it is incredibly important to have a solid understanding and insight into these common trouble areas. A brief overview of each main category is provided below.

11.9.1 Financial Derailers

11.9.1.1 Lack of Financial Understanding

In this section, we will discuss the most common sources of failure for entrepreneurs in relation to finance. Often, an entrepreneur begins their journey with little to no background in finance. This can cause huge misunderstandings and lead to poor decision-making as a leader and within a team. Understanding the financial structure of your business is vital to the long-term success, efficiency, and scalability of your startup. To alleviate some of those misunderstandings, refer to our section regarding finance within this chapter. Along with this, it may be wise to invest in some learning materials such as financial textbooks, small business materials (SBA. GOV), or online courses through websites such as www.Udemy.com. Seeking the advice or assistance of experienced entrepreneurs or individuals with a financial background may also be helpful.

11.9.1.2 Poor Cash Flow Management

The telling quote "cash is king" gives insight into the importance of cash flow management and the success of a business. According to *Entrepreneur Magazine*, cash flow management can be described "as the delay outlays of cash as long as possible while encouraging anyone who owes you money to pay it as rapidly as possible" (n.d.). Poor cash flow can affect your financial health by restricting growth and eventually leading to insolvency of the company. Financial health is not the only thing affected by cash flow management. By not properly managing cash, it can lead to poor relationships between suppliers, customers, and even your internal employees. Suppliers will become frustrated if there are late payments that are made to them, creating a tense relationship. Thus, it is important to have cash on hand to manage payments to remain in good faith with your supplier.

11.9.1.3 Poor Initial Funding Choices

When setting up the foundation of your startup, you need to understand your source of funding. Whether it be an internal or external source, or a mixture of the two, it can determine the financial structure of the company. There is a section within this chapter that focuses on the different types of funding. An entrepreneur should take the time to search and explore various funding opportunities. Becoming knowledgeable on the initial and long-term needs of the business will lead to an effective funding discovery period.

11.9.1.4 Incorrect Investments

An entrepreneur should conduct an analysis of the needs of the organization. This will prevent the entrepreneur from creating unnecessary costs during the early period of the startup process. Related to a topic that has been previously discussed within this text, entrepreneurs often rely on the bootstrapping methodology. Bootstrapping makes use of the company's and entrepreneur's existing financial foundation to build initial products and process. Then the entrepreneur can later make investments and improvements with newly earned profit. Essentially, they should make use of the resources that are available to them at the early development stages of the business.

11.9.2 Leadership Derailers

11.9.2.1 Lack of Understanding of Self

When beginning the entrepreneurial journey, the entrepreneur must have a solid understanding of who they are, how they function, and where their values lie. Key attributes regarding their leadership and management style will be important when making initial company decisions. It is important that the entrepreneur is honest and upfront when developing an understanding of their strengths and weaknesses. This type of insight will be valuable when selecting team members and partners that will be able to assist in areas of weakness. If an entrepreneur does not have a solid understanding and foundation of their personal brand, it will be difficult for the company to develop an authentic brand. The company and its leadership can easily come off as out of touch, impersonal, and more unfortunately... inauthentic.

11.9.2.2 Being Unprepared

While unplanned issues are common among startups, it is vital that an entrepreneur takes steps to protect themselves, their team, and their business from these potential risks. By utilizing risk assessment methodologies, the entrepreneur can become prepared for even the most unsuspected and unfortunate situations. This type of methodology allows the entrepreneur to identify and analyze potential risks that may come up in upcoming situations, both internally and externally in relation to the company. This type of assessment would prepare both the business owner, as well as their team for current and future trouble areas.

11.9.2.3 Lack of Passion

A lack of true passion can be devastating for an entrepreneur and their newly founded business. If the individual is not dedicated to their idea, the startup will suffer. During the process of beginning the startup, there will be many issues and many positive and fulfilling moments. Only true passion will be what keeps them going during those low financial quarters, late nights, and unsuspected issues. They need to be sure that they are in it for the right reasons.

11.9.2.4 Fear and Paralysis

Entrepreneurs must be bold, especially in times of fear and uncertainty. Usually they are characterized as risk takers, which proves very useful for the startup process. When issues begin to arise within the business, entrepreneurs can become fearful and paralyzed with business decisions and with managing their team. It is vital that the entrepreneur be confident in choices that they make. If the entrepreneur allows fear and paralysis to infiltrate their passion and their business it can lead to procrastination and many excuses regarding uncompleted deliverables and objectives. Take steps to become more confident! Create a list of things that must be accomplished and create strict deadlines to continue moving forward. Discuss the harsh realities of uncompleted objectives and deliverables. Most importantly the entrepreneur should remind themselves why they began this journey in the first place and what it truly means to them.

11.9.2.5 Lacking Management Abilities

When doing an analysis of their strengths and weaknesses the entrepreneur should be honest when exploring their management abilities. If the entrepreneur discovers that they have inadequate management skills, they should utilize mentors, partners, and hired individuals to assist with this portion of the company. Prior to hiring in external management assistance, explore team member and partner skill sets. There may be internal skills available that are incorrectly utilized or not utilized at all. A restructuring of duties and roles may be in order for the startup. A lack of skilled management can lead to dissolution within a new company. Consequences include damaged company reputation, poor marketing and product performance, and finally, an uninspired and unhappy team.

11.9.3 Marketing Derailers

11.9.3.1 Not Understanding Your Market and User

An entrepreneur could have the most innovative idea in the world, but if the user base is uninterested and unprepared for the product or service, the business will fail. As an entrepreneur, one of your most important tasks will be to understand your market and, more importantly, your target user base. The startup team should take time to carefully explore the following questions:

- What are the users' "needs"?
- What are the users' "wants"?
- What trends are present within the user population?
- What is important to these users? (Is it price, durability, convenience, etc.)

These questions will begin an exploration of the company's target users and should prompt in-depth conversations between the entrepreneur and their team. The user base drives the success of the company through monetary investments and pledged brand loyalty. Creating a culture in which the user feels appreciated, heard, and the focal point of the brand will ensure long-lasting customer relationships. After long-term relationships are built with the company, it is important to remember that the work does not stop there. The user's opinions should be taken seriously and they should be provided with excellent customer service to ensure brand credibility and user satisfaction.

11.9.3.2 Lacking Storytelling Abilities

Brand development should be a major focus for the entrepreneur and their team during the initial startup process. An acute storytelling ability will help shape the brand as well as the marketing strategy underpinning of the company.

Storytelling allows the entrepreneur to share their passion, vision, and an insight into the roots of the company with the user through the company's brand and marketing. The storytelling methodology creates a humanized and empathetic brand that is both relatable and successful. Creating a genuine emotional connection to the user will create long-term success for the startup.

11.9.4 Product/Service Derailers

11.9.4.1 Lack of Adaptability and Product Adjustment

Because markets and ideas are competitive, it is always imperative that the entrepreneur explore the correct questions when initially beginning their company. During the development process, the entrepreneur must take their brilliant, million-dollar idea, and adjust it for an ever-changing market. The idea that first sparked the business will rarely ever stay 100% intact. The idea will be shaped and refined to ensure success. The entrepreneur and their team should come together and explore the following questions during the product development process:

- Is there a need for the product?
- What differentiates the product?
- What does the product achieve?
- How will this product be used? Is this needed?
- Is there enough demand for this kind of product?
- Are users ready to use this type of product?

After having in-depth discussions about the questions above, the team is able to adjust portions of the product as deemed necessary. While this may be frustrating and hurtful for the entrepreneur, as this idea has been their brain child for a long period of time, it is necessary for the success of the company. Even as the company progresses, the entrepreneur should be aware that the idea will often be changed and manipulated to fit the current needs of the market. If the entrepreneur is unwilling to adjust or adapt his ideations, the company will be unable to experience growth.

11.9.4.2 Overpromising and Poor Outcomes

An entrepreneur needs to be very careful when it comes to overpromising results while achieving poor outcomes. Creating promises that end up being inaccurate and inauthentic will create huge divides between the company and its product users. While the company should create a promise of a respectable and reliable product for its users, the entrepreneur should understand how demand and supply functions. If there is limited supply of the resource, do not oversell just to create initial profit. The entrepreneur should understand the company's resource capacities to avoid poor delivery of the product/service. If the startup gets in the habit of overpromising and under-delivering, there will be issues with quality, customers, and missed opportunities.

11.9.5 Team Derailers

11.9.5.1 Team Creation

The team and partners that are selected will either make or break your entrepreneurial dream. This is one of the most significant aspects of laying the foundation for the company and determining if it will last. Prior to even searching for assistance, the entrepreneur should sit down and conduct an honest analysis of their skills, strengths, and weaknesses. This will showcase the areas in which they will be able to provide the most beneficial energy towards the betterment of the company. The entrepreneur should reflect on the areas in which their skillset was weak or unrefined in building the team. This enables them to seek individuals who will round out the team's skill set as a whole. The entrepreneur should also focus on understanding how differing personalities and leadership styles will function within the team.

11.9.5.2 Unwillingness to Ask for Help

The typical entrepreneur's personality has an emphasis on independence, and they often have a self-driven and self-motivated temperament. Although this type of temperament is useful in many situations that may arise throughout the startup process, at times it can be detrimental to the growth and success of the company. Stubbornness and ego may prevent the entrepreneur from asking for help and admitting failure in their decisions. However, this should not be seen as failure and should instead be seen as developing a network of well-versed entrepreneurial professionals. One person cannot do it all and cannot be an expert in everything. Understanding when to seek help in areas outside of your expertise will create efficiencies, gain, and growth within the startup.

11.9.6 Conclusion

This section focused on the most frequent errors that are made during the startup process. Whether it be financial, leadership, marketing, or team-based, an entrepreneur should familiarize themselves with common shortcomings and learn from past entrepreneurs' experiences. Entrepreneurs can bypass many of these common derailers by becoming knowledgeable of the business, understanding themselves and their team, and by creating a well-rounded network.

11.10 Case Study: Placemeter—Quantifying the World with Computer Vision and IoT

"As demand for IoT technologies grows, entrepreneurs considering IoT solutions should look for ones that might help them achieve a competitive advantage," Rabih Nassar, founder and CEO of IoT solutions firm Scriptr.[5]

11.10.1 Hustle and Bustle of New York Streets

One bright Tuesday morning, Alexandre Winter stepped into the bustling Manhattan area of New York. He had just had lunch with his friend who was visiting New York. From stories about personal lives, they had deviated to business talk and finally to recent happenings in the tech world.

Yet one thing that stood out to him from their conversation was the Internet of Things, IoT, as it is usually called. He knew what IoT meant obviously, but he was fascinated with the many developments and trends in the IoT sphere.

Winter kept on walking the old streets of New York thronging with pedestrians, shops on every sidewalk and corner you turn, and the constant traffic that never seemed to end. He sighed, a mixture of resignation and contentment, as he kept on weaving his way through the mass body of people. He knew New York was where he wanted to be.

11.10.2 Computer Vision

He had been watching the IoT space for quite some time now. He knew there was a whole lot of potential in that area with "a lot more data coming to the world thanks to the explosion of smart objects."[6] With that kind of opportunity, he definitely wanted to be at the forefront of solving that issue and providing solutions to that market.

Right now, he was into computer vision[7] developments, a technology that he hoped to use to pave his way to the IoT side of things.

11.10.3 Customers and Roadside Traffic

As he turned a corner, waiting for the Walk sign to Walk, so he could cross the road with the many people that had clustered around, he looked over and caught a glimpse of customers going in and out of the merchandise store across the road. He wondered how many people could be inside the store. They were tourists and New Yorkers alike, after all its summer, so it is that time of the year. He trudged along and looked at some more stores along the roads, the famous ones and not so famous ones.

As he walked along, he remembered his time in Washington, D.C., some few years ago, when he tried to have brunch and could not get a decent restaurant to take him in. They were all full, and he had to make reservations ahead. He guessed brunch was a serious business in DC. But frankly, it is a problem everywhere, as he thought more about it. Lunch time, rush hours, weekends, and holidays are very busy times for stores and businesses, especially in big cities. And that is when it hit him!

11.10.4 Explosion of Big Cities

Cities have grown and keep growing across the globe, and these cities have businesses and people who live and commute in them. Someone needs to figure out a way to measure what goes on in cities and places, as people interact with these places.[8]

With Big Data, when measured well, people can see when Old Ebbit Grill is swamped, or when Duane Reade is busy and go elsewhere, or adjust the time that they go shopping. Measuring what was going on in these kinds of places, and making it easy for people to see, will enhance the way people interacted with these locations.

11.10.5 Solutions Meets Problem

What if a device could track real-time location traffic, occupancy rate, occupied versus empty tables, wait time, sound level and temperature, average stay, repeat visit, and cross-location visit data through wireless cameras, infrared sensors, real-time and predictive algorithms, and Mac addresses of cell phones?[9] He knew that other people would want to know these things to save their time, efforts, and other resources.

With this innovation, businesses could also have a good idea of where to place a new location based on traffic data and other factors.[8]

Winter knew that he would have to do his research and find out if there was a market for this kind of technology now.

11.10.6 Quantifying the World

After that great eureka moment, Winter decided to start up an IoT business that quantifies modern cities worldwide. In March 2012, Winter founded and became CEO of Placement, Inc, a New York-based startup that quantifies human activity in cities using computer vision and video feeds.[10] This platform would later provide for smart cities, out-of-home advertising, and retail. Winter was able to raise 8 million in venture capital.[9]

Later, in November, Winter brought on his friend, Florent Peyre, to be a co-founder and chief operating officer in the business. It was an interesting journey into this sphere of technology for Winter, seeing as he had had a stint with a technology startup that was into large-scale image searching by content, which reached profitability in 3 years and was successfully acquired by a Japanese company. And his partner, Florent, had also had quite a few business experiences himself and serves as advisor to several New York startups.[11] Increasing their staff team size to 11–50 people was also a great milestone for the startup.[12]

11.10.7 Looking Back

Looking back, before the revolution of better traditional machine learning, Winter believed creating IoT to measure activity in the real world would not have been feasible, as they were just not to that point yet.[8]

Starting with computer vision technologies gave them that momentum and the "one foot in the door" to move into IoTs when the time came. Their computer vision technology was packaged to transform millions of hours of videos into meaningful data to enable a safer, smarter, and more efficient physical world.[13]

Also, harnessing technologies, like the cloud, enabled them to work with computer vision in a way that was previously impossible, which made their business viable.

11.10.8 Looking Forward: NETGEAR

Four years after, placement was acquired by NETGEAR, Inc. Placement and its Board of Directors decided to join NETGEAR because since 2012, their goal was to deliver solution to as many homes, as many companies, and as many cities as possible, and they believed that this acquisition will be a faster path towards achieving their goals.[14]

NETGEAR is a multinational company founded in January 1996, which uses technologies like wired and wireless devices, to deliver innovative products to its clients while focusing on reliability and ease of use. Patrick Lo, chairman and chief executive officer of NETGEAR, believes that "Placemeter will add immense value, and bring commercially proven, computer vision analytics to their smart home security team."[14]

Placemeter's engineering team will be integrated into NETGEAR's Arlo Smart Home Security business and supervised by Pat Collins, senior vice president of Smart Home Products.[14]

11.11 Discussion Questions

1. Outline the design process Alex would follow to start his business?
2. Some business tools help in the planning stage of a business and increase a business's likelihood for success. Identify these tools.

3. What are the business documents or tools Alex would have written up to start his business?
4. What legal documents Alex would need to file to be able to run his business?
5. What type of business structure was Placemeter operating?
6. Describe Placemeter's source of funding.
7. What type of partner was Florent? Briefly describe it.
8. When NETGEAR acquired Placemeter, what do you think happened?
9. Do you think Placemeter should have joined companies with NETGEAR or not? Why?
10. Mention some derailers that may lead to a company being acquired by another company.

Acknowledgments

Special thanks to Rebecca Kizer, Kasia Majkowski, and Lena Tomkins for their assistance in researching and writing this chapter. It would not have been possible or as thorough without their assistance during their time as graduate assistants in the CICS program.

References

1. Sangeeta Bharadwaj Badal and Joseph H. Streur, "Builder Profile 10TM Methodology Report," December 2014, www.gallupstrengthscenter.com/ep10/en-us/getfile?fileName=EP10%2FEP10_TalentDefinitions.pdf&language=en-US
2. Peter Drucker, "Managing Oneself," Harvard Business Review. January 2005.
3. "Design Thinking: A Method for Creative Problem Solving," *IDEO U.* Accessed October 13, 2018. www.ideou.com/pages/design-thinking
4. Michael E. Porter, *Competitive Strategy: Techniques for Analyzing Industries and Competitors.* New York: Free Press, 1980. (Republished with a new introduction, 1998.)
5. Sheila Eugenio, "3 Ways Entrepreneurs Are Making IoT More User-Friendly," *Entrepreneur*, August 13, 2017, www.entrepreneur.com/article/292415
6. Hovhannes Avoyan, "Technology That Sees the World: Welcome to the Future of Computer Vision," *Forbes,* April 26, 2018, www.forbes.com/sites/forbestechcouncil/2018/04/26/technology-that-sees-the-world-welcome-to-the-future-of-computer-vision/#53a0980f33cb
7. Computer Vision: "The initial goal of computer vision was to enable machines to see the visual world and interpret it the way a human would, but AI has advanced computer vision beyond human vision and now machines can see things humans can't, like air quality and temperature"—Hovhannes Avoyan, *Forbes*, 2018.
8. Steven Jacobs, "Placemeter CEO: How 'Computer Vision' Is Making Our Cities Smarter," *StreetFightMag*, January 28, 2014, https://streetfightmag.com/2014/01/28/how-computer-vision-is-making-our-cities-smarter/

9. Internet Software and Services, "Company Overview of Placemeter Inc.," *Bloomberg*, September 02, 2018, www.bloomberg.com/research/stocks/private/snapshot.asp?privcapId=234762000</JoJournal>
10. Crunchbase Inc, "Alexandre Winter," *Crunchbase*, September 02, 2018, www.crunchbase.com/person/Alexandre-winter#section-overview
11. Crunchbase Inc, "Florent Peyre," *Crunchbase*, September 02, 2018, www.crunchbase.com/person/florent-peyre
12. Crunchbase Inc, "Placemeter," *Crunchbase*, September 02, 2018, www.crunchbase.com/organization/placemeter#section-overview
13. Placemeter, "About," *Placemeter*, September 02, 2018, www.placemeter.com/about
14. Press Release, "Netgear Completes Acquisition of Placemeter, An Industry Leader in Computer Vision Analytics," *NETGEAR*, 2016, www.netgear.com/about/press-releases/2016/NETGEAR%20COMPLETES%20ACQUISITION%20OF%20PLACEMETER.aspx, Retrieved September 02, 2018.

Chapter 12

IoT Future Challenges

Rebecca Lee Hammons
Ball State University

Contents

12.1 Introduction

To obtain a fresh perspective on potential future challenges for the Internet of Things (IoT) industry, I interviewed Mr. Arthur Garcia of Tampa, Florida. Mr. Garcia is a former Microsoft MVP for IoT and was recruited to Microsoft based on his experience as an IoT practitioner and systems integrator. Our conversation provides insight into his experience, his areas of interest with IoT, and several areas for research and standards emphasis in the future.

12.2 Interview with Arthur Garcia—June 13, 2018

RLH (Rebecca Lee Hammons): Thank you very much for joining me today, Mr. Art Garcia. I appreciate you allocating your time to talk with me regarding the Internet of Things, an area that I understand you have great specialization in. I was wondering if you could give me some insight into when IoT became a passion for you, and how you entered the field.

AG (Arthur Garcia): I'd say about three or four years ago a colleague of mine was working on an IoT project for a grower in South Florida and he had some issues. And I decided to jump in and help him. And I just kind of got the bug after that, and started building my own devices, working in the Azure space. Because the device is just part of it. It's getting it connected and, as they say, it's the data, Silly. It's not the devices. Getting the things to go from the device all the way into Azure, into the IoT hub, and through all that piece, that part was kind of fun.

RLH: So, putting together the whole infrastructure to support the use of sensors and gathering the data is what interested you.

AG: It's fun to do the soldering and you get nice burn marks on your fingers after a while but putting together all the pieces I think is the fun part. You can solder to get to your devices but typically what you're going to do is buy something off the shelf. Or a company is going to buy some 500 of these units that measure temperature and humidity. They're not going to build them. It's more of how do I use the data I'm getting from those devices? What do I do with it, where do I store it, how do I run it through all the processes?

RLH: That's great. Is this field activity that you did with soil sensors and such your only entrepreneurial experience in implementing IoT or did you have other experiences?

AG: That's the main commercial one. I've done a couple of home projects. I'm working on one where we in Florida have hard water, so we have to use salt machines. It's really a pain in the rear to walk outside and lift the lid to check the salt in the machine. I'm working on a project to measure the salt levels and then send me information every day telling me what the salt levels are. It could become a commercial project.

RLH: I saw an ad for such a device last week in the newspaper. A water softener that sent email alerts to you when it needed attention, or the salt levels are low or such. I was fascinated by that because of the whole smart home thing. I was unaware that anyone was even…I mean, it's a great idea because I hate to take the lid off my water softener to check for salt.

AG: It's not only 'do I need salt?' but take it a step further. Connect it to your Amazon account or to your Home Depot account and place the order for more salt and have it delivered.

RLH: Brilliant! You don't like hauling those 40-pound bags either?

AG: Exactly. Exactly!

RLH: That's a great idea. I like that. IoT has become a fairly broad industry. I went to an IoT conference in Singapore with IEEE back in February and I was dumbfounded at the broad variety of research going on for smart nations, smart cities, smart homes, smart utilities, smart schools. I was curious, do you have an area that you specialize in or are you hoping to remain more of a generalist?

AG: I want to get more involved in a product called Edge, which the computer industry started out, we were distributed and then we went to centralized and now we're going back to distributed. Well, Edge is, instead of having all these IoT devices talking up to the Cloud, yeah, they're still talking to the Cloud but there are instances where you need to do processing at the device much more real-time. So, if you have a machine that you're getting your data from and if that machine temperature spikes above a certain amount, that machine's going to be toast. So, you need to send something to shut that machine down. Typically, what you would do is send the data up to the Cloud and the Cloud would do the processing and say there's a problem and send a command down to turn the machine down. With the Edge devices, all of that happens on the device and then the device reports back to the Cloud to tell it what happened. So I'm going to try to get more into that and that whole Edge industry is pretty interesting.

RLH: Is Edge the name of the concept for a device or is it a brand name?

AG: It's a brand name for Microsoft IoT Edge.

RLH: Oh, okay. Thank you. That sounds interesting.

AG: Other companies are doing something similar; I'm just familiar with the Microsoft offering at the time.

RLH: Well, I think for smart homes and smart cities, that would be a very fundamental, utilitarian use because if you're going to have sensors for fire, for example, you'd like to have suppression technology kick in immediately at that location rather than wait for a round trip to the Cloud to send you an instruction.

AG: Right, and I read something about a year or so ago about Tampa working on a smart grid. And there are other cities working on smart grid technology. And one of them was sound. In high crime areas, if you hear a gunshot, try to figure out the direction it is coming from, to send the police to the right place.

RLH: Wow. That's useful. Is the IoT industry, in your opinion, outpacing the standards and regulations that we've had in place for information and communication technologies?

AG: Yeah, I think the devices came first and now we're starting to think about security. And we're starting to think about security in a big way. When you're doing home systems and you're doing small implementations, it's not a big deal. But when you're doing cities, when you're doing a whole factory, and you're doing things out on the public web, that's when you really need to talk about security. Just simple things like making sure that I've got a secure connection between my devices and the hubs, making sure I can update the devices securely and remotely, we're getting there.

RLH: Do you think we have adequate standards for data sharing amongst a variety of different types of devices, like for emergency services purposes? Do we need create to more standards there?

AG: I think it's still in its infancy. I think we're getting there. But, you know, the IoT industry is only a few years old, if you really think about it.

I mean, back in the 70s, there was a coke machine that was the first, quote/unquote, IoT device, that actually reported back when it was running low on product. And we've come a long way since that to being able to control the lights in my house with my phone. And then we've got the smart refrigerators and all of those other smart home devices; if we're not careful, it could be a big hole for hackers to jump into.

RLH: Agreed. My next question is, do you think that we will see on the horizon, any consumer backlash? This was brought up at the SXSW Conference in Austin in March, they called it 'techlash', against the IoT industry because of the type of IoT issues that are being exploited currently?

AG: I'm not sure about that one. I'd need to do a little research on that one.

RLH: There are also concerns amongst many in the ICT industry about the proliferation of devices that we create that are requiring fairly rapid replacement. A mobile phone is one example that's considered rather throwaway after a couple of years, at least in this country. And not usually maintainable or repairable. IoT devices tend to be designed and developed in that way, as well. Unfortunately, it provides a lot of electronic waste to be handled without many standards in place by manufacturers, for example, as to here's what you do at the end of the lifecycle for this equipment. Do you think that we'll see any special need for people to brainstorm ideas to address this at the manufacturer level?

AG: I think so. At the consumer level, I can buy devices for $20, and if something goes wrong with it, I throw it away and get another one. But when you start talking about manufacturing and when you talk about high volumes, I think we need to work out a way of making these things more sustainable, more componentized. I mean, this piece came out and this piece busted, replace it with a newer model or just replace components and not throw away the entire thing, which adds to all that electronic waste.

RLH: I was also thinking about the possibility that manufacturers could take returns of end-of-life equipment and properly dispose of them as part of their supply chain management rather than have them go into dumps in Africa, for example.

AG: Right. Because a lot of them have all the soldered parts, and the gold in them that can definitely be recycled, it just has to be worthwhile for the manufacturer.

RLH: Okay, so that's an area that one could research and come up with some profitable recommendations.

AG: Right. They're not going to do it if it doesn't make sense.

RLH: Correct. Well, currently only 26% of electronic waste in the U.S. is actually recycled. And with the proliferation of smart devices we could see a lot more e-waste issues to deal with as communities.

AG: Oh, absolutely. What is it? Every year we get a new phone, or every two years we get a new phone.

RLH: Bingo.

AG: The compute power doubles every 18 months.

RLH: Do you have a vision for how we could share data among service entities in smart cities, across smart devices used for things like fire, energy, police, water and sewer?

AG: Yeah, I think with some of the technology that's out there now with the IoT hubs and those devices, in that city, for instance, you'll have millions of devices all over the place, on telephone poles, in hospitals and on buildings. And if we can get a way of taking all of that data and consolidating it, I think the best is to get it into the Cloud and then start doing some machine learning with it and some predictive analytics and some alerting. And just make it so it's not proprietary and make it an open-source standard that all of the municipalities can subscribe to.

RLH: Okay, wonderful. I think that would assist with first responders and all, to have greater accessibility to a lot of data points about a crisis. Let's say Hurricane Harvey and instrumentation in Houston and how cities can use data to respond to a large crisis of that type. I don't think we're going to see fewer of those events in our lifetime. I think there are going to be more of them, and perhaps instrumentation through smart devices will help to get a better handle on what type of a response is needed.

AG: Yep.

RLH: What do you see as the primary benefits of embracing connected devices in our businesses and our enterprises?

AG: In businesses and especially in manufacturing there are huge benefits. For instance, there's an energy company down here [Florida] that I did some work for and they are looking for some monitoring devices to monitor their conveyor belts running from the docks up to the factory. Moving coal and supplies. That's a manual process. If you can put 40 devices in line that measure the temperature and vibration and things like that and have someone monitor that and use predictive analytics and machine learning on that, now you've saved all that manpower which you can use somewhere else in a more productive manner. I think using the IoT devices and using this technology to take care of some of the mundane, repeatable processes—I think that's a great idea. Elevator companies use it to predict when things are going to fail. They'll use machine learning and algorithms to say 'this part will fail in 400 h.' They'll send a technician out there before it's a problem.

RLH: That is a radical improvement. For many devices. For many industries. Definitely. So labor automation is also, besides predicting failure rates so you can address them in advance, labor automation is a fundamental benefit to all enterprises. How about benefits you see when we instrument our homes and cities with IoT? Maybe just start with homes. Is this something you do in your home?

AG: In the homes you can use it to manage lights, manage energy usage—I mean, I think that's the big one, managing energy usage. When I leave I want to make sure that all the lights get turned off. At a certain time, I want the air to kick in. Just managing what my house is doing electricity-wise, heat-wise, even security. With the cameras, you now have a doorbell, and you can see who is coming to your door and you don't even have to be home.

RLH: That is huge. And in our cities, we don't have that many smart cities in the United States. A few are vying to build strategies and implementations, of course most cities use IoT for traffic control, to a degree. I know that's a big benefit for cities. But what do you see in the future for cities in terms of IoT usage?

AG: I think for security, where you have cameras on poles, where you have sensors that measure temperature and sound and things like that. One thing you can do is help predict the weather or even to help with the tourist industry. I'm walking down the street in the mall in Washington. Wouldn't it be nice if I'm walking around and, based on where I am, what streetlight I go near, I get some text or information about what I'm looking at. Just to help that tourist industry would be cool.

RLH: Speaking of tourists, I'm studying this summer, a telepresence robot that was donated to my Human Factors Institute for use by elderly in aging-in-place. And what I learned quickly is that, in cities like Palo Alto, they have these mobile telepresence robots out on the streets already, on the sidewalks, to answer questions for people who are lost or need assistance. I thought that was really great.

AG: Yeah, that's huge. That's a great IoT example!

RLH: Do you see any value in using IoT in K-12 or secondary education? It's not real commonplace yet.

AG: With how technology is today, and how connected everyone is, I think there is some value in starting that conversation with kids and getting them exposed to it. Just to see what's possible. I don't know that we need to put a robot in every classroom but have some sort of IoT devices that can help them with temperature, humidity, little things that can get them used to what it is and what's possible, what it can do. That would be great. And even with some security. Have cameras that alert the teachers when something's going on; I think that would be great.

RLH: Do you think that, in the future, we'll have an even greater need for technology professionals who are able to work with IT infrastructure in support of IoT, for smart cities, smart homes, smart schools?

AG: Oh, yeah. I think the infrastructure engineer and the Cloud engineer is definitely a profession that is not going to go away any time soon. Even with a device, that's easy, it's setting up the whole infrastructure that is important.

RLH: In our program here, a master's program, the Center for Information and Communication Sciences, we do teach Cisco networking and a lot about Big Data, Cloud, security courses and such. But I haven't yet done anything with IoT except to implement it as individual projects for team work or research activities for my GA's. So, I'm curious to have your opinion on whether or not this is the type of course I should create or bring in expert speakers on, the type of job skills people will need to work in IoT.

AG: I agree. I think what they need is exposure to Cloud architecture and the IoT as it relates to the Cloud. You've got your devices and you've got to get data into the Cloud. How do you do that? You use a hub. And now you've got data in there. How do you massage that data so you can actually use it? Streaming analytics and those things. And, okay, I've got the data and it's in a format that I can use, how can I actually make some sense of it? Then you get into machine learning and Big Data, and things like that. All of those are really viable future endeavors, I believe.

RLH: So, it's taking a cross-section of each of those topics and putting them into one course that might be hands-on, where you touch each of these elements through the lifecycle of working with IoT.

AG: Right, absolutely. You start off with…Here's a couple of devices we've got all throughout the campus. Well, now, let's get them all talking with each other. Now let's update them, let's put new firmware out there. You don't want to have people walk to each of the devices and do that, you want to do all of that through provisioning and through the Cloud.

RLH: Okay, wonderful, so I guess there is some value in exposure at this level.

AG: And you could also get into 'How do we talk to these devices?' The traditional way is through Wi-Fi but there's LORAN, there's radio, there's all kinds of other communication formats. And then you start getting into all of the different protocols that are available. MQTT and HTTPS and all the different ways that you can talk to the devices. You can definitely create at least two or three courses out of that.

RLH: Great ideas, that's a blessing! Great outcome for our talk. Do you happen to see, in the work that you do, any generational differences in folks accepting and using IoT?

AG: I see the younger generation embracing it a lot quicker and a lot easier. Folks in mid-life are seeing the value, just not the practicality, just yet. Some are, and some aren't. Like smart refrigerators and things like that, the older generation is starting to get it and starting to see the value. My 83-year-old mother-in-law just got a cell phone, she got a smart phone, and she actually texts, which I think is amazing. I think they're getting there.

RLH: I think that IoT devices can provide tremendous benefit to those that have mobility issues or hearing issues or such, who want to age in place. They haven't designed these products with their user experiences in mind.

AG: Right. There's still a long way to go. The industry is still really in its infancy.

RLH: That's exciting though, to be at the infancy of a new wave of technology. Having been at that point multiple times over the last 40 years or so, in the software industry, it's really nice to see the next wave come on.

AG: I started on punch cards.

RLH: Me, too. Me, too. That's really something. What do you see as the most critical areas to address in the IoT space to assure a sustainable industry segment?

AG: I think we've got to do more in how we're processing the data and starting to show some value. Things like machines, devices, that monitor how things are going on the shop floor. To know when the machine needs to be fixed or parts need to be replaced. I think that predictive analytics is going to be huge for industry. We do it now with things like the elevator companies, but to take that a step further, and have all of the shop floor machines actually talking to each other or talking to a central location and say 'Hey, I've got bearings that are going to go out in 5 h and I need to take action before things go haywire.'

RLH: I would like that on my car. Corporate social responsibility is taking on a larger role in ICT companies in recent years. What types of policies or practices do you think IoT designers and manufacturers should embrace with regards to CSR?

AG: They should keep personal data and people in mind when they're designing these things. Is this something that makes sense? Is this something that's intrusive? With AI, artificial intelligence, yes, AI is great but you've got to be careful about how much you do. You've got to still keep the human part of things in there. And I think they need to keep that in mind when they're designing systems. You know, how far do I want to take this really?

RLH: That's really interesting that you say that because, at the IEEE IoT/Smart Cities conference I attended, they emphasized the need to focus on human connectedness as an outcome; that this was not meant to isolate.

AG: We can do amazing things with technology. And that's the thing—we can. The big question is 'Should we?'

RLH: That's also a great area for future research and policy-making. Especially for smart cities. Do you have a personal vision for how we might be using IoT over the next 20 years, since it is in its infancy?

AG: I don't think we're going to be driving cars very much longer. I think those stories you see in the movies where the cars drive themselves and the cars are flying all over the place; I don't know about flying cars but I don't see us driving very long. Probably in our lifetime and, if not ours, then our kids' lifetime. That's, I think, the big one. It's going to stink because I like driving my car.

RLH: I love a good road trip.

AG: Yeah, exactly. It would be nice to go on a road trip and not be tired.

RLH: That's a fact. I thought about that with the autonomous vehicles. I commuted an hour each way to Indianapolis for 16 years, for work in a software company. And wouldn't it be cool if it was more like a moving gym so I could be working out or walking on a treadmill or doing other things while the vehicle was moving?

AG: That would be cool. I don't think that's too far away.

RLH: And that means self-driving or autonomous trucks are also not too far off in the future.

AG: And, see, that would make a lot of sense. A lot of those drivers are driving a ridiculous amount of hours and some of them are falling asleep.

RLH: So, the downside or the after-effect, as with any labor automation, is that we're going to continue to see a lot of disruption in manual labor activities, that are going to be automated over the next 20 years that were unthinkable a year ago. And how will we help society apply their skills in more meaningful ways to do more value-added activities? And see that future. Painting a picture of that future, a vision, for people who are going to be displaced in their labor, it's going to be critical.

AG: Right. Hopefully there is going to be a lot more re-training, a lot more changing of career paths, I think that's inevitable.

RLH: But people hate change, Art!

AG: Yeah, that's true.

RLH: Getting people on board to change is, like, something else! Although we, you and I, both seem to like it fine.

AG: That's the business we're in.

RLH: Correct. I was wondering, as we close, a question that I failed to ask. If you would give me a little bit about your impressive technology background before we close.

AG: I don't know about impressive, but I started out at University of South Florida as an electrical engineering major. And I had great hopes of becoming an electrical engineer. And then I encountered calculus and physics. Calculus and I didn't get along; physics and I got along fine. But it just happened to be that one of the classes I had to take was Fortran; and I took Fortran on punched cards. I have not looked back since. From there, I had my own consulting business for about 6 years, where it was unique—I was working with fund-raising companies, all non-profits. And I had about 200 or 300 customers all in the metropolitan New York area, helping them with fund-raising software and writing custom software and doing hardware and implementations of that sort. Then I got some Microsoft certifications and decided I'd had enough of working 80 h a week, and I moved back down to Florida and worked as a developer. Worked for companies like Raymond James, Catalina Marketing and I suddenly got hired by Microsoft as an account development manager where I'm responsible for about 7 or 8 clients, helping them with their development processes. I've been doing a lot in Azure. I've got 2 Azure certifications now and I'm working on a third. Doing a lot in DevOps and still dabbling in IoT. I got my MVP in IoT back in August [2017], and still doing some IoT talks and Azure talks, and how—not just the devices—the talks that I've been doing are 'from the device to the Cloud'. The device is the easy part. It's how that data gets used and where all that data goes. I've been working in that space for a while.

RLH: Well, that's fascinating. Your career sort of follows a similar path to my own, over my time in software. … I want to thank you very much for your time today, and also offer you the opportunity to work with student teams on any UX-type of project, innovation in ICT, could be IoT related, or infrastructure—please call on me. … I'd like to collaborate with you on any projects you'd like to do with my students in the future. It's good for the students.

AG: Great!

RLH: Again, thank you for your time! It's been a pleasure!

Chapter 13

Security with IoT

Kevin Keathley
Dean, School of Information Technology

Contents

13.1 IoT Introduction

The Internet of Things, or IoT, is the next evolution of the internet and our interconnected world. Interacting with the internet has become second nature for most of the developed world. Many people, like my daughter, never knew a time without it. Much as every good sci-fi story begins, the IoT is truly centered on machines communicating with other machines. These 'smart' devices communicate over the same networks as do human-to-machine communication—both internal and external private networks, as well as the 'network of networks' or our internet.

What is driving the IoT phenomenon is the promise of improved efficiency and better human experience through connected living. Simple examples include digital assistants like Amazon's Alexa, the ability of home automation adjusting the temperature in our homes, providing security, and even checking our refrigerator contents. Efficiencies are gained in everything from industrial controls to traffic coordination, as well as improved healthcare services and autonomous vehicles. These innovations, as well as future initiatives, improve our lives and the operational efficiencies of enterprises across the globe.

13.2 IoT Growth

Estimates around the growth of IoT vary widely, but one thing that is certain is that the market is huge and growing tremendously. A midrange estimate found at Gartner[1] estimates that the growth in IoT-connected devices will go from 6.3 billion in 2016, to over 20 billion in 2020 as depicted in Table 13.1. Over the same period, IoT spending will rise from $1.3 trillion in 2016, to nearly $3 trillion by 2020, shown in Table 13.2.

Table 13.1 IoT Devices (in Millions)

Category	*2016*	*2017*	*2018*	*2020*
Consumer	3,963.0	5,244.3	7,036.3	12,863.0
Business: cross-industry	1,102.1	1,501.0	2,132.6	4,381.4
Business: vertical-specific	1,316.6	1,635.4	2,027.7	3,171.0
Grand total	**6381.8**	**8,380.6**	**11,196.6**	**20,415.4**

Table 13.2 IoT Spending (in Millions)

Category	*2016*	*2017*	*2018*	*2020*
Consumer	532,515	725,696	985,348	1,494,466
Business: cross-industry	212,069	280,059	372,989	567,659
Business: vertical-specific	634,921	683,817	736,543	863,662
Grand total	**1,379,505**	**1,689,572**	**2,094,881**	**2,925,787**

13.3 Security and Vulnerabilities

13.3.1 Exposing over the Internet

IoT security is concerned with safeguarding connected devices on the internet and preventing unauthorized use or access. Many devices are not designed to be installed with an internet connection. A few years ago I bought a FOSCAM camera that has security features appropriate for an internal network only. For example, the camera uses http protocol instead of https—so if someone were sniffing the network, the password could be detected because it is sent unencrypted. While not an issue if only accessible from the internal network, this is not a device appropriate to expose to the internet. Unfortunately, many use cases involve exposing the camera to the outside world so, for example, you can check on the interior of your home while on vacation. This has led to a lot of these insecure devices being used in a way that is inappropriate for the security features of the device. At a minimum, an internet-exposed IoT device should have restricted network access.

13.3.2 Not Changing the Default Password

The use of noncomplex default passwords brings up another security issue. It is estimated that at least 15% of IoT device owners do not bother to change the default password and that just five common username password combinations will allow access to one in every ten devices.[2] There are websites that live stream up to 73,000 cameras using default passwords.[3] A list of 61 common passwords used by botnet Mirai gained access to as many as 500,000 IoT devices.[4] While there are many security issues related to IoT, the number of devices accessible to the internet that use the default password is one of the most pressing.

13.3.3 Other Vulnerabilities and Pain of Patching

Another issue with many IoT devices is the difficulty in patching the firmware. Many of the devices do not have an easy way to update their firmware. Most computer systems, especially those that can connect to the internet, will 'phone home' and check for newer firmware that often contains fixes to security vulnerabilities. Many of them will also automatically, or prompt a user, to download and update the firmware to improve functionality and security posture. Not only do many IoT devices not have this capability, some of them have no user interface by which to upgrade the firmware. Instead, it must be done by command line making it very difficult, if not impossible, for end users to perform the update. As I discovered with my own device, its firmware cannot be updated unless the device is wire connected—meaning it cannot be updated while on a wireless network. The obvious problem is that almost all at-home IoT devices are connected wirelessly. These factors make updating the firmware very difficult, and as a result, as much as 99% of IoT cameras are running outdated firmware.[5]

13.3.4 IoT Malware History Highlights

- Linux/Hydra—earliest known malware targeting IoT devices (2008). Like most IoT malwares, it contains both a mechanism for participating in a 'Distributed' denial of service (DDoS) attack and spreading to other devices.
- Tsunami—IRC bot that altered an IoT device Domain Name Service (DNS) entry to point at servers controlled by the attacker.
- Linux Darlloz—an IoT worm that spreads using a vulnerability in PHP to access a system and perform privilege escalation using a common credential list.
- BASHLITE—infects Linux-based IoT systems to create DDoS attacks. Uses known default credentials on telnet ports and was thought to have enslaved over 1 million internet-connected cameras and digital video recorder's (DVRs).[6] A predecessor to the Mirai.
- Mirai—one of the most prominent IoT botnet malwares in recent times. Using default passwords and common credentials, Mirai compromised over a half a million IoT cameras, DVRs, and CCTV, and home routers. Mirai has been responsible for DDoS attacks of more than 1 Tbps using well over 100,000 devices.
- Reaper—as of 2018, the next generation of IoT malware uses not only the default and common password methods of its predecessors but also uses at least nine other known vulnerabilities across a dozen manufacturers making it a most prolific spreading malware. It also spreads in a way much more difficult to detect than previous malwares.

13.4 Anatomy of a Botnet

One of the major implications for IoT devices with poorly implemented security is their use as the sources of DDoS attacks on private, government, and commercial internet sites. A group of infected devices that can be controlled as a group is referred to as a 'Botnet'. As we've discussed, some of these botnet networks can contain over 100,000 members. Botnets all contain the two major features and generally have four additional components. The two features necessary to be a botnet include:

1. Bots—or agents (otherwise known as zombies) that are the IoT devices that perform the DDoS attacks when instructed.
2. Command and Control servers (C2s)—the servers used to control the network of bots.

From the basic functions, it's not hard to see how these came to be called botnets. More sophisticated botnets can have these components that add to the capability of the network:

1. Scanners—used to search for vulnerable IoT devices. Can be done by a separate server or by previously infected devices.
2. Reporting server—contains results of scanning from bots or scanning servers.
3. Loaders infiltrate the vulnerable devices and instruct them to download the malware needed to get them under C2 control.
4. Distribution server—where the malicious code is stored to be downloaded by the infect IoT devices.

The method of how these botnets operate is not all the same, but an overview of the flow and interactions is presented here for illustrative purposes. See Figure 13.1.

a. Bots or external scanner servers search the internet for vulnerable IoT devices, looking for open ports by which the devices can be accessed.
b. Once found, the bots or scanners will exploit the system, by means of default username and passwords, or other known vulnerabilities and exploits.
c. The device IP address along with successful exploitation method are stored in the reporting server.
d. Malware loaders then access the IoT device using the IP address and credentials previously stored on the reporting server and download the appropriate malware from the distribution server.
e. Downloaded malware executes and takes control of the device and attempts to escalate privileges using known security weaknesses.
f. The malware may well secure the IoT device by fixing the vulnerabilities used to access the device to prevent future exploits and ensure it remains under this botnets control.
g. If competing malware is found on the IoT device, the malware will attempt to eliminate it again to secure its service for this botnet.
h. The malware performs any remaining configuration to enlist the devices service to the botnet.

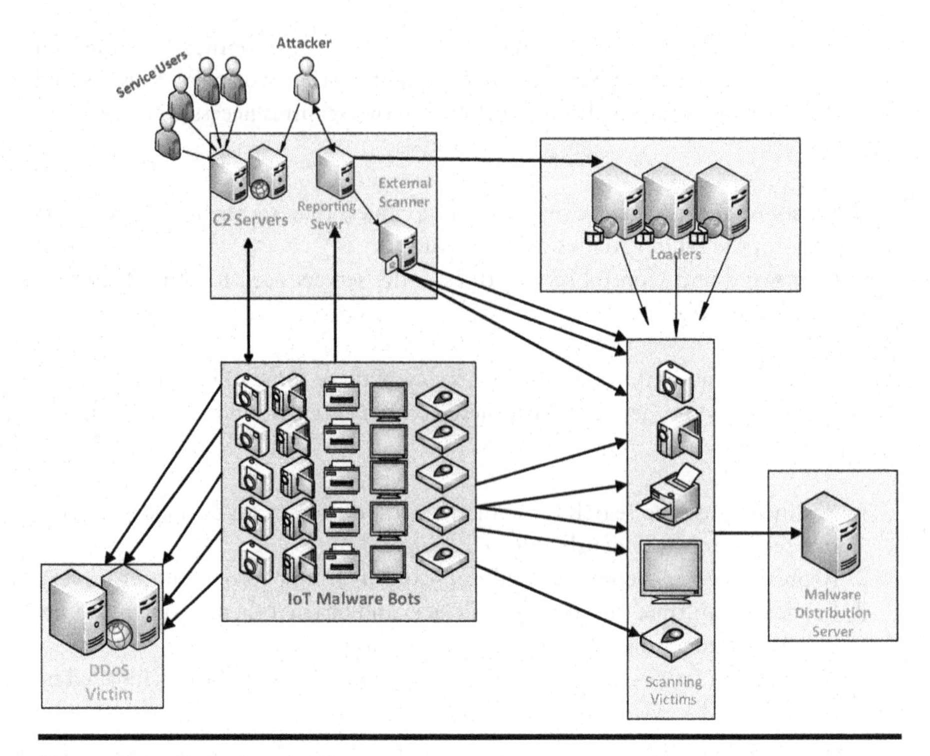

Figure 13.1 Botnet operation.

i. The device, now itself a 'bot', communicates regularly with the C2 servers via an IRC (chat)-based protocol to inform the operator (also known as a 'herder') of their existence and continued service.
j. The malware remains dormant and does not noticeably affect the operation of the IoT device until the operator instructs it to participate in a DDoS attack.

What can be seen clearly from this description is that although the method of accessing the devices is rudimentary, the malware is none the less sophisticated in terms of their operation. They also excel at spreading to other devices, eliminating preexisting botnet competitors and remaining in the communication with the C2 servers.

13.5 DoS and DDoS Attack Basics

DoS—In the most basic form, a denial of service attack is one machine overwhelming the resources of a target machine or server. A well-known open-source tool is called the Low Orbit ION Cannon and can be used to load test systems but also can be used to conduct a DoS attack. This attack method is constrained by the resources available to the attacking system, which many times is not large

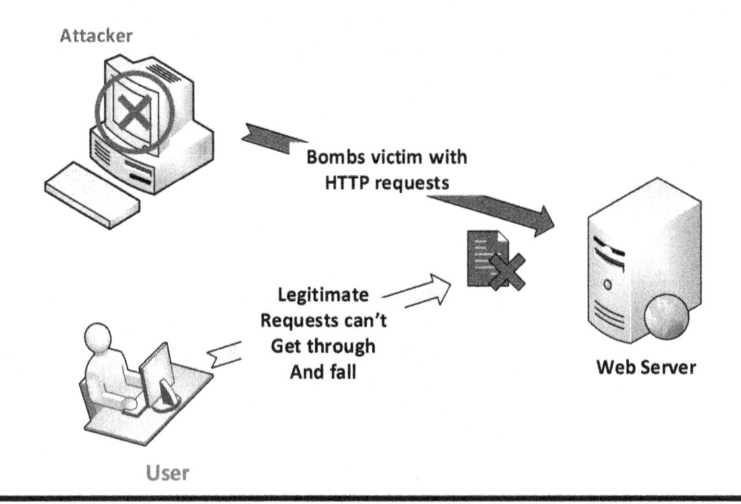

Figure 13.2 DoS attack scenario.

enough or powerful enough to harm the system being attacked. See an example in Figure 13.2.

DDoS—In situations where the attacked webserver has significantly more resources than the attacking system, it is unlikely to prevent legitimate requests from being serviced. This led to methods of 'amplifying' the attacks and gave rise to modern methods including the DDoS. This is not the only method of amplifying the attack, but it is one of the most common methods deployed today.

A DDoS attack uses compromised systems to create a bigger drain on the target victim's resources. As can be seen in Figure 13.3, an attacker sends commands to the compromised systems and they all flood the system under attack creating the desired amplification effect.

13.6 Direct and Reflective Attacks

Some attacks are 'direct' meaning that the attacking machine is sending packets directly to the victim machine, even though it's common to 'spoof' the source address to hide where they're actually coming from. Other attacks are 'reflective' where the address the attacker spoofs is that of the victim. This causes the reflective systems to respond back to the victim server causing a traffic flood, potentially from many systems which can in turn provide another potential amplification to the attack.

In Figure 13.4, the left section labeled (a) shows the path of a direct attack where the attacker machine A sends packets directly to victim server V and in some attacks confounds the victim by tricking it into replying to a third server whose IP address it has spoofed. On the right of this figure is an example of the reflective attack where the attacker machine A sends a packet to the reflective server

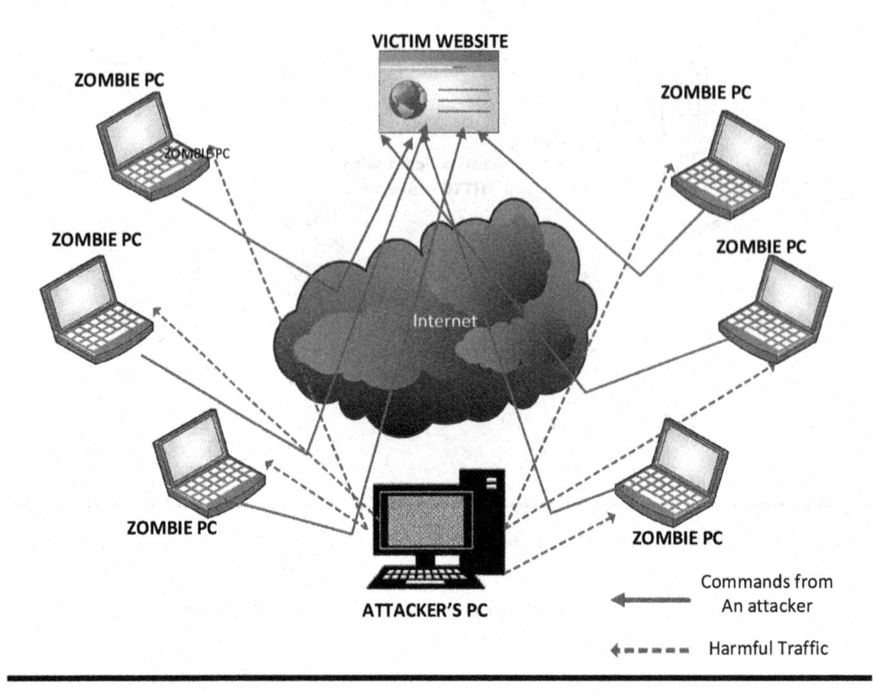

Figure 13.3 DDos attack scenario.

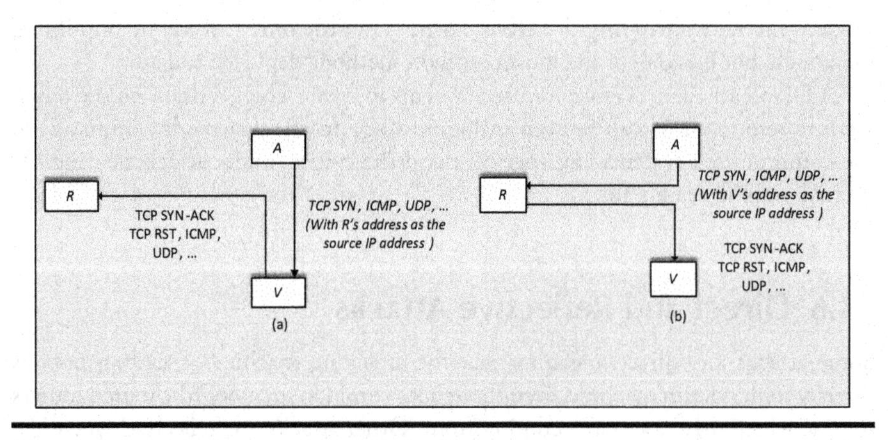

Figure 13.4 Reflective attack scenario.

R while spoofing the source address making it appear it came from the victim server V. Misdirected and half-completed handshakes can be disruptive to the victim machine and in this case do not require compromised systems; it simply relies on the communication protocols of the internet. It's not hard to imagine how the reflective attack can be used to create some of the attack amplification that many modern attacks utilize.

The following section covers a high-level overview of some of the internet's supporting architectures to help our understanding of how attacks use these to their advantage in creating traffic, attack mechanics, amplification, and system resource issues from floods or unconsummated protocol handshakes.

13.7 DDoS Enabling Internet Architectures

DNS is an essential element of the internet. Its primary function is to convert an easy to remember URL name like www.amazon.com to the underlying IP address of that system IE 52.84.1.52. In DDoS attacks, DNS is typically used as an amplifier. Misconfigured DNS servers that are recursive and open can return many times the incoming request in outbound traffic and allow an attacker to spoof the address of the victim server and amplify the traffic flood.

TCP stands for Transmission Control Protocol and relates to the rules used to govern the successful transmission of data over the internet. Features of the protocol include handshakes to set up communications, error correction, and retransmission on failures. In DDoS, the TCP protocol is often used as part of a SYN flood attack. There is a three-way handshake that is part of the protocol where a client sends a SYN (synchronize) request to a server. The server responds with a SYN ACK (acknowledged), and the client sends its own ACK to establish communication. The attacker sends the SYN with a spoofed IP, the victim server sends the ACK to an IP that didn't make the request, so it is ignored, and before this handshake times out, another SYN is sent again. These half-open connections, happening on every port on a server, will eventually cause it to become unresponsive.

ICMP stands for Internet Control Message Protocol and is used by networks to communicate errors and operational information, the most famous of which is the echo or 'ping'. ICMP can be used to perform reconnaissance on systems but is probably more famous in the DDoS world as used in a ping flood. A ping flood relies on the fact that for every ping echo request sent to a server, there is a ping echo reply sent back. The protocol allows specification of the number of attempts and the size of the packet. There is even a command to keep pinging until it times out. This is yet another example of the transmission architecture of the internet being used as part of a network-based DoS attack.

13.8 Botnets, Zombies, and Their Use in Attacks

Now that we've seen how botnets are created and controlled and how DoS and DDoS attacks function, it may be apparent that the 'distributed' part of many DDoS attacks is carried out by these botnets. A key element of DDoS attacks is the ability to direct these 'distributed' systems. This is where zombies, bots, and botnets come into play. Researchers have found that recent DDoS attacks have originated from three types of devices—96% IoT devices, 4% home network routers, and less than 1% were compromised Linux servers.[7]

The definition of 'zombie' in computer science context is a computer connected to the internet that has been compromised. Whether by virus, Trojan horse, or any other exploit, the system can now be controlled by a hacker. Once compromised, a zombie can be turned into a 'bot', short for robot. A robot can be remote controlled to perform some script or repetitive task once set off by the hacker. Botnets are whole networks of these zombie robots that can be controlled by a hacker, sometimes called a 'herder', who orchestrates their activation and scripted tasks. As we've previously discussed, the poor security implementations of many IoT devices is one of the chief sources of bots for these botnet operators.

In Figure 13.5, we demonstrate how a hacker might use a botnet to conduct a reflective DNS attack.

In this example, a botnet herder issues commands to his controller that sets off the network of bots executing their instructions. In this case, the bots are spoofing the IP address of the victim DNS server while querying remote DNS servers, causing them to flood the attacked server with replies it didn't request. The victim DNS server is attacked from regular DNS servers not under control of the attacker, creating a long line of obfuscation from tracking it back. Other features of this attack include amplification from the distributed bots, plus amplification in this particular DNS attack because the DNS replies are typically at least 3–4 times larger than the requests.

13.9 Attack on Krebs Security

On September 20, 2016 an extremely large DDoS attack was committed against at the security-oriented website krebsonsecurity.com, operated by security expert Brian Krebs. There were a few unusual things about this attack and some worry it may be part of a shift in how these kinds of attacks are carried out.

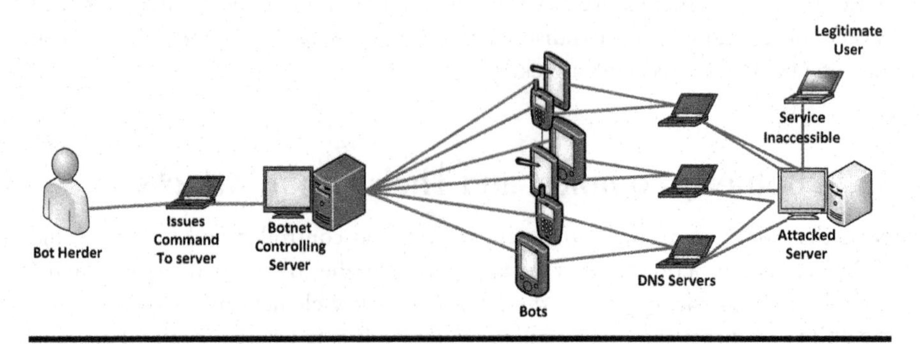

Figure 13.5 Large scale DDos attack.

For one thing, the network bandwidth generated in the attack was extremely large. According to a post on the website the next day, it was approximately 665 Gigabits of traffic per second which was double any previous attack high traffic amount seen by Akamai, Krebs infrastructure provider.[8] While most of the largest DDoS attacks use the DNS reflective method described earlier, this attack did not use any of those tried and true methods, thus making it unusual. Another early indication was that the traffic was GRE (Generic Routing Encapsulation) packets which are usually used to connect two nodes in a point to point communication. If correct, this means that the traffic was probably not amplified at all, which might indicate it was produced in a 'direct' attack by possibly the largest distributed botnet ever recorded, possibly hundreds of thousands of systems.[8] It was also suspected that these large numbers of directly controlled bots were probably a massive number of IoT devices such as routers, IP cameras, and DVRs.

It's important to note that this seemed to be a new and unusual signature for a DDoS. However, some believe we are entering a new era of these kinds of large-scale, direct, IoT-based botnet attacks with new capabilities.

Approximately 2 weeks after the attack on krebsonsecurity.com, a postmortem blog was created by the sites infrastructure provider Akamai. In it, Akamai confirmed that the IoT played a majority role in the attack and that a primary participant had been the Mirai botnet they had been tracking for several months. Akamai noted the majority of these devices were indeed security cameras and DVRs used in small office/home office setups. It also confirmed that this was a 'direct' attack from a large number of devices and that most of them had default or easy to guess passwords.

Given that we are at a very early point in the IoT life cycle, this is troubling because we are likely to see more frequent and larger attacks using this method. Hopefully, better security measures will be integrated into the devices including preset unique passwords as well as ongoing complexity requirements.

13.10 Other Notable Cases

DYN—is an internet performance management company offering product to optimize online infrastructure. They are also one of the leading managed DNS services in the world. In October 2016, they suffered a DDoS attack that was sourced from 100,000 internet-connected IoT devices. It is thought to be from the Mirai botnet, which reached over 1 tb of network traffic in the attack. Many famous websites use Dyn DNS service including Airbnb, Amazon, Reddit, and Spotify. Nearly all of them suffered partial or complete outages as a result of the attack between 11:10 UTC and 17:00 UTC.

U.S. Elections—Both the Trump and Clinton campaign's websites suffered DdoS attacks on Nov 6th leading up to the U.S. presidential elections. Another attack targeted a phone bank service used by both campaigns, TCN. Over a 24-h period, an attacker used varying IP sources and flood protocols to overrun TCN's phone and website network.

WikiLeaks—Nov 7, 2016, WikiLeaks email servers were knocked offline from a DdoS attack lasting almost 24 h. The attack was allegedly initiated in response to leaked emails from the Clinton campaign and released to the media by WikiLeaks. This demonstrated that some DdoS attacks are designed to restrict free speech. Google started an initiative called Project Shield that provides DdoS protection for websites that have 'media, elections, and human rights related content'.

Ransom Denial of Service—RdoS is a newer type of attack that attempts to extort ransom from an organization in order to prevent an attack on them causing a disruption in their business or operations. Most commonly, a small attack to demonstrate the capabilities is followed by the demand for ransom.

13.11 Mirai and Reaper

The largest DdoS attacks recorded so far have been propagated by Mirai, a malware strain that enslaved IoT devices like cameras, DVRs, and wireless routers to create botnets used to bring down some of the world's top online websites. Now, experts are seeing a new strain, named the Reaper, spread via security vulnerabilities in IoT software and hardware.

Reaper is not creating attacks, but apparently content spreading itself and growing its network of bots. In October of 2017, Checkpoint announced they were tracking the growth of Reaper and that as many as 1 million organizations may have been infected.[9]

Unlike Mirai, which exploits vulnerable IoT devices using factory default or hard coded passwords, Reaper uses those techniques along with at least nine known security vulnerabilities across a dozen device manufacturers. While both Mirai and Reaper are worms that are built to spread automatically, the Reaper uses a much more deliberate and stealthy method to enlist new recruits designed to avoid detection from tools on the local network.

It may be important to note that Reaper may be being built not for DDoS attacks but rather could be used for a variety of purposes like a distributed proxy network or as jumping off points to exploit other devices in the internal network. The fact that Reaper is built to coexist with Mirai is interesting to researchers, but it also appears Reaper has the ability to add new functionality. In some ways, this is even more frightening than merely supplanting its predecessor.

13.12 Preventing IoT Infiltration and Mitigating DDoS Attacks

Here are some ways IoT manufacturers can improve their security posture, making it more difficult to infiltrate devices and create botnets.

- Prevent devices from accessing any network addresses other than the private[10] space or the manufacturer's web domain.
- Unique default passwords with sufficient complexity IE ten characters, upper and lower case, and special characters.
- Check in to manufacturer's website to download and apply security updates.
- Consumer groups or government regulation forcing IoT manufacturers to adopt some of these best practices.

There are several ways to mitigate DDoS attacks on commercial or organizational websites.

- Internet Service Providers (ISPs) should follow best practices including Internet Engineering Task Force (IETF) mandating that packets with a source IP address different from its originating network should be filtered at the ingress.
- WAFs or web application firewall services will provide a proxy service that will forward legitimate packets to the protected endpoint but swallow an attempted DDoS attack keeping the actual site operating.
- CDNs or content distribution networks like Akamai or Amazon Cloudfront can also mitigate DDoS attacks by distributing the web endpoint making it much more difficult to disrupt the normal website operation.
- Insurance can be a viable option for many commercial enterprises who need to be protected from business disruption.

13.13 Other Security Risks—Not Just DDoS

It was mentioned during the discussion about Reaper that instead of being used as a source of bots for DDoS attacks, it could be used for other purposes like infiltrating the internal network where the IoT device is connected. A great number of devices live on an internal network but have internet access as well. This fact makes not just the device vulnerable, but the internal network it's attached to is vulnerable as well.

A very unique and interesting case was discussed at a Wall Street Journal CEO conference where Nicole Egan, the CEO of Darktrace, told the story of hackers using the smart thermometer in a casino's lobby fish tank display to steal a high-roller database.[11]

Because of the confidential nature of the client and data, the 'who' and 'when' were not elaborated. However, it is yet another example of an IoT device, connected to both the internet and the internal network that was used as a gateway to access and retrieve sensitive data.

A lot of time was spent in this case study discussing the use of IoT devices as bots to deploy DDoS attacks on other internet targets. There is another threat vector to data on your internal network that may include appliances and home automation that most people wouldn't consider a serious threat. Unfortunately, they would be wrong.

13.14 Conclusion—Do I Really Need One?

- Patch your IoT devices.
- Adhere to best practices and change your password.
- Don't put things on the internet unless you really need them to be—really.
- Mitigate your commercial interest with DDoS protection services.

13.15 Case Study

13.15.1 Overview of the IoT Capability Models

The IoT is already changing how we interact with our environment and there is little doubt that these changes will continue far into the future. From smart home appliances to industrial controls, IoT is growing every day in pervasiveness and capability. The most visible IoT from a consumer perspective comes in the form of smart and connected products. While we'll discuss several classes of IoT devices, this case study focuses on examples of consumer IoT devices that are available and popular in 2018.

The majority of smart and connected consumer products are mechanical and/or electrical, physical devices that are incorporated into the home. Over the last several years, many of these devices have become 'smart', including on-board software, some processing power, and, more than likely, a sensor of some kind. These units may have a user interface and functionality designed to enhance the service capabilities or provide some level of personalization for users.

As smart devices become connected, primarily through the use of Wi-Fi networks, their capabilities increase substantially. Connectivity allows for remote monitoring, control, and service of these devices. This makes them even more useful for consumers, and the ongoing analysis of product usage patterns allows the device manufacturers and service providers to expand the products' capabilities and optimize existing performance and processes.

The next level of evolution is for these smart and connected devices to work together as a product system. The integration of various components into a system allows for enhanced features and operation including optimization across the entire system. More complex systems perform real-time analytics that may include predictive algorithms to expand product and services across the system on the fly.

The future of IoT lies in the systems described above to interconnect with other disparate systems sharing relevant information to automate and coordinate advanced capabilities like machine learning and predictive analytics. It is not unreasonable to expect this kind of next-generation processing capability and interconnectedness to be truly transformational to human interaction with our environment.

13.15.2 Consumer IoT Examples

There are currently several connection standards to allow home-connected devices to interoperate. Zigbee and Z-Wave are a couple of the most popular but have historically been competing standards. However, as of 2018, there is some convergence in hardware that supports both standards, including units from manufacturers like Samsung SmartThings and Wink 2. There is an ever-expanding range of connected devices from numerous manufacturers, but most consumers interact with smart home gadgets through just a few interfaces. Many will have software that runs on a tablet, laptop, or smartphone, but the majority will also support one of the major systems for voice recognition and virtual assistants—Google Home and Amazon Echo, also known as 'Alexa'. These two companies are currently dominating the market for the voice recognition hardware that allows consumers to interact with their devices using natural speech. Because of how convenient speech interaction is, it will likely continue to be an important interface to consumer use of IoT devices. Following are a few broad categories and specific examples of IoT devices in use today.

13.15.2.1 Voice Recognition and Virtual Assistants

Amazon Echo (aka Alexa) is a Wi-Fi-enabled virtual assistant developed by Amazon used in a variety of hardware footprints. The Echo is the original full-size unit with seven microphones and a full range speaker to enhance music playback capabilities. The Echo also comes in other configurations including the Dot, a smaller footprint unit without the full range speaker; the Show which includes a 7″ LCD screen; and the Spot, which is similar to the Dot but with a 2.5″ circular screen. All of the Echo devices are activated by the keyword 'Alexa', so it is common for them to be referred to as Alexa.

Google Home is a virtual assistant and smart Wi-Fi-connected speaker developed by Google. Google Home also has different hardware configurations including the original Home, the Mini, and the Max. The original Home has a full range speaker and is a cylindrical device with a similar footprint to the Echo. The Mini is the small footprint Home unit about the size of a flattened baseball with all the same functionality, aside from the full range smart speaker (similar to the Dot). The Max went the route of major sound with stereo speakers including subwoofers and connectivity that includes a USB type C and standard analog audio connectors. It seems a certainty that LCD screen versions of Google Home will show up soon. Google uses the activation keywords of 'Hey Google' or 'Ok Google' to summon the virtual assistant.

13.15.2.2 Security and Convenience

Cameras are clearly one of the major IoT device categories in the market. There's an amazing number of cameras deployed in homes around the world. Many of these were designed to be used only on the inside of a network, but consumers want to view them when they are not home making it common to find them exposed on the internet. The lack of security controls appropriate for external exposure to the internet has made them vulnerable to becoming part of robot armies used in DDoS attacks. Keeping these on the inside network and changing default passwords is a best practice and always a good idea. There is a class of camera engineered to be viewed from the outside from manufacturers like Nest and Amazon Cloud camera. Instead of allowing traffic in from the internet (a dangerous threat vector), they only reach out to a known endpoint that controls access and abstracts the actual camera from the internet. These usually involve a subscription service but are inherently more secure due to the nature of the traffic origination being outbound only, from the inside network.

Smart thermostats are another very popular smart home device. Programmable thermostats have been around for a long time, but the newer generation can be controlled by your smart phone or your voice using one of the virtual assistants. Modern units can learn the patterns of your usage to become predictive about how you want the environment set and when. These 'learning' thermostats can support remote temperature sensors and adjust to time of day, day of week, and even the season of the year. Since this one control is responsible for about half of a home's energy use, efficiency gained from a 'smart' one can pay for itself in under 2 years.

Home automation includes regular LED and color-changing light bulbs; motion sensors that can be used as an event trigger; control programs that can create environmental 'scenes'; automatic door locks that can control access to any part of the home; power plugs that can be used to turn on or off virtually any electrical device; sensors for smoke, carbon dioxide, and moisture that can alarm or trigger remediation; and even doorbells with built-in camera that allows you to interact with someone on your front porch, no matter where you may be physically located. Home automation is an example of a connected product system as the devices can work together to provide a single solution. For example, you could set your status to 'coming home' and you're your home automation system might unlock doors, turn on lights, and set the house to your desired temperature. The same technology in 'theater mode' could turn the TV and sound system on, dim and change light colors, and initiate your favorite streaming service. These are just a few examples of how the devices may act as one system to create convenience in managing the home environment.

13.15.2.3 Other Consumer Smart Connected Devices

Refrigerators. Yes, there are smart refrigerators! And they come in handy! Besides controlling the temperature and having a 'virtual message board', another great feature is being able to peek inside your fridge real time while grocery shopping to see what you may or may not need. Combine this with a light and camera in your pantry, and the days of making shopping lists could be over!

Another great use case for smart connected appliances is the ever-so-convenient Crock Pot. Set it in the morning before you leave for work and you can monitor it all day long and adjust the cooking time and temperature remotely.

Robot vacuums have been around for a few years now, and they are an interesting use case. Vacuuming is a chore that needs to be done often, and it involves a lot of physical motion as part of the work. Because it requires significant movement throughout the environment, these devices are some of the most 'robotic' consumer IoT devices operating today. These mechanical sweepers can work together to provide better coverage and perform analytical geo mapping to understand and navigate the complex environment they must operate in throughout the home.

Though larger and more powerful than robotic vacuums, robotic lawnmowers function similarly in terms of operating on a schedule with a complex ability to learn the geography and terrain of the area they cover. These beefy, motorized weed whackers only recently made the jump into mainstream consumer use.

13.15.3 What Will the Future Hold?

One important level of maturity for IoT capability models is having smart and connected devices work together as a product system. As discussed previously, some of these IoT devices and systems are already operating in this manner with home automation and other integrated services. The expansion of these capabilities will continue, as well as their extension across product systems to even further integrate the whole system, and enhance both the capability and the user experience. One thing seems certain that IoT will become a larger part of how we interact with our environment and the many objects that exist within it.

13.16 Discussion Questions

1. Why would you not allow external internet traffic inbound to your in-home cameras?
2. What are the advantages for cloud camera services? Disadvantages?
3. Which of the connected products or systems do you think are most useful?
4. What do you think the future will hold for IoT and consumer connected products?

References

1. www.gartner.com/newsroom/id/3598917
2. www.theinquirer.net/inquirer/news/3012365/15-per-cent-of-iot-devices-owners-dont-change-the-default-password

3. www.tripwire.com/state-of-security/latest-security-news/73000-security-cameras-viewable-online-due-to-use-of-default-passwords/
4. www.csoonline.com/article/3126924/security/here-are-the-61-passwords-that-powered-the-mirai-iot-botnet.html
5. www.csoonline.com/article/2224469/microsoft-subnet/hacks-to-turn-your-wireless-ip-surveillance-cameras-against-you.html
6. https://threatpost.com/bashlite-family-of-malware-infects-1-million-iot-devices/120230/
7. www.multivu.com/players/English/7911451-level-3-threat-research-malware/
8. https://krebsonsecurity.com/2016/09/krebsonsecurity-hit-with-record-ddos/
9. https://research.checkpoint.com/new-iot-botnet-storm-coming/
10. https://tools.ietf.org/html/rfc1918
11. www.maplecasino.ca/blog/casino-database-hacked-through-fish-tank/

Chapter 14

Legal Issues with IoT

Gerald DeHondt
Ball State University

Contents

14.1 Introduction

Technology moves at the speed of light, and regulation moves at a sluggish pace. Difficulties in regulating technology have always relied on a cumbersome process, fraught with red tape, to determine what is legal and what is allowed. Laws are typically enacted *after* a change occurs, in response to a precipitating event or situation. The advent of the Internet of Things (IoT) has led to much confusion within the legal community and related legal matters (Kim, 2016). Technical areas inevitably create a high degree of conceptual confusion to nontechnical persons. As a matter of scale, Gartner, Inc. estimates that 4.9 billion IoT devices were in use across the globe in 2015; this is quadruple the number of automobiles in use. They further estimate the number of "Things" to be 20.8 billion worldwide (Kim, 2016).

Like most evolving technologies—standards, practices, and regulations follow behind the actual innovation. Regulating technology is akin to hitting a moving target that travels along an inconsistent path, very common with emerging technologies. This chapter will discuss some of the challenges in regulating an area just after conception, discuss evolving legislation, and identify potential areas for developing legislation.

The "IoT" refers to the ability of everyday objects to connect to the Internet and to send and receive data (GSMA Association, 2016). This connectedness is for the benefit of users, individuals, and society. These connections would include Internet-connected cameras that allow users to post pictures online with a single click, home automation systems that turn on the front porch light when you leave work, and bracelets that share with your friends how far you have biked or ran during the day. Certainly, the number and types of devices connected to the Internet will only increase as developers figure out ways to make devices talk to each other in ways that are helpful to us both at work and as everyday consumers (Mighell, 2014). In 2010, the number of these "things" connected to the Internet surpassed the number of people accessing the Internet, and experts estimate that by 2020 there will be 50 billion connected devices (GSMA Association, 2016). This can be viewed as a significant tipping point in the growth of the Internet and communication. The purpose of the Internet is to enable communication between individuals and provide benefits by connecting us to these things when we are away. This infrastructure has grown to include not only those persons that need to communicate but also the devices and "things" that are communicating on behalf of these persons in the form of garage doors, thermostats, refrigerators, and any host of once solitary devices. The infrastructure used to provide this communication—on behalf of these persons—has grown to such an extent that it exceeds the participants in this communication.

14.2 The Need for Regulation

Taking all of this into account, what are the regulatory necessities for the IoT or should these devices even be regulated? The European Union (EU), U.S. Federal Trade Commission, and the National Institute of Standards and Technology (NIST) have all weighed in on the practical legal issues surrounding the IoT (Kemp, 2017). It is important to note the global nature—and necessity—of discussions between and among national organizations. IoT cannot be localized, and it is a challenge for nations to deal with this new concept because almost every whisper in one country is collected and sent to the country that is providing the service (AboBakr and Azer, 2017). This multinational aspect demonstrates that IoT does not affect users in a single country, but their data may be sent across borders as part of normal data processing activities.

The Federal Trade Commission, in their recent workshop on IoT, noted caution about proposing regulation of the IoT sector, given its importance to innovation in America. They further noted that this could mark the beginning of a regulatory regime for a new set of information technologies that are still in their infancy and recommended that policymakers should exercise restraint and avoid the impulse to regulate before serious harms are demonstrated (GSMA Association, 2016). In effect, enacting legislation once a particular harm has been identified. This last point is most serious to note in that regulation tends to follow harm; it is only after a malady has occurred that legislators take interest in resolving a conflict. Corollary to this is that legislation and regulation will shape the industry, beyond the natural evolution and guidance provided by trade associations and working groups. Legislating IoT too early may lead to the stifling of intellectual and economic growth in the sector and prevent such evolution as necessary to allow fruition of benefits.

Alternatively, the Global Privacy Enforcement Network (GPEN) stated that about two-thirds of devices surveyed failed to adequately explain to customers how their personal information was collected, used, stored, and disclosed (Kemp, 2017). Almost three-quarters failed to show how personal data could easily be deleted off the device. Regulators find this lack of consumer information particularly concerning and may be an area for regulation. Other areas that may be of particular interest to regulators and legislators are identified in Table 14.1.

Table 14.1 Inherent Challenges of IoT Devices

Lack of control and information asymmetry	Device connectedness results in personal data generation, storage, and communication over which the user has no control.
Quality of user consent	The user's consent to the processing of data carried out by IoT devices must be informed. Many times, the user is not informed of the data processing carried out by an IoT device. In these instances, consent cannot be granted as it is not properly informed.
Secondary use and repurposing	Big Data analysis techniques may lead to device data obtained for one purpose being used for another purpose for which no consent has been given.
Aggregation of data	Different devices may reveal specific aspects of user habits, behaviors, and preferences in an intrusive manner.
Remain anonymous	Limitations on the possibility of anonymity when using IoT devices or services.
Security risks	Physical constraints with the device may lead manufacturers to prioritize IoT device performance or cost over user security.

Adapted from 2016 Global Privacy Enforcement Network (GPEN) Privacy Sweep Opinion.

Aside from legislation, there will be the day when Fitbit data will be requested as discoverable during litigation and be just as useful—and commonplace—as e-mail, instant messages, or social media postings (Mighell, 2014).

Further consideration notes that in the absence of specific regulation, self-regulation, and best business practices—that are technology neutral—along with consumer education serve as the preferred framework for protecting consumer privacy and security while enhancing innovation, investment, competition, and the free flow of information essential to the IoT (GSMA Association, 2016). Furthermore, self-regulation has tended to work well to ensure consumer privacy and foster innovation, and industry has a strong track record of developing and implementing best practices to protect information security.

Given the unique nature of certain industries such as healthcare and financial services, they are unlikely to wait for broader protections and best practices to develop governing the industry. Instead, these areas are likely to develop their own rules and standards regarding the IoT, along with the extension of consumer protection regulations (Kemp, 2017). Within the next 15 years, most medical care services will be delivered virtually, resulting in a dramatic growth in IoT technologies. Along with this will come an explosive growth in the amount of data held within these devices (AboBakr and Azer, 2017). Data held on medical devices will most likely come under the purview of the Health Insurance Portability and Accountability Act (HIPAA).

It is also important to consider how these devices are being monitored. What are we doing to monitor the use of the same devices that are monitoring us through tracking our whereabouts and movements through GPS and video monitoring? As stated initially, it oftentimes takes a period for the law to catch up to technology that moves at an ever-changing pace. Presently, many of the legal issues relating to the IoT deal with the data that these devices provide (e.g. Mighell, 2014; Kim, 2016; Walker, 2014). Using IoT devices for theft, breaking and entering, stalking, and other similar crimes would be covered under current statutes.

One legislative area requiring specific attention is the use of this data in ways not originally intended. Can information retrieved from fitness devices be used by insurance companies to rate policyholders and assign premiums? Would this represent a violation of privacy guidelines or laws? It is common for health insurers to provide discounts to customers for living a healthier lifestyle and charge reduced rates for nonsmokers, losing weight, and other healthy lifestyle changes. These actions are periodically monitored to ensure continued compliance and the insured remain eligible for discounted rates. These metrics and monitoring methods are hardly intrusive and serve merely to verify continued adherence. Fitness trackers that monitor numerous biometric data—data that may be unknowingly provided to companies without explicit approval from participants—provide significantly more information on daily behavior and routines than an annual checkup and would be seen as significantly more intrusive. Will laws be enacted that would prevent insurance companies from demanding this information from their customers?

First, insurance companies would have to request—or demand—this information before potential backlash could occur. In the absence of these demands, legislation would be unnecessary until it grew to a large enough issue.

14.3 Data, Privacy, and Security

Privacy matters are intertwined with cybersecurity and data breaches (Kim, 2016). There is no single law or standard that provides a uniform set of rules governing all the issues related to security, safety, and privacy. Due to this, a majority of legal challenges may be resolved on a case-by-case basis; developing a set of precedents and case law in lieu of—or until—legislation is passed governing this. Kim (2016) continues that this is the more likely scenario as the real-world problems are not well defined as far as the technical aspects are concerned.

Consider the Illinois Biometric Information Privacy Act of 2008, which states that no private entity can gather and keep an individual's biometric information without prior notification and written permission (Graham, 2017). The Illinois law allows for private citizens to sue companies that collect their data without meeting the requirements of prior notification and permission. Although the legislation is almost a decade old, it is only recently where it has been used in a slew of lawsuits against tech giants Facebook, Google, Shutterfly, and Snapchat, with consumers claiming their biometric information was handled illegally. These cases would be the first time a judge will interpret the law to see its applicability in a broader context.

Recent legal cases also demonstrate the ability of IoT devices to help solve crimes. In a murder case, police seized the defendant's smart speaker on the theory that it may offer evidence of what happened the night of the murder. A search warrant was then served on the speaker's manufacturer for recordings made by the speaker during the time of the murder. This search warrant was initially challenged by the manufacturer, although the challenge was made moot by the defendant's permission to turn over the recordings (Zatz et al., 2017). In an arson case, pacemaker data demonstrated that a patient's heart rate barely changed during the time of a house fire. A cardiologist testified at trial that it was "highly improbable" that a man in his condition would be able to escape the fire without elevating his heart rate. Zatz et al. (2017) continue that IoT device cases present interesting issues with regard to First and Fourth Amendment rights and privacy issues.

Auto insurance companies are using IoT devices to monitor their policyholders' driving habits. Customers may voluntarily install the auto insurance application in their vehicle as a means to receive safe driving discounts. In this instance, customers are voluntarily and knowingly allowing the insurance provider access to information on how they drive, where they drive, and other information that is beneficial in setting rate schedules. The voluntary and overt nature of this interaction provides customers with the benefit of potentially lower insurance rates.

It allows auto insurance companies to better target customers with safer driving habits, in turn lowering premiums for this group.

Fitness devices are primarily used as a personal device to provide activity information solely to the user, on their daily routines, eating habits, bioinformatic information, and similar to help track their healthy lifestyle. Any interference or dissemination to an outside entity may corrupt the beneficial process of an individual attempting to lead a healthier lifestyle and receive feedback on their progress.

14.4 Working with What We Have

Experts recognize that the IoT is in its infancy, and legislation at this stage may be premature and unnecessary at this time (GSMA Association, 2016). Further, extending legacy regulations based on outdated technologies may serve to produce more harm than good.

As such, legislative activity in the IoT space is just getting underway and investigating what may be necessary. The U.S. Congress recently passed legislation (United States. Cong. House. 115th Cong. 2nd sess, 2018) directing the Secretary of Commerce to conduct a study and submit a report on the state of the Internet-connected devices industry in the United States. In 2017, Senator Mark Warner sponsored legislation to provide minimal cybersecurity standards for IoT devices purchased by federal agencies (Warner, 2017). This bill has been referred to the Committee on Homeland Security and Governmental Affairs and awaits further action. Both items highlight the early stages of government understanding of this rapidly evolving technology. Similarly, the legislation introduced by Senator Warner builds upon the current understanding of the cybersecurity space and applies it to IoT.

Regulation occurring too early may limit the full capacity of this revolutionary technology. It will also be difficult to predict the direction this evolving technology will take and enact relevant regulation. Given the length of legislative cycles, relevant laws may be outdated as soon as they are enacted as the developing technology moves quicker than the legislative process. Specific risks of this technology, primarily due to privacy and security, may be dealt with through existing laws. Regulation enacted too early may miss the mark and only serve to confuse matters, possibly even stifling the growth and potential benefits of this technology. As most industries grow, initial self-regulation leads to industry consortium that set standards and practices for convention. Out of this governance will typically grow legislation that serves to protect consumers once the industry and technology have stabilized. Beyond this, there is recognition of the need for strengthening data security and breach notification legislation at the federal level (GSMA Association, 2016). The GSMA also notes that existing, well-established privacy laws and regulations around the world are sufficient to ensure that IoT services align with consumer privacy expectations. An extension of this could also include legislation requiring certain security standards for IoT devices, similar to the legislation sponsored by Senator Warner mentioned above. This point also touches back to Chapter 7 in that the IoT system is only as strong as its weakest link. A breakdown anywhere in the system could serve as an open door to provide entry into the entire system.

Given the evolving nature of both the technology and the regulatory environment, companies will have to prepare their solutions and implementations to match a set of laws that are yet to be determined (Walker, 2014). At this stage, much of these regulations will be governed by the contractual relationships negotiated between the vendor and customer. Still, data protection, privacy, and security will be at the forefront of these standards and negotiations. Much of these areas are already covered by existing legislation and contract law. Tarouco et al. (2012) suggest that the best way to deal with security and privacy challenges is to implement a framework of key principles by legislators in an international level, with the aid of the private sector to implement detailed regulations.

The EU Commission has been investigating IoT for several years and focuses its efforts on the loss of privacy and data protection (Walker, 2014). Not surprisingly, their report concluded that the IoT should be designed from the start to meet suitable and detailed requirements that underpin the right of deletion, right to be forgotten, data portability, privacy, and data protection principles. These are critical components of the EU's General Data Protection Regulation (GDPR). Domestically, privacy and security are also critical components of legislative investigation.

14.5 Summary

As a revolutionary technology, the IoT continuously redefines the connected environment; both in what society needs and what is possible. Along with these expanding possibilities come new questions about what is right, what is ethical, and what should be legal. This environment raises never-before-considered questions about data, privacy, and utilitarianism. There are presently laws on the books relating to each of these areas *based on current technology*. IoT will redefine what is possible and, so too, raise new questions about whether we should fulfill all our capabilities.

As a global medium, this technology, data, and communication will cross international borders potentially making it bound by various—and possibly contradictory—laws in each of the jurisdictions it crosses. Presently, these evolving technologies are regulated by existing laws; laws meant to govern current technology. IoT is still in its infancy, and possibly the best solution is to define regulation by what is important now; using existing laws and regulations regarding data, privacy, and security to govern this space. As this technology matures, we will see what is possible in the IoT and the according questions, debates, and necessary restrictions. Corollary to this is the challenge of regulating future society with current restrictions. Consider the viability of governing present-day society with laws in place during the early 1900s. Much of our lives and what is taken for granted had not been considered possible during that time. At present, we cannot predict the impact the IoT will have on our future lives. It is through this lens of impact that we can more intelligently enact the legislation necessary to govern this new environment.

Another consideration in the regulatory space is restrictions will dictate how the technology can and will grow. Legislators cannot move with the speed needed to govern how the technology evolves, although restrictions may limit what the technology is able to do. In this instance, industry and trade associations will self-regulate—those with the most knowledge in this evolving space—as they are the most suited to make these decisions. These decisions and standards placed by the necessary trade associations can be used to guide legislative efforts once the IoT technology and environment are mature and stabilized.

14.6 Case Study: Health Insurance Fitness Tracker

John is a healthy, fit, 55-year-old male. He is a nonsmoker, occasional drinker, who eats healthily, and works out at the gym three times a week. He bicycles with his family once or twice a week and walks the family dog on a regular basis.

New this year as part of his company-provided health insurance, it is mandated that he wear a fitness tracker to verify his healthy and active lifestyle. This is in addition to the annual checkups his insurance provider already requires. On a weekly basis, the fitness tracker uploads bioinformatic data to his insurance provider to verify his activity levels, time of day he goes to bed and wakes each morning, his eating habits, and a plethora of other information that his insurance provider uses to verify adherence to his healthcare policy.

The fitness tracker is a requirement for his enrollment in the Healthy Living Wellness Plan of his healthcare insurance. Should he choose not to wear the tracker and provide this information to his insurance company, he will automatically be enrolled in the Healthy Trying Wellness Plan with premiums about 20% more than the Healthy Living Wellness Plan. His insurance company, and his employer, sees this information as vital to ensuring a healthier workforce, lowering healthcare expenses, and maintaining decent margins for the health insurer. John does have concerns about the healthcare provider selected by his employer and their mandate that he provide excessive information on his routine and daily activities. Many of his co-workers have spoken with Benefits Administration about the coverage provided and the providers need for what they view as intrusive information requirements.

John does not have to provide this information to the company's insurance provider, he and his co-workers will simply be enrolled in the Healthy Trying Wellness Plan with the requisite higher premiums. Having been selected by his employer as the company's health insurance provider, the employer stands behind the policy negotiated and provided to their employees.

14.7 Discussion Questions

Should this type of insurance mandate be allowed? Is this legal or reasonable? Is it too intrusive? What if the information gathered on employees was shared with the employer by their health insurance provider? Would this represent a violation of privacy laws as excessive monitoring by an employer? Would this be considered outside the bounds of reasonable information necessary to perform a person's job? Would the health insurance provider face liability if the bioinformatics information was stolen from them?

Discussion Question 1—Future Legislation. The EU, U.S. Federal Trade Commission, and the National Institute for Standards and Technology have all weighed in on the issues surrounding the IoT and proposed caution in regulating this evolving and emerging area. What area(s) of the IoT will have the greatest need for regulation over the next 18 months? The next 2 years? Where particular caution should be exercised so as not to restrict the evolution of this technology?

Discussion Question 2—IoT Liability. Who holds liability when an IoT device malfunctions? As example, current thinking is that self-driving vehicles will prioritize the safety of vehicle occupants over the safety of those outside the vehicle. Considering a pedestrian stepping into traffic, the autonomous vehicle will prioritize passenger safety over pedestrian safety. Will the liability for actions reside with the owner of the autonomous vehicle as it is their car; the manufacturer as prioritization of safety is based upon faulty logic or a "manufacturing defect"; or the pedestrian who was jaywalking?

Discussion Question 3—Legislators and Trade Associations. Who should take the lead on defining and developing legislation and regulation that will govern the IoT? Should the legislators be given the lead to develop, discuss, and legislate what is legal or illegal in the IoT environment? Should this be left to Trade Groups such as the Federal Trade Commission and NIST to put in place Best Practices and Industry Standards that will govern how the IoT evolves and develops? Legislative action carries the weight of law and is more imposing but slow moving and possibly outdated by the time enacted. Industry Standards can carry the impact of law yet will move and adapt quicker to a changing environment. Who should be responsible for taking the lead in this evolving environment?

References

AboBakr, A. and Azer, M. 2017. IoT Ethics Challenges and Legal Issues, *2017 12th International Conference on Computer Engineering and Systems (ICCES)*, Cairo, 233–237.

Graham, M. 2017. What's Next: Illinois Biometrics Lawsuits May Help Define Rules for Facebook, Google, *Chicago Tribune*, January 17, 2017.

GSMA Association. 2016. Comments before the National Telecommunications and Information Administration. GSMA Response to US Department of Commerce Consultation on IOT.

Kemp, R. 2017. Legal Aspects of the Internet of Things, *Kemp IT Law*, June, 2017.

Kim, I. 2016. The Internet of Things: A Reality Check for Legal Professionals, *Law Practice Today*, January, 2016.

Mighell, T. 2014. Web 2.0: The "Internet of Things" in Law Practice, *Law Practice Magazine*, 40, 3.

Tarouco, L., Bertholdo, L., Granville, L., Arbiza, L., Carbone, F., Marotta, M., and de Santonna, J.J.C. 2012. Internet of Things in Healthcare: Interoperatibility and Security Issues, *2012 IEEE International Conference on Communications (ICC)*, 6121–6125.

United States. Cong. House. 115th Cong. 2nd sess. Discussion Draft. Washington: GPO, 2018. U.S. House of Representatives Document Repository. Web. 7 July 2018., retrieved online at: https://docs.house.gov/meetings/IF/IF17/20180522/108341/BILLS-115pih-TodirecttheSecretaryofCom.pdf

Walker, K. 2014. The Legal Considerations of the Internet of Things, July, 2014, retrieved online at: www.computerweekly.com/opinion/The-legal-considerations-of-the-internet-of-things

Warner, M. 2017. Internet of Things (IoT) Cybersecurity Improvement Act of 2017, *S.1691-115th Congress* (2017–2018).

Zatz, C., Meadows, J., Aradi, L., and Mathis, P. 2017. Recent IoT Device Cases, Data Law Insights, retrieved online at: www.crowelldatalaw.com/2017/07/recent-iot-device-cases/

Chapter 15

IoT: A Business Perspective

Thomas Harris

Ball State University

Contents

15.1 Introduction

Businesses are created to meet the benefit not only of the owners but to provide needed goods and services that all people need. There exists a symbiotic relationship between the business and the customer in that both provide something the other needs or wants. The customer gets the goods and services needed while providing the business the means to continue operation. While some not-for-profit organizations exist, most businesses are established to make a profit. A major driving force for business profitability and sustainability is a high level of productivity of a quality good or service. The higher the productivity level, the more profitable the business typically becomes. All businesses perform the tasks necessary to produce the desired output by creating processes that enable the firm to achieve its end result. These processes are the steps of activities that must be performed to accomplish some specific task. Business processes exist in every aspect of business such as financing, operations, marketing, accounting, or whatever other business task must be done. The easiest way for most businesses to increase productivity is to focus on either efficiencies or effectiveness or both. Any technology that can potentially increase either is soon adopted into the business's processes. Any process that can be made more efficient, and therefore less wasteful, can add to the overall profitability to the firm.

This trend can be seen throughout the ages. For example, consider the impact of Gutenberg's printing press. This technology completely revolutionized the world and made education possible to the masses because of the ability to produce written word much quicker. The genesis of the Industrial Revolution was the invention of the steam engine. This technology enabled businesses to produce at a much higher level than was ever possible before. This new form of energy allowed for the development of factories that could produce much more output and enabled the emergence of powered tools. This trend has continued as automation has become commonplace in many business areas. Automation simply refers to using technology in place of human or animal effort. The term automation first appeared in the 20th century and is oftentimes attributed to Henry Ford's use of the assembly line. This innovation created an amazing amount of efficiencies and effectiveness by using standardized parts and jobs to manufacture automobiles. This led to the concept of mass production of affordable output, and it enabled Henry Ford to obtain a fortune while providing the masses with transportation. His concept of the assembly line has now become commonplace in business manufacturing (Figure 15.1).

Automation comes in different forms and applications. Throughout the 20th century, more and more innovations have been adopted in business applications to improve the levels of productivity. The use of automation within the information sector of business started in the early teens and twenties of that century. This was facilitated by the use of what was known as unit record equipment, more commonly known as punched card processing. This innovation led to the development of data centers in many large corporations in the late 20s, 30s, and through the 1940s. World War II led to the development of automatic electronic computing devices to provide information processing power for many necessary applications in the military. This led to the development of what we know today as electronic computers. The advent of the commercial computer in the 1950s began a major

Figure 15.1 "Automated assembly line." (Source: ShutterStock.com.)

trend of automation with the use of information processing within business. By the 1960s, computers were becoming fairly commonplace in larger corporations and automated data processing centers were becoming a norm. As innovations in electronics were being established, computers were becoming smaller and more powerful. During the mid-1970s, as a result of the space race, microcomputers first evolved. The term personal computer (PC) first evolved in the early 1980s, and at that time, computers started to become commonplace in business regardless of size. It soon became evident that a stand-alone microcomputer was not terribly efficient in that all information had to be entered via some type of magnetic media. Soon after PCs became commonplace in business, the use of networking quickly followed. By the end of the decade of the 80s and into the early 90s, the use of highly networked computers was emerging. By the mid-1990s, the Internet had evolved and almost all businesses regardless of size were using computers to automate their information needs.

The Internet of Things (IoT) within business, generally referred to as machine-to-machine processing, started to evolve shortly after the Internet became a reality. The advantage of having machine-to-machine communication with little or no human intervention allowed faster and more accurate information processing. Multiple types of data sensors have been recently developed, and these sensors are capable of being connected to a network, allowing the concept of machine-to-machine utilization to expand greatly. As better, smaller and more energy-efficient sensors and monitors have evolved, the use of IoT in business has become exceedingly important.

15.2 IoT in Manufacturing

15.2.1 Robotics

When most people consider automation, they often think of the loss of jobs, but in many cases, this is not necessarily correct. Automation typically is used to increase productivity in job areas where the work can be either monotonous or dangerous. This type of work lends itself easily where machines can take over the drudgery that many humans face, thus enabling people to obtain better, more creative types of work. Toward the end of the 20th century, many businesses saw the large growth in the use of robotics in the manufacturing area. The term robot generally refers to any machine that can carry out a fairly complex set of tasks. Oftentimes, this may be programmed by using computers to control the operations of the robot.

Robots allow for computer-controlled work processes that are either highly repetitive or potentially hazardous. For example, robots have been used very successfully in painting mass-produced products such as automobiles. The advantage of the robot is that hazardous fumes from paint will not affect the machine. Robots have been employed in many of the welding jobs that are required in certain types of manufacturing. Not only do the robots perform this sometimes-hazardous task, but they can also be set up so that they do not have to stop to move while welding as humans do. This stoppage creates hot spots in the weld which can weaken it. Therefore, the robot can produce a higher quality weld than might a human. Robots have also been used in production lines to perform mundane highly repetitively tasks. The advantage being that machine does not get tired, does not take breaks, can be in dangerous or unhealthy conditions, will work in less than comfortable environments, and can work in the dark. All of these attributes of robotics allow them to produce products that are of a high quality with a highly efficient level of production. By attaching sensors to these robots, data from each process can easily be monitored and recorded (Figure 15.2).

15.2.2 Radio Frequency Identification Tagging

Radio frequency identification technology, which is commonly referred to as RFID, has made a tremendous impact on the advance of automation in business. RFID tags are known as AIDC technologies. AIDC stands for "Automatic Identification and Data Capture." AIDCs automatically identify objects, collect and analyze the data, and then send it to computer systems with little to no human interaction. The concept for this technology was developed shortly after World War II. However,

Figure 15.2 "Robots welding." (Source: ShutterStock.com.)

it wasn't until the 1960s that the technology was advanced enough to start using commercially. These tags are similar to barcodes in that data from the tag or label is stored into a database utilizing radio waves. Any number of items can have one of these tags attached to them.

There are several advantages of using RFIDs but the most notable one is the fact that the tags can be read out of sight and not have to be scanned like a barcode. The tag is an integrated circuit composed of protective materials that holds the pieces together and shield it from the environment. The material used depends on where the tag is being used. These tags come in a variety of shapes and sizes and can be passive or active. Passive RFID tags are the most popular. They are smaller and less expensive to implement. Passive tags must be activated by a reader to transmit data. Active tags, on the other hand, have on-board power, such as a battery, which allows them to transmit data constantly. There is a total of three components that involve these tags. The tag is one of them and the other two are a reader and an antenna. The antenna is used to transmit the data to the reader, which can also be called an interrogator. The reader converts the radio waves to a more useable form of data. This data is transferred through a host communication interface then on to a host computer system where it is stored for later use (Figure 15.3).

RFID tags began to be used commonly in the 1970s, especially in tracking inventory levels in manufacturing. The use became more commonplace as the cost of these tags became more affordable. Toward the end of the 20th century, these tags were regularly found on commonly sold items in retail stores for theft

Figure 15.3 "A typical RFID tag." (Source: ShutterStock.com.)

protection. If the tag had not been deactivated by the sales clerk as the item was leaving the store, the tag would set off an alarm that would indicate the tag had not been deactivated. This led to a major decrease in in-store stock shrinkage. By the turn of the 21st century, RFID tags could commonly be found in inventory management, logistics tracking, and supply-chain management.

The integration of inventory tagging and robotics has been incorporated in many different industries to automatically pick up items from inventory. This process was done by workers who would have to physically locate products and load them into bins to either be shipped or put into production. There was potential error in dealing with people doing this work. Individuals often failed to meet the deadlines for shipment, the wrong product could be selected or a myriad of other potential problems. With IoT, when an order is placed, the robot receives the product number, locates the product within the warehouse, and then loads the product onto a pallet or into a bin, all within a matter of minutes.

A major manufacturing issue is to assure that the machines that are being used to create the products are always in good working order. IoT can be used to assist in the maintenance of factory equipment. Sensors are available that can be attached to the machine as it is being used in a factory. These sensors determine the condition of the machine and can predict when the machine will need servicing and what parts on the machine would need to be replaced. This allows the company to schedule its maintenance time in such a way that machine downtime will be greatly reduced in production.

In manufacturing, IoT could be utilized through production flow monitoring. In digitally connected factories, IoT would allow for complete monitoring of production lines from the beginning process through the packaging of final products. This meticulous supervision of processes in real time would provide information that could help make adjustments in operations reducing operational costs substantially. The close monitoring of lags in production can help eliminate waste and unnecessary processes in the workflow. This would help smoothen the entire manufacturing process and save time and money in the production costs.

15.3 IoT in Marketing and Retail

The process of marketing is a data-driven, highly necessary aspect of business. The role of marketing in the business environment is to assure that the firm can attract customers that will consume the product or service that is offered. The issues of concern in retail are the same that are in marketing. Since the concept of marketing is about attracting customers, it is important that marketers understand many issues of consumer behavior. The amazing connectivity of the Internet and of the IoT enables marketers to do many things that would have been impossible before. The concept of the connected customer is being very heavily applied in marketing because of the ability to track customers to determine where customers shop and their various shopping habits. This enables marketers to adjust advertising to fit the specific needs of specific categories of customers.

Customized ads and banners have proven effective in online shopping and have provided an edge over brick-and-mortar stores. Retailers are trying to catch up by implementing smart displays within their stores that will personalize offers and promotions. Sensors, RFID tags, and beacons can collect data from customer interactions within the store. These displays will analyze the data with machine learning solutions. The data will then be turned into extra information that will be pushed out to the customers. Customer reviews, recommendations, and special offers are just a few examples of this information. Another way stores can compete with online shopping is through mobile apps. Data from online browsing histories would be collected and analyzed to provide useful information to customers with the app. There would be special offers and would drive up the loyalty programs, creating a more personalized experience.

One of the major problems that retail businesses often must deal with is stock-out or dead inventory. IoT can assist in this issue. Stock-out is a major problem for retail businesses. If a product is not available when the customer comes in for that item, that customer will leave the store and find it elsewhere. The chances of that particular customer coming back to the retailer who had the stock-out are then greatly diminished. Likewise, dead inventory, or stock that is not moving, is a major problem in that shelf space is being utilized for a product that is not selling. Not only is the cost of inventory of that item a problem, but the potential loss of sale that that space could be used for is doubly troubling. Smart shelves and RFID can be used together to determine in-store inventory levels (Figure 15.4).

If a shelf level gets low, monitors can alert the inventory system that that particular product needs to be restocked. This information can then be immediately

Figure 15.4 "Reading a smart shelf." (Source: ShutterStock.com.)

delivered to management. Likewise, inventory levels of those items that are not moving can be identified and tracked automatically through these types of systems. This information would then allow managers the ability to make faster decisions regarding the dead stock.

Retail stores would have a difficult time operating without their in-store staff. Part of their duties is to identify customers in need of help and tend to their needs, making them an important factor of retail. Employees can be the difference in improving sales and conversion rates. Since they can only watch so many customers at once, IoT can be used to assist. The constant presence of a sales associate can be misinterpreted by the customer and is often considered offensive. IoT can solve this problem while not disturbing their experience. Motion detectors can be used to see if a customer has been standing in one spot for too long. Facial expression logarithms can be used to see if the customer looks confused. A sales representative can be notified by a mobile or smart watch app. This way shoppers aren't kept waiting or feeling offended by an employee. With the IoT, retailers may optimize their in-store staff while ensuring the customer has the best possible experience.

Global positioning systems (GPS) and smart phones may be used to track the location of potential customers. With GPS mapping, it is possible for marketing to determine specifically where a customer is located and customize advertisements to be sent specifically to that individual. If the individual is close to or in a restaurant, it is possible to entice the customer or potential customer to enter that restaurant, eat, and then rate that particular establishment. This can all be done with a minimal amount of human interaction.

With the ability of the many sensors that monitor sales data, by using data analytic methods, marketers can create a much better view of customer behavior. It is extremely important for marketers to have a complete and total understanding of what customers are desiring or needing. With the ability to obtain data from social media, combined with data collected from retail sales, marketers can now generate much information about consumers' likes and dislikes. This information will give the marketers a much better concept of how to approach current customers and how to identify potential customers for their product or service. In turn, this ability to connect to the customers leads to generating more relevant data that can be used to develop better marketing campaigns. The result is that firms will be able to get richer, fuller customer engagement, and a more loyal customer base.

The use of IoT marketing is only limited to the imagination of the marketing individuals. With all of these data becoming available and with the current data analytic processes, marketing departments should be able to generate a holistic, customer-centric view of the consumer. For businesses to succeed, it is imperative the customer experience is an end to end given and that marketing needs to work with other divisions to remain relevant.

15.4 IoT in Healthcare

The IoT has a myriad of applications in healthcare that could benefit patients, families, and physicians alike. The potential of the IoT in healthcare is ever expanding.

From the individual perspective, there are three major aspects of IoT that may be used to advantage. Most of these uses of IoT involve wearable monitoring devices. By using these devices, individuals can monitor their health in a number of different ways. You can track your sleep habits, your exercise, heart rate, and other vital signs. By individuals having these data available, they can generate information that can help them live a healthier lifestyle. These physical trackers are designed to help people sleep better, eat a better diet, and become more active (Figure 15.5).

These trackers have helped many people become healthier by providing estimations of these physical traits. A major use of IoT for individuals is providing tracking for family members. As individuals age or become infirm, there is a likelihood that they might need more healthcare services in an emergency situation. IoT devices can be used to monitor individuals who are at risk of becoming ill or falling and then notify family members or emergency providers when and if needed.

Another use of IoT and individual healthcare is to use these monitors to provide information directly to the family physician. Monitors for blood pressure, blood glucose levels, heart rate, and other potential health issues can be sent directly to a physician's office or hospital. These monitors can alert healthcare providers when instant care is needed. Imagine having a monitor on your body that can determine

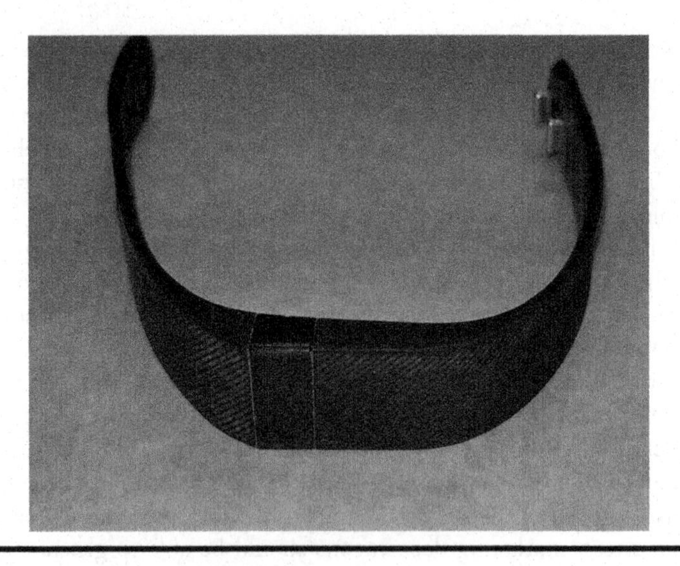

Figure 15.5 "Wearable health tracker." (Source: ShutterStock.com.)

if you are about to have a heart attack. This monitor could notify you immediately and, at the same time, contact emergency medical services to come to your aid and notify a hospital's emergency room staff to prepare for your arrival. This technology has the potential of saving numerous lives.

From a physician's standpoint, IoT offers the potential of engaging patients in a much more personal manner. While electronic health records to help physicians better understand their patients have been around for a while, the use of wearables and other electronic communication devices can generate immediate information that the physicians can use to determine how best to work with their patients. Diagnosis of illnesses can be improved because of the vast amount of current data about the patient that is available. As patient care is one of the major responsibilities of physicians, the use of IoT can have a major impact in that more personalized and specialized care can be afforded to their patients than ever before.

IoT has had a major impact on how hospitals go about their business. Some hospitals are using the IoT in healthcare to keep the tiniest patients safe and healthy while others are using the technology to keep track of inventory. The diversity of the use of IoT in hospitals is nearly as widespread as the number of hospitals that exist. IoT monitoring, which has become commonplace in many hospitals, allows for a much more efficient use of nursing staff. The use of these monitors enables fewer nurses to closely monitor more patients in a highly effective manner. Hospitals often lose many patients every year due to the fact that hand hygiene standards are not being met. For years, hospitals have tried to find ways to check on the sanitation of their employees' hands. With IoT, there can be a badge worn on the shirt of the employee which reacts to a monitor set up close to the patient's bed. This will track the status of hand hygiene and automatically transmit information to a server that may be accessed from the hospital's dashboard. This allows hospitals the opportunity to actually track employees' hand hygiene in a real-time environment.

Many hospitals are using RFID tags for inventory management to track the use and quantity of drugs available. This greatly reduces shrinkage of inventory and allows for patient medication to be administered in a much less error-prone manner. This automated dispensing of medications has made prescription filling much easier, allowing doctors and nurses to attend to those of most critical need. With these smaller processes being taken off the hands of nurses and doctors, the flow of processes within medical institutions can become much smoother, giving a better experience for both the employees and patients. Along with all the benefits to employees and patients comes the added benefit that the hospital saves money by cutting down on issues such as errors, time being spent doing simple processes, and reducing the stay of patients and the frequency of their hospital visits.

15.5 IoT in Logistics and Transportation

The field of logistics deals with the movement of products from place to place. This area of business has become exceedingly important as the concept of supply chain management has evolved in the past few decades. Supply chain management is the interorganizational linkage of businesses that supply necessary inputs for production from the acquisition of raw materials until the finished goods are sold to the consumer. The concept behind the supply chain is that businesses needing the inputs to produce whatever they need for manufacture can place inventory orders electronically and immediately receive information regarding delivery. This electronic movement of information allows for businesses to receive inventory levels just in time. One of the major issues of creating efficiency in business is reduction of inventory levels. Therefore, the concept of getting inventory at the point when needed is extremely important in the creation of more efficient businesses. The supply chain simply expands the use of this information not just between a supplier and manufacturer but with the supplier's supplier and so forth. This linkage or chain has commonly been referred to as the supply chain. Information going upstream, that is, from the needed inventory to the supplier, can continue throughout many organizations until it is at the source of the raw material. Downstream, that is, from supplier to consumer, information regarding delivery time and quantity being shipped can be provided. This movement of information allows for those businesses in the supply chain to be able to plan for inventory levels and delivery dates in a most efficient manner. This practice has become so commonplace that, in many areas of business, competition is now not so much business versus business rather supply chain versus supply chain. Since the supply chain relies on the tracking of the movement of materials, logistics, and warehousing, IoT has become a necessary standard in this industry.

Logistics are designed to help control and manage the deliveries of product. Any element that can make the process of delivery more efficient will assist in the profitability of the logistics company. Since information regarding orders and delivery dates can be transferred immediately, it is imperative that the logistics aspect is included in the information flow. Since almost all inventory can be tagged using RFID or some other technology, it becomes easy to automatically track where the inventory is in movement. Location tracking and environmental sensing are two great components of the logistics and supply chain. It is important to understand exactly where and what the shipment is and when it will be received.

The trucking industry is one of the areas most impacted by IoT, with monitors available to place throughout the trucks. These monitors can check things such as length of runtime, engine efficiency, time between maintenance, and also detect any potential problems that may occur. The use of logbooks, which has long been required by the Interstate Commerce Act, can now be tracked electronically, which no longer enables truckers to cheat on the number of hours driven. With the use of GPS, trucks can be located automatically, anywhere at any time. The use of GPS in the trucking industry has become essential. It is imperative to minimize time without a load to increase the efficiency of a truck. With the movement of electronic information, it is possible for trucking companies to schedule drop off and pickup loads within a minimal amount of distance. The GPS system can guide the truckers to the most efficient route available, thus minimizing loss load time.

The development of autonomous vehicles is having a major impact on the transportation industry. Vehicles are now being equipped with computer-controlled technologies, through the use of cameras, GPS, and sensors, which can drive and control the vehicle without human intervention. The potential impact of this development is that someday we are expected to have transportation that is completely controlled by automation, not humans. The obvious advantage is that, with computer-driven vehicles, the likelihood of accidents is greatly reduced because human error due to inattentiveness, distraction, or tiredness will not be an issue. This concept will undoubtedly have a major impact on the trucking and the taxi cab industries. There would be no need to have downtime with autonomous trucks because the driver is unavailable to operate the vehicle. Likewise, in cities having autonomous vehicles that could be summoned from a smartphone and automatically paid for through that application, the need for the traditional taxi cab would be eliminated. Autonomous vehicles could immediately drop people off at their desired location and could go self-park, if they were owned by individuals. The potential of autonomous vehicles is incredible (Figure 15.6).

Another use of IoT in transportation is the development of smart infrastructure. With the sensors available, it is possible to create bridges and highways that can self-monitor. A bridge sensor could determine the condition of the bridge and automatically notify highway maintenance whenever attention to the structure would be necessary. Sensors in highways can determine the condition of the road and automatically report this data to anyone who needs it. This data could be used to

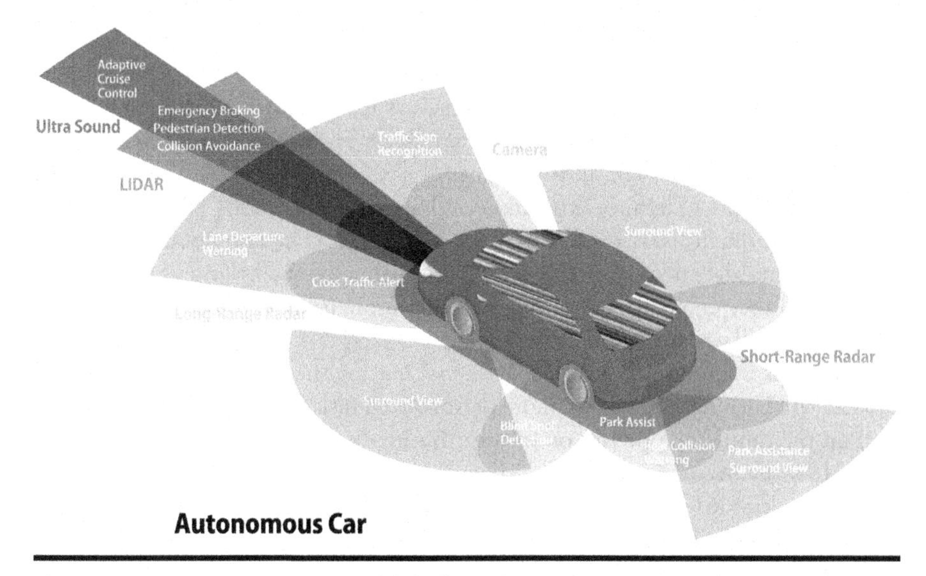

Figure 15.6 "Autonomous car with indicated sensors." (Source: ShutterStock.com.)

automatically regulate speed limits on highways. When conditions are poor, speed limit signs connected to these sensors could automatically indicate a reduction in the speed limit. Conversely, when road conditions are good, with fair weather and light traffic, speed limits could automatically be increased. Highway data could also be used to reroute traffic in cases where there are accidents. All of this data would greatly increase the efficiency of highway traffic management.

15.6 IoT in Finance and Accounting

The areas of accounting and finance are both majorly concerned with the use of money and tracking of money in a business. The need for information within these areas is critical for the operations of any organization. Accountants rely heavily on information and data collection to do their job accurately and efficiently. With IoT, accounts will have information readily available on a regular basis. As the use of IoT devices in accounting evolves, accountants will have better communication with all people within the organization and the resulting information will help top managers move the firm in a more effective manner. IoT devices will bring clarity to the data and will move accounting processes into a real-time realm. The use of IoT throughout the business will remove the human error factor from data collection. The realm of auditing is often about checking the veracity of the data recorded. The accountant can, with IoT, focus more on the source and the use of the data. Cost accounting, the estimation of how much it costs to produce something, will be greatly impacted by devices. Cost accountants will have immediate access not only to the most current inventory levels but also the most current pricing of those items. With this information, cost accountants will be able to do their job much more effectively than in the past. With instant access to information, accountants can work much more closely with the financial officers of the corporation to maintain control over the flow of cash in the business.

A major aspect of finance is insurance. One of the areas of IoT that has become popular with the insurance industry is the use of sensors in automobiles that automatically send data about the driver's driving habits to the insurance companies. Many insurance companies are using this as a promotion to provide discounts to drivers that will adopt these sensors (Figure 15.7).

As IoT is adopted, when someone is looking for the right car insurance, the process will become easier and a more personalized task. The insurers could simply pull up the profile of a person or an organization looking to be insured that contains all of the necessary data about the potential insured, automatically compiled by a network of IoT devices. This profile could easily and readily tell the company everything they need to know about this group or individual. It will improve the accuracy of the decision process for the insurers.

Agents would no longer need to be the ones who decide or place the blame on someone when an accident happens. These companies will have the ability to know immediately when someone they insure either caused an accident or is the victim in an accident. This decision-making process would be simplified immensely. The company could see the subjects' driving history, their lifestyle, and the way they handle their financials. This would help insurance companies know that they are going to be making the right choice by investing in the group or individual looking

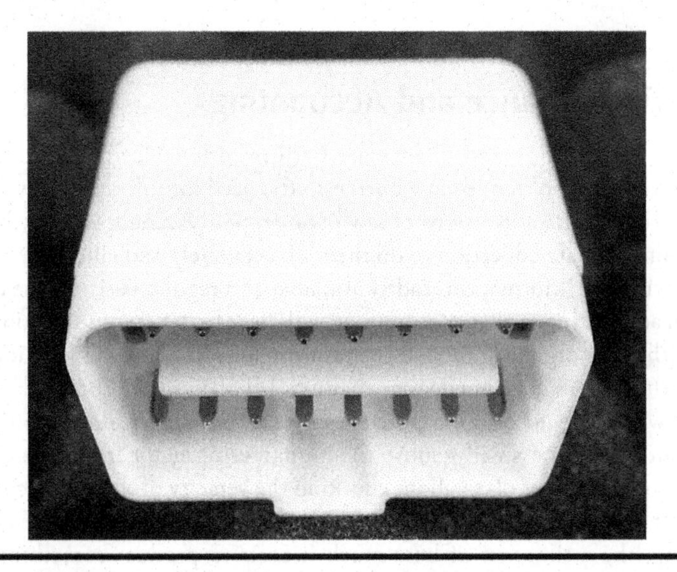

Figure 15.7 "Automatic sensor for car's computer." (Source: ShutterStock.com.)

to be insured. These decisions could also ultimately help all insurance companies increase reliability and profitability for years to come.

There is also an important impact on commercial insurers with IoT innovation.

In commercial insurance, applying sensors on shipping containers and transport vehicles could provide insurers with the opportunity to enhance shipping insurance coverage. The ability to better detect and model risks due to theft or damage could move the pricing of these products from an actuarial exercise to one that better assesses risks and losses in real time.

Customers and consumers can use their smart phones or other devices with internet streaming capabilities to access any desired financial data from their banking institutions. This could be as simple as checking an account, transferring money, or actually speaking with a member of the bank. The banks using this new technology can anticipate the needs of customers through the data collected. This allows the banks to provide a complete and full view and history of that person's finances in real time. In turn, the banks may offer solutions and advice that can help customers make the financial decision that is going to be the best for them in the long term and possibly in the short term. This development is going to make banking and checking financials much easier for years to come. Banks believe that this will help improve the overall customer loyalty and satisfaction. The banks know that the instant availability of individuals' financial information could also help boost their business as they will have real-time information available for loan and investment decisions.

15.7 IoT in Agriculture

Precision farming covers everything from livestock operations to the growth of crops. The feature that differentiates precision farming or precision agriculture from normal agricultural operations is the use of smart Information Technology in different gadgets like robotics, control systems, sensors, automated hardware, autonomous vehicles, and variable rate technology. The requirements of fast Internet access, portable devices, and reliable and affordable satellites for the purpose of displaying the field conditions are vital for the appropriate operations to be executed in precision agriculture. Furthermore, precision farming is one of the most popular applications of the IoT, and different farming firms and corporations have optimized their operations as it provides more efficient results. It can deliver a return in the form of investments for farmers.

There are a multitude of uses for the IoT in the field of agriculture such as the ability to sense the soil moisture and nutrient requirements and monitoring and scheduling of water usage for optimal plant growth. IoT can help assess the need for different fertilizer types based on the soil chemistry, investigate the right weather and time for planting and harvesting, and provide weather forecasting of the weather. When farmers can have access to technology, which can provide them with weather forecasts, information on the standard of the soil, the cost that needs to be invested, and live updates on the agriculture processes, they can schedule their agricultural activities according to clear requirements. This can help a farmer anticipate and address any problems to help them produce the best yield possible.

The use of greenhouses helps yield the best batch of fruits, vegetables, and crops regardless of the external temperatures or conditions. Greenhouses are made for the efficient production of edibles in a controlled and appropriate environment, which cannot otherwise be produced. In such processes, manual intervention is known to be quite risky, as it has the potential to produce errors, which can further result in production loss, energy loss, and labor cost. However, in the modern era, through the applications of IoT, 'Smart' greenhouses can be developed, which allow for the integration of artificial intelligence that can take over the control of the maintenance of the greenhouse as per its requirements (Figure 15.8).

This eliminates the possibility of human interactions that might lead to error. Inside a smart greenhouse, sensors are integrated that can measure the status of environmental requirements of the plants and then modify the environment of the greenhouse accordingly. A cloud server can be used for the monitoring, and applications can access the greenhouse with the help of IoT. This helps in limiting the need for constant assessment of the greenhouse and also limits the need for manual interactions with it. The integration of IoT in greenhouses can deliver to the farmer's regular live information of the state that it is in, such as the pressure, temperature, humidity, and light levels. The sensors can have the authority to control the functions of the greenhouse. It can open windows and turn the lights off or on, control a heater, or switch on a fan; all these and other functions may be controlled by cloud servers through Wi-Fi signals.

Figure 15.8 "Automated greenhouse." (Source: ShutterStock.com.)

Since agriculture and livestock work together, large farm owners can, with the use of wireless IoT applications, monitor the health and the well-being of their cattle. The data that is collected can help assess the animals that are unhealthy so that they can be taken out from their herd, so as the disease that a certain animal suffers from does not get transferred to other animals. The IoT assists by saving time and resources in livestock management, with the integration of IoT in ear tags, which can help in determining the disease from which a farm animal is suffering.

IoT also can be integrated into collars, which can feed us live data of livestock whereabouts and also helps in monitoring the movement of the animals on farmland, which can help save on the labor costs of ranchers (Figure 15.9).

IoT can provide cow monitoring services to farm owners. One popular service is the monitoring of the cows that are in labor and about to give birth. When the water breaks, the cow has a battery-powered monitor that is expelled, and this sends alert information to the manager of the herd or the rancher. During the time that the herd is giving birth, the sensor allows the farmers to be more focused to potential issues within the herd.

To manage live monitoring of the farms, some companies are manufacturing Unmanned Aerial Vehicle (UAV) or what are more commonly referred to as drones. This can help the farmer successfully obtain live data on the weather conditions, climate changes, and live coverage of the field. Drones further assist farmers in scheduling and planning their farming activities in a more secure manner against any potential disaster. Drones are extremely useful in agricultural sectors to boost and assess different cultural practices (Figure 15.10).

Figure 15.9 "Monitoring animals with sensors." (Source: ShutterStock.com.)

Figure 15.10 "Drone surveying." (Source: ShutterStock.com.)

Drones are being used to provide operational information about the crop's health, irrigation, crop analysis, crop spraying, plantation, and soil and field assessment. The use of drones in the agricultural sector is highly beneficial and consists of crop health imaging and integrated GIS mapping, is user-friendly, does not consume a lot of time, and also has the potential to rapidly increase yields. With the immaculate high performance and live data collection and processing, drone technology is bound to give a new high-tech makeover to the industry of agriculture. Drones are being used to collect important data with the use of sensors for imaging, mapping, and surveying of agricultural land. These drones help in accessing live field and crop data, they are the user-friendly product of IoT, and work by just a simple set of instructions provided by the farmer. With the use of these data provided by the drone, farmers can perform a variety of tasks such as assess the plant or crop health, perform plant counting or plant height measurement, map canopy cover, map field water posing, create scouting reports, measure stockpiles, measure chlorophyll, assess nitrogen content in wheat, map drainage, map weed pressure, and predict yields. Furthermore, the drones can also provide services of multispectral thermal and visual imagery and can work regularly after maintaining their settings.

Food storage and security are another useful application of IoT for farmers. Food storage needs the same level of care as that provided to plants during their growth phase. To maintain the healthy and fresh state of fruits and vegetables, it is essential for them to be delivered to the market in an appropriate manner.

The IoT monitors help in analyzing the state the fruits are stored in, and the temperature is maintained in favor of the fruits and vegetables. It further provides us with the record maintenance and the time of delivery. It is essential to store milk under the best-suited environment for the purpose, which would be storing it in a cool environment. To ensure that milk is contained in the best environment, an IoT application is used to monitor the milk-storage chilling temperature, lactose measurement, plant wash monitoring, etc. It is essential for milk to be stored in its desired condition or else a huge loss of revenue can occur for the farmers.

Pests are one of the biggest threats to the crops on a farm. They portray a huge potential for damage to farms crops and fruits. The farmers usually tend to spray pesticides over their crops and fruits but, eventually, even that damages the fruits. The agricultural industry has been looking for an alternative to pesticides. The IoT has played an important role to replace pesticides and helps agriculturists in monitoring, identifying, and controlling the pest growth. There are many sensors that assess insect behavior and the seasons in which they are more active, and through this information, farmers are able to take measures to anticipate and apply any counter measures for the reduction of any major loss.

A balanced amount of water supply is essential for crops to remain healthy and also for seedlings to grow. Every crop requires different care when it comes to providing water: rice seeds are placed into a flooded land, whereas wheat need sufficient rainwater during ripening. Anything with more or less quantity can temper with the quality and the quantity of the harvest. With the help of IoT in ground monitors, farmers can better investigate the need for water in a certain field. IoT also assists in comprehending the topography or resources of water that are available in a certain area. IoT-based technology also helps and guides to ensure the best and efficient use of water is sustained. This can be obtained through remote controls. Through either automated control or manual control, the farmer can schedule and perform irrigational activities in fields, which can help significantly in cutting off water loss.

Another innovation of automation in farming is the use of GPS to actually steer the farmer's equipment. By having monitors and cameras mounted to the tractors and combines that the farmer uses, this information, along with information from a satellite, then may be used to determine exactly where in the field the equipment is located. This enables the actual equipment to take control of where and how the equipment is moving through the field. This also enables a much more accurate ability to plant and harvest crops both efficiently and effectively.

Farming and agriculture are high risk activities, at each turn, an error can have the potential to ruin the whole field of crops or fruits. To address the risk of losses and emerge with higher yields, smart agriculture technology using IoT-based portable devices and gadgets can be a huge boost for farmers and the food industry. Mobile applications can be used to analyze the farms and produce efficiently. Smart sensors work by checking the soil requirements such as its acidity, moisture content, and humidity. They also check the health of the crop, anticipate pest attacks, health issues and problems in farm animals, especially dairy animals. Smart GPS, drones, and robotics are used for effective monitoring and controlling of farms and livestock. These are some of the latest technologies that can help respond to problems faced by farmers in conventional farming. The data can be further used to anticipate any future problems in agriculture, making farmlands smarter, and more productive.

15.8 General Impact of IoT in Business

The IoT is having a tremendous impact throughout all areas of business. The examples cited are just a few of those areas. There are many more that could be discussed. IoT has been used in research and development departments, human resources, communications departments, and many other areas. The IoT, as it is evolving, is regularly finding more and more applications. The growth of IoT in business is showing no signs of slowing and will only continue to increase at a near exponential basis as the ability to improve processes throughout the business organization becomes reality.

As technology advances, efficiencies and effectiveness are built into business processes and within their respective organizations. Through the years, there have been many technologies that have had a major impact on business; however, few have had the impact on businesses that we have recognized during the last three decades with the advent of affordable computing, the Internet, and now the tsunami of the IoT. These trends have forever changed how business processes are done. All of these changes that have been described have occurred within the last two decades. In many ways, the IoT is in its infancy. As long as business processes can be continually improved by the adoption of these technologies, there will be no slowing of the advances of machine-to-machine technologies. It would take much more space than this chapter affords to describe the many ways in which IoT has impacted business in the last 15 years. What has been described is just a sampling of the many applications that are currently available using IoT technologies.

Information is all sensory based. When we consider how we process information as individuals, information takes the form of visual, audible, textual, touch, and olfactory (taste and smell). With the myriad of sensors that are currently available and are being developed, it is not improbable to consider that, in the future, we will have sensors that can deal with all forms of information. The future of IoT in business, therefore, is essentially unlimited. The only limitation is the imagination of humans as they consider how to automate their multitude of business processes. The IoT is an ongoing venture that will continue forever because of the constant flow of new ideas created by people. The one aspect of the IoT that it cannot do independently is to accomplish the generation of new ideas. This will require the human element of IoT in business.

15.9 Case Study: Smitty's Golf Emporium

Alex Smith created and is the primary owner of Smitty's Golf Emporium. His business includes a driving range and sells golf equipment. In addition to the business, Smith gives golf lessons as a teaching pro. He has been in business for about 20 years and has been fairly successful; however, he is concerned that he is not doing as well as he might. Currently, the only information systems that he uses are a computer-based cash register that is not point of sale equipped and an office computer that he uses for financial spreadsheets and basic accounting. Smith knows that some form of automation would probably help increase his bottom line, but he feels that he is technically challenged in the area of information processing.

The driving range aspect of the business operates year-round since 20 heated booths are available. The driving range offers buckets of balls, either small or large, for a single visit or a regular customer may purchase an annual membership that allows for unlimited use of the range. Smith feels this part of the business is working fairly well, but he is unsure if the amount of revenue that is being generated by memberships covers the costs. He needs to be able to track how much usage each member incurs.

While he is comfortable with his ability to give good golf lessons, he realizes that there is always room for improvement. He currently is using the traditional method of observing his student's swing and, based on what he sees, determines the best approaches to improving their games. Smith has read that there are now technologies that may be attached to the golfer's glove or clubs and cameras that could generate additional swing information that might assist in the lessons. He is concerned that some of the teaching pros near him might be using some of the devices that he is not employing.

The retail side of the business is what is most worrisome to Smith. He has a fairly extensive inventory of golf equipment. That includes clubs, balls, bags, clothing, and other golf-related merchandise. He does not have any form of electronic tracking in place for inventory control and does not know to what extent, if any, there is inventory shrinkage. Smith feels that his retail sales are good but is unsure how to best measure and meet his customer's desires for the products he carries.

Considering these details, how might Mr. Smith be able to employ various forms of IoT assistance to make his business more profitable?

15.10 Discussion Questions

1. Why is the concept of increased efficiency and effectiveness of processes so important to all business productivity and how might IoT provide for these increases?
2. Hospitals are constantly looking for ways to lower costs. How could IoT be used, in ways not mentioned in the text, to make them more cost efficient?

3. Manufacturing automation is extremely important for maximum productivity. Research and consider alternative applications of IoT that might be used in manufacturing.
4. How do you perceive that IoT may change our automobile driving habits?
5. Logistics involves the movement of things from one point to another. The use of IoT has had a major impact on the field in the areas of vehicle and load tracking. Examine and expand on the many ways that these impacts affect logistics.
6. Inventory costs are a major concern in most businesses. How might IoT be utilized to reduce inventory costs?
7. The term "smart farming" is used to define using automation to improve agriculture output. How important is this concept to society as a whole?
8. As was mentioned in the chapter, there are many other current and potential uses of IoT technologies in business. Search the Internet for other applications that you find important and interesting.

Bibliography

"4 Innovative Internet of Things Examples in Retail." Beaconstac. October 27, 2015. Accessed April 24, 2018. https://blog.beaconstac.com/2015/10/4-innovative-internet-of-things-examples-in-retail/.

Aleksandrova, Mary. "IoT in Agriculture: 5 Technology Use Cases for Smart Farming (and 4 Challenges to Consider)." Eastern Peak. June 07, 2018. Accessed April 21, 2018. https://easternpeak.com/blog/iot-in-agriculture-5-technology-use-cases-for-smart-farming-and-4-challenges-to-consider/.

Bello, Juan Jose. "IoT in Retail: 7 Real Examples." Ubidots Blog. August 15, 2018. Accessed April 4, 2018. https://ubidots.com/blog/iot-applied-retail-real-examples/.

"Big Data and Business Analytics Revenues Forecast to Reach $150.8 Billion This Year, Led by Banking and Manufacturing Investments, According to IDC." IDC: The Premier Global Market Intelligence Company. March 14, 2017. Accessed April 25, 2018. www.idc.com/getdoc.jsp?containerId=prUS42371417.

Clark, Tracy. "How the Internet of Things Will Transform the Insurance Industry." Argo Group. November 07, 2017. Accessed April 20, 2018. www.argolimited.com/how-internet-of-things-will-transform-insurance-industry/.

"Deep Dive: The Future Customer Experience-AI and IoT in Retail." Coresight Research. 2018. Accessed April 25, 2018. www.fungglobalretailtech.com/research/deep-dive-future-customer-experience-ai-iot-retail/.

Desai, Nisarg, and IDG Contributor Network. "IoT in Agriculture: Farming Gets 'Smart'." Network World. April 18, 2018. Accessed April 23, 2018. www.networkworld.com/article/3268971/internet-of-things/iot-in-agriculture-farming-gets-smart.html.

Drinkwater, Doug. "10 Real-life Examples of IoT in Insurance." Internet of Business. October 10, 2016. Accessed April 26, 2018. https://internetofbusiness.com/10-examples-iot-insurance/.

Eha, Brian Patrick. "Four Ways the Connected Car Will Change Banking." American Banker. January 30, 2017. Accessed April 24, 2018. www.americanbanker.com/news/four-ways-the-connected-car-will-change-banking.

Gubbi, Javavardhana, et al. "Internet of Things (IoT): A Vision, Architectural Elements, and Future Directions." NeuroImage. February 24, 2013. Accessed April 19, 2018. www.sciencedirect.com/science/article/pii/S0167739X13000241.

"Internet of Things Global Standards Initiative." ITU Committed to Connecting the World. 2018. Accessed March 29, 2018. www.itu.int/en/ITU-T/gsi/iot/Pages/default.aspx.

Marous, Jim. "Should Banking Build an Internet of Things (IoT) Strategy?" The Financial Brand. November 21, 2017. Accessed April 18, 2018. https://thefinancialbrand.com/63285/banking-internet-of-things-iot-data-analytics-payments/.

Ranger, Steve. "What Is the IoT? Everything You Need to Know about the Internet of Things Right Now." ZDNet. September 19, 2018. Accessed April 23, 2018. www.zdnet.com/article/what-is-the-internet-of-things-everything-you-need-to-know-about-the-iot-right-now/.

Ray, Brian. "5 IoT Applications In Logistics & Supply Chain Management." AirFinder. January 5, 2018. Accessed June 10, 2018. www.airfinder.com/blog/iot-applications-in-logistics.

"RFID Tags Market." Transparency Market Research. Accessed March 20, 2018. www.transparencymarketresearch.com/pressrelease/us-rfid-tags-market.htm.

Ryan, Luke. "How The Retail Industry Can Benefit From The Internet of Things." Mokriya Blog. October 03, 2016. Accessed April 24, 2018. http://mokriya.com/blog/how-the-retail-industry-can-benefit-from-the-internet-of-things/.

Staff, 24/7. "How the Internet of Things Is Improving Transportation and Logistics—Supply Chain 24/7." Supply Chain 24/7. September 9, 2015. Accessed April 20, 2018. www.supplychain247.com/article/how_the_internet_of_things_is_improving_transportation_and_logistics.

Svaidi, Amirul. "The Internet of Things (IoT) in the Retail Industry—Evolutions and Use Cases." I-SCOOP. 2016. Accessed April 23, 2018. www.i-scoop.eu/internet-of-things-guide/internet-things-retail-industry/.

"What Is RFID and How Does RFID Work?—AB&R®." AB&R. 2018. Accessed April 24, 2018. www.abr.com/what-is-rfid-how-does-rfid-work/.

IoT CASE STUDIES

Case Study 1

Agile Development for IoT Applications: Lessons Learned from a Case Study on Hydrants

Timon Heinis, Johannes Heck,
Filippo Fontana, and Mirko Meboldt
ETH Zurich

Contents

Does Agile Development Support IoT Application Development for Industrial Goods?

While digital technologies offer a high innovation potential, developing Internet of Things (IoT) applications can be challenging for industrial manufacturing companies. Most of them are struggling to identify solid value propositions as they are following a technology-oriented approach (Fleisch, Weinberger, and Wortmann 2015). Thus, their development projects become lengthy and costly. Examples of such companies can be found in the Swiss manufacturing industry which represent the second biggest export industry in Switzerland. They face inhibitors in the domains of organization, business, technology, and industry. More specifically, their organizational inhibitors are a lack of a clear (user-centered) digital operations strategy, a lack of in-house expertise and skills, and no integration of physical product and software development processes. Moreover, their business inhibitors are insufficient information and uncertainty to predict demand and revenues, issues with monetizing the current business model, and difficulty in identifying market opportunities (Heinis, Hilario, and Meboldt 2018). Despite market excitement about the potential of IoT, concrete examples of mainstream adoption with sustainable results are scarce (SAS Institute Inc. 2016).

An agile development approach promises to reduce both the overall development time and costs as well as associated risks by focusing on early value delivering from a user perspective. Originating from the domain of software development, "agile development" as a methodology aims at solving complex problems with high uncertainty by splitting the whole work to be done into increments, starting with the most valuable and critical one from a user perspective. Before proceeding with the next increment, the current work increment—a potentially shippable product—is tested and evaluated. Thus, the risk of rework (in time and costs) due to potentially necessary change propagation based on false assumptions strongly decreases compared to the traditional "waterfall development process." In recent years, the agile development methodology spills over into hardware development (Johnson 2011). Thus, developing an IoT application with an agile approach would be a great case study for an industrial manufacturing company on how to handle a complex challenge with high uncertainty in a time and cost-sensitive environment.

This case study illustrates how agile development techniques facilitate identifying promising value propositions and how to develop an IoT application for an industrial manufacturing company. A hydrant, for example, is a good produced by an industrial manufacturing company. Hydrants are installed all over urban areas, and their purpose is to provide a high amount of water to fight fire. From an engineering perspective, their robustness is achieved by a simple design and a rugged cast metal construction. They have been chosen as the topic for this case study due to the following reasons: Traditionally a fully mechanical product, their Swiss manufacturers have no prior knowledge in the information and communication technology (ICT) sector. However, both big Swiss hydrant manufacturers tried to digitalize their products. While Hinni announced their acoustically sensing water

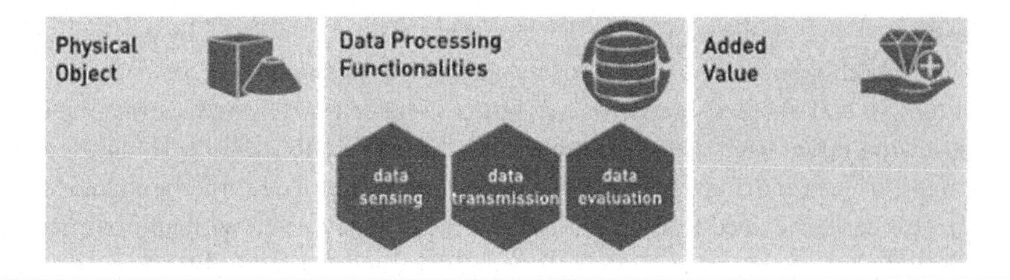

Figure CS1.1 IoT application canvas with three main blocks. (Heinis, Gomes Martinho, and Meboldt 2017.)

flow module "Lorno" 2005, VonRoll announced a software-based "Internet of Water (IoW)" program. While IoT applications generally comprise three building blocks (Figure CS1.1), i.e. physical objects, data processing functionalities, and value adding application, especially the data processing can be split further into data sensing, transmission, and evaluation (Heinis, Gomes Martinho, and Meboldt 2017). In this case, the hydrant as physical object was set as the starting point, so that we could focus on identifying the added value and developing the corresponding data processing functionalities. The following illustrates how we addressed these building blocks with the agile development concepts of need finding, job story prioritization, iterative development, rapid prototyping and testing, as well as retrofitting.

Applying Agile Development for IoT Applications

This section discusses the agile design activities applied in this case study. Design activities can be divided into four distinct phases—Discover, Define, Develop, and Deliver—using the double diamond model (UK Design Council 2015). In agile development though, such phases do not occur in a linear process but iteratively, over and over again, and, in reality, even in unstructured micro iterations. Nevertheless, we used the four phases to structure the narrative of this section.

Discover: User-Centered Need Finding for Promising Value Propositions

> Discover—Look at the world in a fresh way, notice new things and gather insights. (UK Design Council 2015)

To identify potential value propositions of IoT applications, developers must first of all focus on human needs and pains. The identification of target users and their needs is called *need finding* (Leifer and Steinert 2011). The outcome of an IoT application is basically semantic information about the physical world generated automatically and independently of human input (Ashton 2009; Atzori, Iera, and Morabito 2010). The beneficiary of semantic information is in most cases a human being. Thus, human needs and pains are an essential starting point to develop value-adding IoT applications. The development should be started by gathering all possible stakeholders in the context of a specified physical object or an area of interest. From all stakeholders, gathered key target users can be selected. Surveying and interviewing target users as well as modelling them by establishing personas, empathy maps, or user journey maps are proven methods to isolate underlying needs.

In this case, the need finding is based on the city of Zurich with approximately 8,000 hydrants. Four target users have been identified—a fire fighter, a municipal maintenance technician, a building contractor, and a municipal administration employee—and the mentioned need finding methods were applied. For each target user, a key job, a related pain, and the corresponding need resulted from the need finding (Table CS1.1). By this, four domains are identified outlining potential value propositions of the IoT application.

User stories support the derivation of clear solution features from the high-level value propositions outlined. User stories describe the functionalities of a

Table CS1.1 User Jobs, Pains, and Needs Outline Potential Value Propositions of the IoT Application

Target User	*Job*	*Pain*	*Need*
Fire fighter	Consume water to extinguish fire	Hydrant defects may remain undiscovered for a long period of time due to low inspection frequency (2years)	Increase disposability of hydrant infrastructure
Municipal maintenance technician	Inspect hydrants to guarantee working hydrant infrastructure	Inspection effort not resulting in the detection of defects (90% of inspected hydrants do not show any defects)	Reduce inspection effort
Building contractor	Consume water to construct buildings	Logistic effort to get equipment necessary to measure water consumption from municipality and related rental costs	Reduce effort for water consumption
Municipal administration employees	Administer and billing of water consumption to distribute costs fairly	The registration and billing of water consumption is manual and paper based	Reduce administration effort

solution—the IoT application in this case—desired by a target user (Blomkvist 2005). As high-level requirements, user stories should be testable and independent from the technical implementation (Beck 2000). User stories follow a certain format: As a "someone," I need "something" because of "some motivation." The underlying motivation allows one to better estimate the value of a certain requirement. This is necessary for the subsequent prioritization of work increments. As IoT applications deliver semantic information, promising user stories demand "something" that users want to know.

In this case, 37 user stories are identified. Examples of the 19 most valuable stories are listed in Table CS1.2.

Table CS1.2 Most Valuable User Stories Identified

Index	*User*	*Need*	*Motivation*
U1	Municipal administration employees	Wants to know who has last inspected the hydrants	To make the inspection process traceable
U2	Municipal administration employees	Wants to know what the results of inspections are	To better plan maintenance work
U3	Municipal administration employees	Wants to know if the hydrant has been used correctly	To make sure that there is no water left in the hydrant
U4	Municipal administration employees	Wants to keep as much as possible of our present infrastructure	To keep the cost at a minimum
U5	Building contractor	Wants a hydrant that is easy to use	To keep the chance to make mistakes while using a hydrant at a minimum.
U6	Municipal maintenance technician	Wants to know if one has done all inspection steps correctly	To improve the inspection quality
U7	Municipal maintenance technician	Wants a system that makes it easy to identify the hydrant	To not manually register hydrant ID
U8	Building contractor	Wants to know the amount of water that is drawn each time	To keep costs under control

(Continued)

Table CS1.2 (*Continued*) Most Valuable User Stories Identified

Index	*User*	*Need*	*Motivation*
U9	Municipal maintenance technician	Wants a report of the key inspection findings of today	To have an overview of today's inspection quality
U10	Municipal maintenance technician	Does not want to use the notebook	Because it is not handy
U11	Municipal administration employees	Wants a flexible and intuitive software to easily add, edit and delete hydrants	To easily register changes
U12	Building contractor	Wants to know how much water he has drawn	To keep an overview of the costs
U13	Municipal administration employees	Wants to know where and when water is drawn	So the consumers can be billed accordingly
U14	Municipal administration employees	Wants to have a list of all hydrants including details	So we can manage them
U15	Municipal maintenance technician	Wants to keep the input about inspection and maintenance complete but as simple as possible	To not lose valuable time
U16	Municipal maintenance technician	Wants to know a hydrant's position	To minimize searching time
U17	Municipal maintenance technician	Wants to know if the hydrants have been used correctly	To faster identify faults
U18	Municipal administration employees	Wants to know when a hydrant has been last inspected	So we can start maintenance in case it is necessary
U19	Fire fighter	Wants a working hydrant at all times	To efficiently fight fire

Define: Setting Up a Backlog, Allowing the Solution to be Delivered Incrementally

> Define—Develop a clear creative brief that frames the fundamental design challenge. (UK Design Council 2015)

A new method that we call information variable mapping (IVM) supports the identification of the right components for data processing (sensing, transmission, and evaluation) based on user stories. This method is necessary, as the process of combining several physical components cannot be informed by simply and consecutively working through the user stories—as the agile methodology suggests. Ideally, the relationship between components, information variables—i.e. the semantic information obtained by the data processing functionality of an IoT application—and user stories would be a one-to-one relationship. For example, one sensor delivers one information variable which then satisfies one user story. In reality however, these relationships may appear with one-to-many, many-to-one, or even many-to-many characteristics. For example, the information variable(s) obtained by a sensor can satisfy multiple user stories, or in turn, multiple sensors and other components (for example, machine learning classifiers) are required to provide one information variable. Therefore, understanding the relationship between user stories, information variables, and technical components is essential.

In this case, the IVM resulted into five sensors, two data transmission channels, and one web application for data evaluation and presentation. Some of the relationships are illustrated in Figure CS1.2.

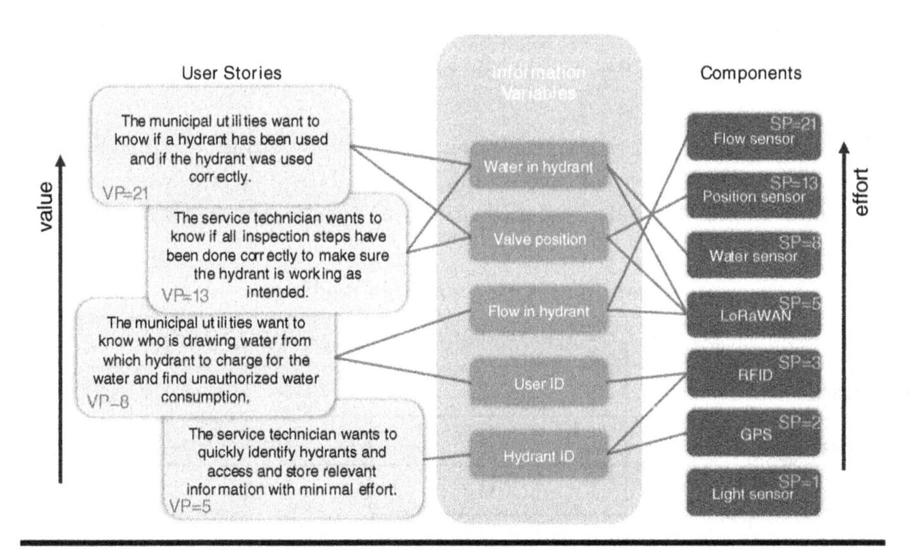

Figure CS1.2 IVM to connect user stories and components.

A prioritized backlog of information variables is the basis for the incremental development of testable functionality and of a potentially shippable product. The prioritization is done following the idea of the first agile principle: early and continuous delivery of value (Beck et al. 2001). Early delivery of value can be achieved by prioritizing the information variables which deliver high value for little effort. The value and the effort cannot be estimated directly for the information variables but indirectly by mapping value from user stories and effort from components using the IVM relationships. Adapting established agile estimation techniques (Cohn 2005), the value and the effort are not estimated absolutely but relatively by creating a ranking based on comparison. Once a ranking is done, so-called value points (VPs) can be assigned to the user stories and story points (SPs) to components using the Fibonacci sequence, for example. By summing up VPs and SPs for every information variable using the IVM relationships, the bang for the buck (BFTB) can be calculated for every information variable.

$$\mathrm{BFTB} = \frac{\mathrm{value}(\mathrm{VP})}{\mathrm{effort}(\mathrm{SP})}$$

In this case, the highest prioritized information variable is "user id" followed by "water in hydrant" and "valve position" (Table CS1.3).

Develop and Deliver: Create Testable and Potentially Shippable Added Value

> Develop & Deliver—create, prototype, test and iterate solutions or concepts and finalize and produce product increments (UK Design Council 2015)

Work on possible solution designs for one information variable at a time, in time-boxed sprints. The concept of time boxing aims at subdividing large work packages into short process intervals. So-called sprints are special time boxes that divide a

Table CS1.3 Prioritized Information Variable Backlog

Information Variable	*Accumulated VPs*	*Accumulated SPs*	*BFTB*
User id	8	3	2.67
Water in hydrant	34	13	2.62
Valve position	34	18	1.89
Hydrant id	5	5	1.00
Water flow in hydrant	8	26	0.31

whole project into a series of time boxes with constant duration. Each sprint starts with a planning session to determine features of the next product increment and ends with a retrospective to evaluate the built increment and to reflect on learnings for the next sprint (Varma 2015). The purpose of these sprint meetings is to incorporate feedback from several players (such as the product owner and stakeholders) early on.

In this case however, only one 2-week sprint was conducted addressing the information variables "user id" and "water in hydrant." After the first 2 weeks, only a fraction of the sprint goal was attained because the development and the testing of physical components had delayed the progress. The delay was mainly due to a missing testing infrastructure at the beginning of the project. Instead of ending the sprint, reflecting the impediments, and setting up a following one (as the agile methodology suggests), the duration of the current sprint was extended. Later on, time boxing was not in focus anymore, leading ultimately in neglecting the initial sprint concept (with 2-week sprints) for the rest of the project. The idea of working on the solution for one information variable at the time was maintained, and reflection on the development and testing progress was done in weekly meeting.

Promote quick, early, and iterative testing of components through rapid prototyping. A hardware solution feature might need more time to develop than a software one. Therefore, rapid prototyping is essential for hardware components to minimize the difference in development pace for hardware and software. Allowing testing of solution increments—finished or unfinished ones—is key in agile development and essential for successful sprint completion. The learnings from those tests can then be fed back into the design. The concept of rapid prototyping trades both quality and physical characteristics for the reduction of manufacturing time and costs (Hodge 2016). Rapid prototyping and testing always have to come together, to get the most learning within a short period of time.

In this case, we applied the fastest available manufacturing process for rapid prototyping of hardware components. This process was fused deposition modelling (FDM), which is basically 3-D printing of thermoplastic materials (we used the material acrylonitrile butadiene styrene (ABS)). For rapid prototyping of sensing and data transmission on the edge, the Arduino microcontroller "Lorauino" was used. It is part of a development kit for IoT applications that Swisscom (a Swiss national ICT provider) offers to promote their newly announced nationwide LoRaWAN network. Lorauino enabled the controlling and testing of sensors (especially for users who are not experienced with hardware programming) and the data transmission to the web application. For fast software development and testing of the cloud API and the web application Django—a high-level Python web development framework—and PyCharm—an integrated testing and development environment—was used.

In this case, the original idea for rapid testing was to build a test rig with several degrees of freedom. One challenge was the test rig design to enable rapid testing on a real hydrant provided by the city of Zurich. It was not possible to make it watertight because of the high amount of seams in this concept. Therefore, rapid testing in real conditions—inside the hydrant—could not be conducted until the fully integrated sensor module was finished. However, most components such as the moisture sensor and the encoder could be tested outside of the hydrant—under the tap, for example. The possibility of testing components right after producing/prototyping them led to an early understanding of whether the chosen design was applicable or not. This is the main reason why rapid testing contributed a lot to this case.

The concept of retrofitting allows upgrading of physical objects to IoT applications by adding components to already existing components without changing these existing ones. This is especially important when an integral redesign of a physical object would require a lot of resources or when it is not at the end of its life cycle yet. The focus again is on the development speed, i.e. to bypass a redesigning of the whole (physical) product by just fitting in new IoT functions without changing physical components. If it is possible to identify and validate the added value of the IoT application, one can still redesign the physical components.

In this case, the costs of replacing all the existing physical components with IoT applications would have exceeded the added value of the additional IoT functionalities. Thus, one challenge of our retrofitting was the lack of precise blueprints of the existing hydrant in combination with the inaccuracy of cast metal parts. Especially, the form-locking parts of the encoder and the geometry of the sensor ring had to be tested beforehand to ensure the right fit. The integration of the whole module could be better realized with fewer design constraints.

What was Delivered and What was Learned

This section presents the IoT application delivered and discusses the main learnings the case studies.

NERO: An IoT Application for Smarter Hydrants

The IoT application NERO emerged from this case study. The IoT application comprises three main components, i.e. the NERO sensing add-on, NERO cloud API, and the NERO web app (Figure CS1.3). It is possible to retrofit existing hydrants with a NERO sensing add-on, regardless of its manufacturer (Figure CS1.4). The NERO Add-on measures usage, water flow, and detects errors and transmits this measurement data via LoRaWAN to the NERO cloud API. The NERO web app enables several users to interact with the hydrant in an application-specific manner via customized user interfaces. Users can connect directly to the hydrant via radio-frequency identification (RFID) using a mobile phone or tablet (Figure CS1.5), for example, to register a water supply or call up maintenance information. The NERO sensing add-on was implemented as working prototype for one hydrant. The NERO cloud API and the web app are running live on an online test server and can be accessed with various devices over the internet.

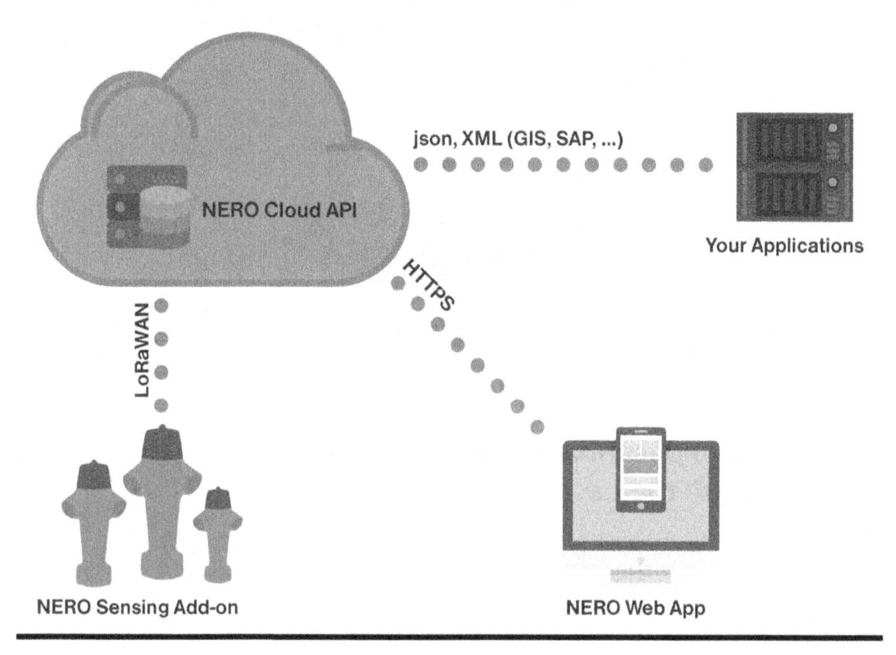

Figure CS1.3 The three components of the NERO IoT application.

Figure CS1.4 NERO sensing add-on can be retrofitted between upper and lower part of hydrant (FDM prototype).

Figure CS1.5 NERO web app accessed from mobile phone after scanning RFID chip on hydrant.

Managerial Implications: Learnings Which Emerged from the Case Study

This case study illustrates how an agile approach was applied for the development of an IoT application. The IoT application is based on an industrial manufacturing good and demonstrates how IoT technology can be used to add value in an industrial setup.

The strong user focus and the extensive need finding activities facilitated the creation of clear value propositions. The case study thereby corroborates the existing literature stating that a focus on user needs is essential when building the IoT (Hui 2014). We strongly recommend one conduct need finding activities by applying, for example, the presented methods for developing IoT applications. This approach will lead to the implementation of solution components that provide an added value for targeted users.

The agile approach suggests to incrementally develop the solution based on prioritization of user stories by estimating value and effort of each increment (Cohn 2005). This simple concept was not directly applicable in this case study. Estimating each user story's development effort was not possible due to the complex and multidimensional relationship between user stories and technology components. However, our agile estimations based on the introduced artifacts *information variables* with support of IVM allowed us to define and prioritize development increments.

While the existing literature on agile development recognizes the importance of testing (Blomkvist 2005), the importance of a comprehensive testing infrastructure for the physical components of the IoT application was underestimated in this case. To test, for instance, the sensing add-on in the hydrant, large amounts of water had to flow through the hydrant. Providing this water and sealing all components were challenging and led to many setbacks during the project. We learned that an appropriate testing environment is essential for rapid testing activities. We also recommend one allocate enough resources on the development of a testing infrastructure at the beginning of an IoT project.

The implementation of physical components generally takes longer than of digital ones. Even though this challenge is known (Heinis, Hilario, and Meboldt 2018), we underestimated the strong effect on the applicability of agile development methods in this case. It was not possible to deliver tested solution increments within 2-week sprints, and thus, the concept of time-boxed sprints with constant length was abandoned rather early. Moreover, the maturity of the digital components was usually higher than the one of the physical components. Nevertheless, the concept of incremental and iterative development was still an essential element of the development process but not with a fixed iteration duration. We recommend to actively provoke iterations, comprising early validation and testing with users.

Although challenges occurred while applying agile methods for the development of this IoT application, we consider "agile development" as a valuable approach for the identification and development of value-adding IoT applications.

Acknowledgments

Special thanks go to Thomas Benedek, Sebastian Klotz, and Samuel Blösch for their efforts in developing the IoT application NERO for smarter hydrants. Additional thanks go to Livio Tolomeo and his colleagues from the water supply administration of Zurich, Switzerland for their inputs during need finding and for providing a hydrant.

References

Ashton, Kevin. 2009. "That 'Internet of Things' Thing." *RFiD Journal* 22 (7): 97–114.

Atzori, Luigi, Antonio Iera, and Giacomo Morabito. 2010. "The Internet of Things: A Survey." *Computer Networks* 54 (15): 2787–2805. doi:10.1016/j.comnet.2010.05.010.

Beck, Kent. 2000. *Extreme Programming Explained: Embrace Change.* Boston, MA: Addison-Wesley Longman Publishing Co., Inc.

Beck, Kent, Mike Beedle, Arie van Bennekum, Alistair Cockburn, Ward Cunningham, Martin Fowler, James Grenning, et al. 2001. "Principles behind the Agile Manifesto." http://agilemanifesto.org/principles.html.

Blomkvist, Stefan. 2005. "Towards a Model for Bridging Agile Development and User-Centered Design." In *Human-Centered Software Engineering—Integrating Usability in the Software Development Lifecycle*, edited by Ahmed Seffah, Jan Gulliksen, and Michel C. Desmarais, 219–44. Dordrecht, The Netherlands: Springer. doi:10.1007/1-4020-4113-6-12.

Cohn, Mike. 2005. *Agile Estimating and Planning*, 1st ed. Upper Saddle River, NJ: Prentice Hall PTG.

Fleisch, Elgar, Markus Weinberger, and Felix Wortmann. 2015. "Business Models and the Internet of Things (Extended Abstract)." In *Interoperability and Open-Source Solutions for the Internet of Things: International Workshop, FP7 OpenIoT Project, Held in Conjunction with SoftCOM 2014*, Split, Croatia, September 18, 2014, Invited Papers, edited by Ivana Podnar Žarko, Krešimir Pripužić, and Martin Serrano, 6–10. Cham, Switzerland: Springer International Publishing. doi:10.1007/978-3-319-16546-2-2.

Heinis, Timon B., Carlos Gomes Martinho, and Mirko Meboldt. 2017. "Fundamental Challenges in Developing Internet of Things Applications for Engineers and Product Designers." In *DS 87-5 Proceedings of the 21st International Conference on Engineering Design (ICED 17) Vol 5: Design for X, Design to X*, Vancouver, Canada, 21–25, August, 2017, edited by Anja Maier, Stanko Škec, Harrison Kim, Michael Kokkolaras, Josef Oehmen, Georges Fadel, Filippo Salustri, and Mike Van der Loos, 279–88. The Design Society.

Heinis, Timon B., Jan Hilario, and Mirko Meboldt. 2018. "Empirical Study on Innovation Motivators and Inhibitors of Internet of Things Applications for Industrial Manufacturing Enterprises." *Journal of Innovation and Entrepreneurship* 7 (10). doi:10.1186/s13731-018-0090-7.

Hodge, Shayne. 2016. "A Rapid IoT Prototyping Toolkit." IEEE Internet of Things. https://iot.ieee.org/newsletter/january-2016/a-rapid-iot-prototyping-toolkit.html.

Hui, Gordon. 2014. "So You Want to Build an Internet of Things Business." Harvard Business Review.

Johnson, Neil. 2011. "Agile Hardware Development—Nonsense or Necessity?" www.eetimes.com/document.asp?doc_id=1279137.

Leifer, Larry J., and Martin Steinert. 2011. "Dancing with Ambiguity: Causality Behavior, Design Thinking, and Triple-Loop-Learning." *Information Knowledge Systems Management* 10 (1–4): 151–173.

SAS Institute Inc. 2016. "Internet of Things Visualise the Impact."

UK Design Council. 2015. "The Design Process: What Is the Double Diamond?" www.designcouncil.org.uk/news-opinion/design-process-what-double-diamond.

Varma, Tathagat. 2015. *Agile Product Development: How to Design Innovative Products That Create Customer Value*. Berkeley, CA: Apress.

Case Study 2

Automotive IoT

Alex Peczynski
Ford Motor Company

Contents

Introduction

Ever since the automotive industry began around the 1860s, it was an industry that had the whole world enamored with its complexity. It started with the transition from steam-powered engines to gasoline-powered automobiles and the industry never looked back. In the beginning automobiles were predominantly produced in Europe and were hand crafted, one by one. This was so until Henry Ford came along and changed the game. He was able develop a mass production process that allowed for just about anyone to be able to purchase an automobile. The assembly line was the beginning of many advancements made in the automotive industry. From there, a vast number of features were being added to such as different engines, interior, paint colors, and options. Advancements in automobiles created one of the fiercest global competitions between automotive manufacturers where the market leader was king. Competition from firms in the United States and overseas spurred an automotive revolution that included the best technologies of the times (Rae and Binder 2018).

Fast forward to the beginning of the technological revolution where computers were first being developed and computer chips are starting to be embedded into cars. The computer chips in cars support the basic functions of the different modules in an automobile. Some examples of these basic functions are the instrument panel cluster (IPC), antilock braking system (ABS), and restraint control module (RCM). The more and more that technology was developing outside of the automotive industry, the more it was being directly inputted into automobiles (Belford 2017).

Since the increase in technology has become even more intertwined with society and how we interact with it, it has also become even more intertwined with the automotive industry and its cars. With every new vehicle today, it is standard that the vehicle has sensors for many different parts and purposes. The sensors and actuators allow for the devices to receive input and act upon it. Without the use of sensors, one would physically have to check tire pressure, turn on airbags in a crash, or tell the lights to come on when it is getting dark outside. The technology in cars and their sensors have been taken even farther in today's lineup. Sensors and actuators are becoming even smarter and have opened the space for the automotive internet of thing devices to enter.

The Internet of Things, or IoT, can be defined as many different things, but when it comes to automobiles, it is about the sensors, actuators, and technology that connects cars to users and even cars to other cars. These sensors are much smarter in their capabilities and functions than ever before. We now expect that our phone will connect via Bluetooth in every car and tell us when we need to make a turn. This is all because of the smart environment that has been developed with IoT. Even though the true "connected car" is not yet a distinguishable selling point, it is becoming more and more popular. Below, you will learn about three main areas in which the IoT is affecting the automotive industry and specifically automobiles themselves (Ninan, et al. 2015). This case study reviews three main areas where IoT and the automotive industry have come together to produce smart cars with cutting-edge technology that will change how we feel about going from point A to B.

Connected Cars

In a study conducted by Gartner, it is predicted that by the year 2020 more than 250 million vehicles will be connected. This means that the number of vehicles with installed connectivity units such as Ford's SYNC 3, or GM's OnStar, will be on the road. With the increase of in-car connectivity units there is also a trend predicted that 67% of vehicles worldwide will be connected and consumers will spend twice as much money for this in-vehicle connectivity. In-car connectivity units are exactly what they sound like (Ninan, et al. 2015). They provide a platform for the driver of the automobile to connect their smart phone to the vehicle and perform an array of functions. With Ford's SYNC 3, there has been the addition of Apple CarPlay and Android Auto. These are two separate car connectivity platforms, one by Apple, Inc. and the other by Google. These platforms allow for users to connect a smart device via USB and be able to display a version of their phone. Apple CarPlay allows for users to link multiple apps such as Pandora, Waze, and Spotify for easy use. Furthermore, it does not just allow for users to play music through the car, but the user can talk to the car and use voice commands. An example would be clicking a button and Siri asks what she can help you with. One can ask, "Find me the nearest restaurant." Next, the in-car connectivity system would use Apple Maps and shows you possible restaurant locations. The functions of the connected car are not limited to just Apple CarPlay or Android Auto.

Other companies such as progressive use IoT connectivity to assess how someone drives and provides an insurance quote based upon real-time data. Progressive created a device that streams your driving information to their servers where the data can be analyzed and insurance rates can be determined. Progressive Snapshot plugs straight into your OBD-II port (Vehicle Communication Module Port) where it can access information about your car such as location, acceleration rate, speed, stopping rate, and other data points. This provides Progressive with a comprehensive "snapshot" of how one drives. Progressive Snapshot even goes further by providing an app to help you with certain trip details, trip logs, and driving tips. This is a smart extension of IoT based in-car connectivity where your automobile can connect with devices outside of itself (Hunt 2017).

Automobiles are becoming more connected and smarter than ever. This is where another realm of the connected car comes into play. When modules, such as the infotainment system, need to be updated, but are already out of dealership control, how are software and firmware upgrades done? The over-the-air software updates are a form of IoT that enable this to take place. This means that the automobile is connected to the internet through its main infotainment system. Next it receives a notification. It is like the pop-up one receives on their phone when the newest operating system has been released. With over-the-air updates, no longer do the owners have to spend time going to a licensed dealership for a software update (Keith 2018). Owners can leave their car in the garage, let it update while connected to the internet, and spend their time as they want. With the development of wireless technologies such as 5G, over-the-air updates will become increasingly popular and effective.

Artificial Intelligence

Many artificial intelligence (AI) applications can be found in a car. The most popular and talked about is the premise of self-driving cars. Companies such as Toyota are going so far as to invest $1 billion in research and development funds to develop autonomous systems. Some of the other big names in the AI self-driving arena are

Tesla, Google, and Uber. Before a completely autonomous car hits the road, there will have been advances with AI in developing driver-assisted AI features. This means that that the car is not fully autonomous but has features that are dependent upon smart sensors and actuators. Some of the functions are as follows:

- Automatic braking
- Collision avoidance systems
- Pedestrian/cyclist alerts
- Cross-traffic alerts
- Intelligent cruise control
- Parking assist

Another arena in AI that is currently being developed by auto manufacturers is the cloud-hosted intelligence realm. The use of cloud platforms has seen a large increase in the last few years and especially in this industry. General Motors and IBM's Watson supercomputer have partnered to try some very special features. They are trying to create a platform of features that are developed by AI. One might wonder how a car can utilize a cloud-based AI platform. This means that the automobile has these functions that at any one moment can be accessed by the car from the cloud and acted upon. In Figure CS2.1, you can see the cloud being accessed by the connected vehicle. Some of the ideas that are being developed on the cloud platform are as follows:

- Locating gas stations and making it so the driver can pay without getting out of the vehicle
- Identifying and suggesting restaurants that are similar to what the driver normally goes to
- When approaching stores on a route, reminding the driver of needed household items
- When approaching a restaurant, automatically preorder the food of choice
- Finding open parking spots (Figure CS2.1)

AI is an element of the connected car but not necessarily connecting the car to the user. AI can connect your car with other cars. Through AI, vehicle-to-vehicle communication is being developed by taking the sensor data on each car and communicating it to many others. For example, if two people were trying to merge into the center lane on the highway and both were in each other's blind spots, the cars would be able to communicate with each other and notify each that they were trying to merge into the same lane. Therefore, through a command sequence, AI could notify the driver of the issue and it could wait for the lane to clear and then merge. The ability to have vehicles communicating with vehicles outside of human acknowledgment could be a large benefit to safety. Also, this could allow for the aid in developing autonomous driving features or functions (Lyudmyla 2018).

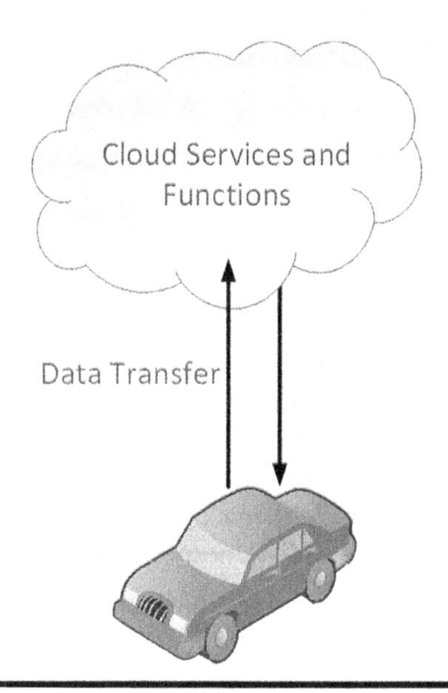

Figure CS2.1 Cloud connectivity.

Big Data Analysis

Having a connected car is a thing of now and produces more data than you probably imagine. It is said by the IHS Automotive 2015 study that Big Data from the connected car industry will produce more than $14.5 billion or more in revenue by 2020. This money can be made from creating applications that utilize the data or selling the data for business purposes. In a connected car, the sensors collect valuable information not only from the car but also from the user. The data that is typically collected about the car is the vehicle diagnostics, maintenance, wear, and defects. Each serve as important information to manufacturers or dealerships to see how the car is functioning and what can be done better. Furthermore, this data can also be important to other outside companies such as aftermarket producers or top tier suppliers. Top tier suppliers supply different modules or parts to automotive manufacturers. Having data that shows product life and performance can be crucial to future development of parts or services. On the other hand, the data collected by the connected car about drivers is even more sought after in today's market. Information that is included is the location, usage patterns, driving history, and more. The interesting part about both types of data collected in the connected car is that it can be in real time. Therefore, while one is driving, the data can be collected and utilized. If there is data collected and analyzed that is critical about a certain automobile part, the system can let the driver know it is about to fail. Imagine if you could schedule an appointment at a dealership before a problem occurred because of this real-time data stream. This can go far into the preventative care of automobiles in the future. Big Data collected from smart connected cars will continue to be a growing industry because of its size, necessity, and value (Dorsey 2013).

Conclusion

As IoT and the associated functions become popular, IoT will be increasingly incorporated into vehicles. The development of fifth generation, or 5G, networks will soon support the amount of real-time data throughput and drastically increase the ability to connect vehicles to vehicles, and car to driver. Along with Big Data and the connected car come the innovations that are cutting edge that add to the user experience and influence the purchasing of cars. The smart car of the future is almost here, waiting behind the closed doors of a test garage.

References

Belford, Geneva G. 2017. *Encyclopædia Britannica*. April 20. Accessed August 4, 2018. www.britannica.com/science/computer-science/Computers-in-the-workplace.

Dorsey, Jim. 2013. *IHS Markit Says*. November 19. Accessed August 4, 2018. news.ihsmarkit.com/press-release/country-industry-forecasting/big-data-drivers-seat-connected-car-technological-advance.

Hunt, Janet. 2017. *The Balance*. June 5. Accessed August 4, 2018. www.thebalance.com/progressive-snapshot-review-4141266.

Keith, Barry. 2018. *Consumer Reports*. April 20. Accessed August 4, 2018. www.consumerreports.org/automotive-technology/automakers-embrace-over-the-air-updates-can-we-trust-digital-car-repair/.

Lyudmyla, Novosilska. 2018. *Ignite*. July 30. Accessed August 4, 2018. igniteoutsourcing.com/publications/artificial-intelligence-in-automotive-industry/.

Ninan, Simon, Bharath Gangula, Matthias von Alten, and Brenna Sniderman. 2015. *Deloitte United States*. August 18. Accessed August 4, 2018. www2.deloitte.com/insights/us/en/focus/internet-of-things/iot-in-automotive-industry.html.

Rae, John Bell, and Alan K. Binder. 2018. *Encyclopædia Britannica*. August 2. Accessed August 4, 2018. www.britannica.com/technology/automotive-industry#ref65765.

Case Study 3

Butler University and Academic Scheduling: A Platform for IoT

Ruth Schwer and Timothy Roe
Butler University

Contents

Introduction

The Internet of Things (IoT) broadly describes how data collected from sensor-enabled computing objects can be shared and analyzed to gain efficiencies or create new opportunities. Corporations have used IoT technologies to gain logistical efficiencies with manufacturing. Higher education has watched this trend with interest, as a combination of changing expectations, and imperiled funding forces many schools to reevaluate how they do business. A university campus offers a stable physical environment and predictable patterns of activity, and IoT technology might benefit the institution in many ways, from reducing complexity to gaining insights into ways to control costs, improving products and services, and creating new lines of business. However, getting clarity on the goals and buy-in from stakeholders can be challenging, especially in the higher education environment. Given that IoT technology allows for increased data collection and systemic controls, is it possible to use this technology to expand the services offered to students without increasing costs or adding unreasonable risks? As higher education confronts its rigid dependency on tuition dollars, could the IoT help it meet increasing technology expectations without adding to the bottom line?

Many U.S. colleges and universities have learned from the mistakes of the past. In *College (Un)Bound*, Jeffrey Selingo notes that in the period from 1999 to 2009, colleges and universities were growing their population, growing their footprint, and growing their tuition.[1] During this "Lost Decade" as he calls it, a strong sense of traditionalism and rosy predictions for the future shaped an "all-is-well" attitude among higher education decision makers. In 2009, online and open source education forced a new conversation about tuition costs and a reevaluation of the purpose and mode of postsecondary education. Universities found a new spirit of innovation in their quest to demonstrate value to students, family, and alumni.

Butler University is at the forefront of the new face of higher education. A campus of just over 5,000, Butler is located in urban Indianapolis, Indiana, and boasts the title "Most Innovative School" in the Regional Universities Midwest rankings by *US News and World Report* in 2018.[2] Butler president, James Danko, states that "Innovation has become an important differentiator among universities, as we work to create learning experiences that enhance student engagement and success." The culture of innovative practices at Butler University extends beyond the teaching and learning environment into the operational units of the institution also, and this is especially evident in its IT department.

Butler makes significant expenditures on technology and the personnel to support it. From classroom learning spaces to financial management, nearly every sphere of activity at the university is supported by some type of technology. The infrastructure for these applications on Butler's physical campus is a continuously evolving creation that's expected to be reliable and secure. Like many schools, Butler IT is adjusting to the changing landscape of technology support as more applications that were hosted on premises are migrating to vendor-hosted services. Growing pains are to be expected, and the change makes different demands on personnel. While technical skills are still prized, the ability to communicate with vendors and coworkers is even more important. When support analysts have no direct, physical access to the server or database, words become the only way they can effectively resolve problems. This is especially true for IoT technology due to the interconnected systems and vendors it is built upon. Skilled management of vendor relationships is a key capability necessary for successful implementation of IoT technology.

Technology demands on campus have been fueled by consumer technology experiences, such as mobile devices and smart appliances. The Pew Research Center reports that 95% of teens 13–17 years have a smartphone.[3] Students come to campus quite experienced with using their mobile phones and social media to connect with their peers; the college transition, then, forces students to connect with the university through official e-mails and a campus hierarchy of colleges and departments. While posting a photo to Instagram is a simple, seamless experience, reviewing their class schedule on a mobile device is not. Keeping track of a given semester's class meetings, assignments, and messages from classmates and instructors can quickly become overwhelming for students.

The Project

Amidst the technology proliferation that the Butler IT department manages, a student mobile app was one item that was missing. As early as 2012, Educause research indicated that students were eager to communicate and study using their mobile phones, and although Butler IT created some mobile-ready content, a cohesive mobile student experience proved extremely difficult to turn into a reality.[4] Some vendors simply didn't offer a robust mobile app frontend that could satisfy multiple, common self-service goals of students. Efforts to custom build a mobile app faced difficulties ranging from political resistance to security concerns.

In 2015, Butler, was looking for technology opportunities that would help the university leverage its existing IT investments into sustainable gains in capacity and functionality for the campus. In December, 2015, Butler joined in discussions with Indianapolis-based High Alpha, a leading venture capital firm and studio, about creating a mobile student engagement tool. What started as a promising concept demo evolved quickly, and in June, 2016, Butler became both an investor and a customer of a new company, ClearScholar, whose software-as-a-service product was a new student engagement mobile platform.[5]

Over the summer months of 2016, Butler IT staff collaborated with ClearScholar personnel to incorporate data from its Customer Relationship Management (CRM) and student information system as well as other content. The Butler app, built on the ClearScholar platform, debuted on campus as a limited access release in fall, 2016.[6] A preselected group of 100 Butler students were welcomed into the process of app development with a release party and special, insider status. Their feedback and ideas for development directed some of ClearScholar's functionality. As the buzz surrounding the app spread through the student body, eager students sought out the app and signed up to be "waitlisted" to gain access to the product. Access to the Butler app was expanded to all students in January, 2017.

The Butler app provided a unified, curated navigation path for students in a mobile, context-aware application. It also allowed the student to select and personalize the content they see, and it used the phone's position in space to aid way finding on the physical campus. From the university's perspective, these interactions with the app created usage data. Administrators could monitor the app's usage and send messages to subsets of students who met complex filtering criteria.

A vital part of any system is the people. In fall, 2017, ClearScholar was used to effectively augment communication regarding welcome week activities: during this annual event, Butler welcomes students back to campus, offers special orientation events for those new to campus, and generally prepares to engage in another year of student learning. With regular doses of relevant content, created and maintained by campus staff, the app's introduction met with higher than anticipated usage rates. (Please see the accompanying chart for more information.)[7] In fall, 2018, a change in welcome week planning leadership led to less emphasis on Butler app content. With less communication going through the app, active user rates declined.

Arguably, the most popular feature of the app is an optical barcode version of the student ID card that could be used to purchase food at the student union, borrow books, register attendance at cultural events that satisfied an academic requirement, and check in to the Butler Health and Recreation Center. While the traditional plastic card with radio frequency identification (RFID) tag is also accepted for these uses, students immediately embraced the app version of their ID card, as they were far more likely to have their smartphone with them when they needed campus services. Early on, some students realized that it was easy to take a picture of the ID and share it with friends, which extended access to the Health and Recreation Center and other campus amenities to individuals without a valid ID. Needless to say, this wasn't the kind of collaboration that we hoped to foster with the app, but it was a valuable learning experience, as it highlighted the desirable limits for automated identification. Having staff randomly check that the ID photo matched the person was enough to deter misuse.

The Butler app also provides a panic button, prominently displayed, that places a call to on-campus police. This augments the kiosk emergency phone system (known as Blue phones) already in place on Butler's campus. From that initial idea grew another feature currently in development for the ClearScholar platform: known as Aware, it adds the mobile device's current GPS coordinates and the student's identity to the telephone contact with campus police. The information is sent directly to police dispatching systems, so that officers can recognize the student in need once they are on the scene. For institutions trying to combat sexual assault and campus crime, Aware offers a novel tool that follows an IoT use case scenario.

Ultimately, ClearScholar was acquired by Civitas Learning in June, 2018, and Butler became a preferred customer with stock ownership in the company. The journey from venture studio to startup to acquisition was swift, as high tech tends to move much faster than higher ed can. Butler's involvement with ClearScholar became a touchstone for solving problems and extending services by embracing new relationships outside of academia.

Risks and Reactions

More than a few campus administrators had questions and reservations about ClearScholar's intended use and implementation. The app started more campus discussions about data privacy and security, campus monitoring, content creation, vendor relationships, and accessibility that helped educate decision makers. Long-standing questions about how the Federal Education Rights and Privacy Act (FERPA) should apply to modern student data and records resurfaced. Administrators reevaluated their own expectations of third-party vendors when handling student data, in light of the university's increasing number of such relationships. Socializing the risks and benefits of new technology in a way that allows those risks to be discussed and managed is a tricky balancing act. Nevertheless, these discussions are a necessary part of forming a strategy poised to shape future choices about IoT deployments. As consumer devices continue to evolve, expectations will likely continue to rise: having a strategy in place will help Butler keep pace.

IoT frequently raises concerns about risk, particularly with regard to data privacy and security. The General Data Protection Regulation (GDPR), the EU data privacy laws, underlined those concerns when its compliance deadline date arrived in May, 2018. Because GDPR applies to European subjects no matter where they are in the world, U.S. universities who accept international students from EU countries could be held accountable for their data practices for those students. International students are an important source of educational demand that American universities can leverage to increase enrollments. Universities who can wrestle with the risks introduced by the GDPR will be in a better position to maintain their tuition-dependent budgets for years to come.

GDPR draws a clear distinction between data controllers (an entity that owns user data) and data processors (those that analyze or transform user data) and creates clear expectations for both roles. However, IoT technology blurs the line between these roles and creates an environment where giving specific consent for data collection can be difficult to design. Can a student walking through the university's campus choose not to accept being surveilled on wireless security cameras?

When the utility of an IoT device is dependent on its ubiquitous presence and use, then building privacy and security into the system from the beginning becomes mandatory. Designing with privacy in mind is extremely important, both to protect data and to satisfy GDPR requirements. For Butler, having ClearScholar collect and process data on its behalf was similar to other applications hosted by vendors for the university. However, this vendor's data practices will receive more scrutiny as the IoT makes new features possible and the app collects more data.

IoT and the Future

With the Butler app established as a primary mode of communication for students, faculty, and staff, adding capabilities inspired by IoT technology will create valuable opportunities to make the student experience better. In the future, a transformed campus with a constantly connected ecosystem of devices could be monitored and controlled through the app. The Butler app's streamlined reporting will offer decision makers real usage data collected quickly, regardless of the system or sensor from which the data originates. In turn, the university could use the data to contain operational costs, and the intangible gains of new functions offered are just as compelling. The university is starting to imagine the possibilities for personalized student interactions, which might include items from the following lists.

Redefining the Student Experience

- Wellness and fitness tracking is made more robust when students are able to purchase food, monitor calories, and track fitness activities through point-of-sale devices and workout equipment monitors.
- Internet-connected cameras let students monitor the waiting lines in the cafeteria or the availability of the free-weight area in the gym at any given moment.
- Navigating campus becomes simpler, even when construction or maintenance activities block traffic or parking areas temporarily.
- University-owned resources such as library books or recreation equipment are checked out to students without a checkout station. Students simply walk out of the geo-fenced area with a tagged item; later, the app reminds them to return the resource or assists in locating lost objects.
- A network of room occupancy sensors validate campus buildings are empty when personnel must evacuate a location for safety reasons.
- Students participate in campus life by adding to the augmented reality data about the campus, which becomes visible only in the ClearScholar app. Changes to the campus, fond memories, or classroom gamification activities are recorded in space and time by student authors.

Improving University-Provided Services

- Users can request help from any university service department and schedule an appointment time to meet service personnel from IT, HR, or campus maintenance.
- Real-time video tutoring is provided upon request through the app, linking students with questions to faculty or other students for help.

- Heating, ventilation, and air conditioning (HVAC), lighting, or classroom mediation equipment failures trigger repair alerts for the appropriate staff member, complete with failure diagnostic information.

Campus Sustainability

- Administrators make decisions to optimize utilities in campus buildings based on near real-time data. Rooms stay warm or cool regardless of outside temperature fluctuations.
- Soil monitors ensure landscaping crews know when water or fertilizer is necessary and reduce or prevent chemical runoff in public waterways.
- Easy access to reserve ZipCars or call ride-sharing services through the app decrease the number of cars students bring to campus—and the prices they pay for parking.

Summary

Butler's journey through collaboration and development of the ClearScholar app laid a foundation on which to build. As students continue to come to campus with more IoT devices and higher expectations, Butler now has a chance to keep pace. Although the technology and its implementation can be complex, it still holds great promise in creating a campus that's even more valuable to the students and community it serves in the future. The insights gained through the IoT data of the future will drive value in new and innovative ways.

Chart

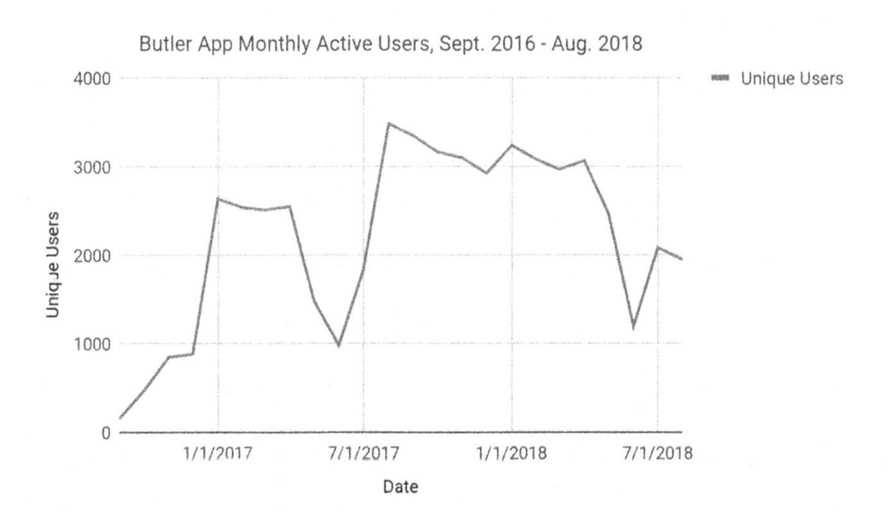

Butler App Monthly Active Users	
Date	*Unique Users*
9/1/2016	150
10/1/2016	461
11/1/2016	845
12/1/2016	878
1/1/2017	2,631
2/1/2017	2,535
3/1/2017	2,507
4/1/2017	2,544
5/1/2017	1,490
6/1/2017	973
7/1/2017	1,827
8/1/2017	3,480
9/1/2017	3,345
10/1/2017	3,163
11/1/2017	3,096
12/1/2017	2,922
1/1/2018	3,238
2/1/2018	3,076
3/1/2018	2,967
4/1/2018	3,064
5/1/2018	2,460
6/1/2018	1,193
7/1/2018	2,081
8/1/2018	1,946

Acknowledgments

Many thanks go to Ruth Schwer, Timothy Roe, Rob Hartman, Pete Williams, Mary Reiman, and Brandon Quarles from Butler University for their contributions to this case study.

End Notes

1. Jeffrey Selingo, *College (un)bound: The Future of Higher Education and What It Means for Students* (Las Vegas, NV: Amazon Pub., 2013).
2. "Butler University Fact Book." (Butler.edu, 2018); "How Does Butler University Rank among America's Best Colleges?" (U.S. News & World Report, 2018).
3. Monica Anderson and Jingjing Jiang, "Teens, Social Media & Technology 2018 (Pew Research Center: Internet, Science & Tech. May 31, 2018); Pearson Student Mobile Device Survey: College Students, 2015.
4. Kyle Bowen and Matthew Pistilli, "*Student Preferences for Mobile App Usage*" (Louisville, CO: EDUCAUSE Center for Analysis and Research, 2012).
5. "Student Engagement Platform," *ClearScholar*.
6. Kate Harrison, "ClearScholar Launches New Butler University Student App" (TechPoint, April 2018).
7. ClearScholar, "Butler Monthly Active Users, 2016-18." Clearscholar administrator report, 2018.

Bibliography

Anderson, Monica, and Jingjing Jiang. 2018. "Teens, Social Media & Technology 2018." Pew Research Center: Internet, Science & Tech. May 31, 2018. www.pewinternet.org/2018/05/31/teens-social-media-technology-2018/.

Bowen, Kyle, and Matthew D. Pistilli. *Student Preferences for Mobile App Usage* (Research Bulletin). Louisville, CO: EDUCAUSE Center for Analysis and Research, September 25, 2012. www.educause.edu/ecar.

"Butler University Fact Book." Butler.edu, 7 June 2018. www.butler.edu/oira/fact-book.

Harrison, Kate. 2018. "ClearScholar Launches New Butler University Student App." TechPoint. April23,2018.https://techpoint.org/2016/08/clearscholar-butler-university-student-app/.

"How Does Butler University Rank Among America's Best Colleges?" 2018. U.S. News & World Report. www.usnews.com/best-colleges/butler-university-1788.

"Pearson Student Mobile Device Survey: College Students 2015." Harris Poll. June 2015. www.pearsoned.com/wp-content/uploads/2015-Pearson-Student-Mobile-Device-Survey-College.pdf.

Selingo, Jeffrey J. *College (un)bound: The Future of Higher Education and What It Means for Students*. Las Vegas, NV: Amazon Publishers, 2013.

"Student Engagement Platform." 2017. *ClearScholar*. clearscholar.com/.

Case Study 4

Connectible Me: On Using the Internet of Things to Collect Points for Wellness Tracking

Jared Linder
Indiana Family and Social Services Administration

Contents

Introduction

This piece started out as a case study on how government was using the Internet of Things (IoT) to create a transformative landscape of connected devices that were working together to create a more seamless client experience and to improve data collection and government efficiency. As defined, IoT can be elusive to specifically pinpoint by definition (Wortmann & Flüchter, 2015) but loosely can be used to explain the methods by which physical objects, "things," can be used in an interconnected fashion to facilitate connectedness between users and physical objects (Weber & Weber, 2010). We can observe a few current instances of this: our Indiana Department of Transportation is exploring using drones to analyze traffic patterns and to monitor roadways, bridges, and possible damage in areas where traditionally crews would need to take heavy machinery and cranes to monitor and inspect structural integrity or damage (Dorsey, 2018). Our Bureau of Motor Vehicles offers kiosks where users can renew vehicle registrations and perform simple transactions that do not involve the assistance of a live person (Indiana Bureau of Motor Vehicles, 2018). And our natural resources and wildlife divisions have been collecting data through remote devices for years. As a state, we are continuing to work to embrace the concept of IoT.

The agency I work for, the Indiana Family and Social Services Administration (FSSA), helps administer health and social services benefits to as many as one out of every five Hoosier citizens on any given day (Kaiser Family Foundation, 2018). To do that across 92 counties and in 10 regional change centers, we have set up computer terminals for client self-service and multifunction devices for workers to upload and scan paper documents to be attached to client records. But we have yet to make much more headway into the IoT as we presently stand. FSSA has at least a few projects we are exploring like electronic visit verification for home health services (CMS, n.d.) and automatic check-ins for children in day care, but those are still in the preliminary planning phases. Those use cases of course are evident: tracking to see if providers of services are in fact on location where needed or verifying that children are on-site and accounted for during their time in day care. To prepare for possible projects like this, we have taken preparatory steps (Kranz & IDG Contributor Network, 2018) as a state through our Indiana Office of Technology to focus on improving our network strength and throughput, to increase wireless capability across our campus and remote buildings, and to focus on the security of that network (Ammar, Russello, & Crispo, 2018), which is now heavily supporting thousands of devices over both public and private connection points on a daily basis. That network availability and security are both top of mind for the state Chief Information Officer's office and efforts to enhance that network are ongoing as well as increasing.

This seemed a good start but not enough to write about yet. Then I glanced at my fitness tracker on my wrist to check the time. It was an hour off because I had apparently not yet synched it to its native mobile application since returning that day from business travel. I opened the fitness app on my phone, swiped down to start the synch process, and watched all the data start to populate: steps, hours slept, minutes in motion, and most importantly to me at that moment, the time of day: all the relevant data I would expect. And then, I suddenly realized that I had become the primary node on my own network of devices, apps, wearables, and other collected data that all talked to me and influenced my day, my habits, and my thinking. I had personally become constantly connected: an internet of me, if you will.

Plugging In to the Network

A note on the text: this piece will by no means go into technical architecture or discussions on the future of fog computing as an intermediary technical methodology to manage thousands of connected endpoints (Sarkar, Chatterjee, & Misra, 2018). It will also not address how latency can be an issue with the constant connection to, and chatter from, thousands of controllable devices (Petrakis et al., 2018). This piece will, however, talk about how I have learned to embrace the fact that, with the appropriate control over my personal devices and the apps that translate data from those devices or connect to other apps to create synergies of data collection, I can more thoroughly monitor and control the endpoint that is me. This does slightly stray from the traditional meaning of IoT by altering the endpoint under analysis but, as of this writing, smartphone users worldwide are estimated to over 5 billion (Statista, n.d.). That volume itself cannot be underestimated as a reasonable exploration for study as well as for establishing baselines and knowledge bases of what all that collected data means: a kind of diagnostics for humanity.

Debate continues regarding if smartphones even constitute IoT devices given that they are computing devices and not physical endpoints (Duffy, 2014) but, for the purposes of this exploration, a smartphone is used as a device that can connect to and across IoT devices and display the data that those sensors collect and aggregate in order to present that sensor data in a usable way. I will argue that the smartphone is in fact part of the IoT ecosystem when it is used as the user interface and dissemination device for collected data. I also argue that secondarily, since smartphones contain sensors like GPS antennas, gyroscopic sensors, and movement analyzers, that smartphones need to be considered IoT devices given that these features are integrated into smartphone hardware.

Let me explain how this started. Like many employers, my workplace encourages employees to participate in a wellness program. It is simple really: get a few thousand points collected over the course of the year and one can qualify for a different tier of pricing for health insurance, including lower monthly deductible payments and higher employer contributions to health savings accounts. So, how do you get points? Get a health screening, visit a dentist, maintain healthy cholesterol, and most importantly, log steps. Honestly, how hard is it to get steps? So, immediately, the principles of gamification apply in this situation, and since studies show people tend to care more and try harder when challenged with playing a game (Cotton & Patel, 2018), I was immediately on board. Personally, I know that the competitive streak in me is activated the moment someone dares me to win, so this was easy to get started. I bought a fitness tracker for my wrist and downloaded the corresponding app. This is probably where I pause and say my smartphone is my best friend and the enabler of all this data and activity collection. This is the platform where all the needed apps that connect to wearables and other hardware required to conduct my data collection experiments take place.

A smartphone and good Wi-Fi or cellular access make all this possible. This setup is basically my IoT bring your own device approach: BYOD IoT, I suppose.

While I was out walking to get points, I began to realize that this fitness tracker had the possibility to change my habits and possibly my life. First, it measured heart rate. What I started to notice was that my heart rate was terribly high, in my opinion. It would be in a reasonable range on days where I would jog for a few days in a row and try not to eat fast food lunches. It would be substantially worse when I had nights where I would stay up late studying or working (more on sleep in a minute) and would take at least a few days of regular sleep, workouts, and eating regimens to get back in a more normal range. I am not a performance athlete, so I have never really cared about my heart rate. I am not sure I really do now, but it was the first metric I saw and paid attention to that captured more complicated metrics than simple steps, which oftentimes could be so inaccurate that an occasional hour on the lawn mower could often translate erroneously into miles of physical movement. And, although I appreciate the points, steps alone did not feel like a reliable metric.

Continuing to explore my possible capture of new data, I then noticed that this sleep tracker measured sleep and stages of sleep. Interesting but, at first, I was not sure that I really cared about this metric either. When I started over the course of a long summer to notice I really felt groggy and unrested in the mornings, I noticed that my sleep tracker was displaying several dozen periods at night where it was showing I was awake. That is not something I would have guessed on my own or realized without seeing the data. I noticed that these two metrics—sleep and heart rate—appeared somewhat correlated. This prompted me to go see a doctor (more points collected by me for that checkup, by the way) to make sure nothing was wrong. As it turns out, like many Americans I have pretty solid allergies and other interesting chronic sinus issues, but hey, easily fixable and now under better control. Now I sleep better and feel like I can typically track those metrics and adjust habits as needed. This is important because I was now relying on this fitness tracker, my smartphone, and a fitness app as part of my daily routine for wellness (Huang, Chen, & Kwon, 2018). I had been tricked into monitoring myself! And it was working!

Now that I had started to see some need for these self-diagnostic measures, I began to dig deeper into both the wellness app and the fitness tracker app to see what else I could collect to get precious points. Some ways were easy: run a 5k, upload a picture of the race bib, and collect a few hundred points. Or start a fitness habit and take the stairs for a month, prove it by uploading a picture, get 20 extra points. Sleep 7 h a night, get a few points. And it continued: log a food diary, get 5 points a day. This of course resulted in me downloading a separate food tracker app and using it consistently. That food tracker also had a barcode reader so part of the fun was, and continues to be, scanning in different food labels to see if the database contained the item and if it could provide correct label and nutrition information. Surprisingly, it seems to be consistently accurate most of the time.

Expanding My Network

I should have expected it—logging food got a bit techno-creepy: I started to get advertisements and article posts that looked representative or similar to the things I was logging. Log a granola bar and a banana, get an advertisement for yogurt and a blog post about how to make easy breakfast overnight in a thermos. Log a piece of steak and a beer, get a story about how to steam vegetables to lose weight. The trade-off was that now I knew if I was hitting my current caloric intake target and balancing my macros. Importantly, I was still getting those daily food points counted in my fitness tracker. I was creating a pretty accurate picture of me, at least through my movement and consumption habits. At that point, I could have probably produced a report and drawn a relatively accurate graph of how I was spending my days.

I mentioned previously that I was using the fitness app for running 5 kms but what I needed in addition to the steps counter was an accurate GPS and course tracker to monitor running distance and performance and to see if I was improving as a runner both in distance and in speed. And, like any addiction, I had a strong desire to push it as far as I could. After my first run, I noticed a difference in the output between the two trackers, so I downloaded two more running GPS apps to triangulate the accuracy. I have no idea which one is most accurate but "guess what?" They all have gamification concepts built in such as rewards, stars, noted wear on shoes, and social media presence. I kept them all and still run with at least three tracker apps open at the same time. The only drawback is they all try to play music simultaneously and, due to the constant GPS signal, require substantial usage of my phone's battery.

During a long summer, the temperatures got high enough that running and tracking and getting steps in was not enjoyable so I found myself spending more time inside both at the office and in general. After several weeks, I noticed that again my heart rate was rising, and my sleep was down substantially. Also, I was not earning the type of daily points I had gotten accustomed to; psychologically, I started to feel a bit defeated. I had been trained to look at and care about these metrics, and I was not where I expected to be. I needed to try something else. I started using a rowing machine, given that it was an inside activity, but it was not adept at logging steps. Sure, I was exercising, which was good, but I wanted to have better diagnostics on physical activity through the translation of points—after all, that was the goal: get as many points as possible. I noticed that my fitness tracker had an active minute's metric that was keeping score during that time but it was not yet translating that output to overall progress on the wellness app (that has since been fixed and now works well).

To counter this point's deficiency, I did some research on other fitness tracking devices and found that a few bicycle manufacturers make cadence and distance monitors that attach directly onto a bike's frame. I was in. I immediately bought a bike, installed a tracker, and downloaded a bike tracker app. I then synched that app's output to my fitness tracker app and immediately I was seamlessly back to logging points on my wellness app. I was now leaving work at reasonable times to get 10–12 miles in each day to move the needle on the points. It could not have been easier, but while testing out this new gear, I had a running app on simultaneously and noticed that the distances were not identical. After some research, I downloaded another respected bike tracking app and started using that one as well. My data collection methods had probably become a bit overkill at this point. Although at the time of this writing the author has not tested it, that second app has video output displaying one of many world-famous courses and can vary resistance and incline position of the bicycle based upon course conditions when paired with an indoor smart trainer that connects to the app as well. Basically, you can use the app to control the resistance on the bike, which reports the news through the cadence monitor back to the app. Beautiful.

Accepting My Place on the Network

This examination might seem more about apps and phones than devices and things, but I am currently using my smartphone to do things like intentionally bypassing physical devices in favor of digital paradigms: skipping a physical step if you will. For example, my current most used digital manifestation of a previous physical device is using my credit card stored on my smartphone. With a simple thumbprint, most card readers know that my credentials combined with the device code it stores serves as a new physical manifestation of that card. This creates a new view of me as the cardholder that only exists when in possession of those digital tokens. This of course then involves creating new security protocols that work to ensure the device and transaction on the network are in fact valid and secure. Again, instead of me monitoring a "thing," the "thing" is in fact monitoring me.

Since this analysis has slightly sidestepped or deviated from the typical discussion or exploration of the IoT, I will note one recent example that helps tie all these experiences together: Amazon Go. Amazon Go is a retail store where users make no actual monetary exchange for goods and do not wait in any checkout lines, instead using a workflow where customers simply select intended items for purchase and leave the store with those items in hand. The products themselves are not the connected things; however, they are known to the store itself. I recently went to a Seattle location and believe it is the future of not just retail sales but product logistics, surveillance, and ultimately both employment and customer satisfaction. By downloading a simple app and attaching my stored credit card mentioned previously, I was able to walk in with a simple tap of a phone barcode; grab a bison jerky and a kombucha; and walk out with no product scanning, no money changing hands, and no waiting in line. In fact, they have a reverse greeter (a goodbyer?) on staff who stands near the exit and whose job it is to encourage you that it is in fact okay to leave. It is quite unnerving the first time to simply take an item and walk out of a store without performing any kind of validation of purchase of that item.

While marveling about this technology inside the store, I scanned the barcodes of the labels into my food tracker app, so I could get points. This example is a tremendous shift in the model of what IoT can be. I would argue that the entire store is the "thing," whereby Amazon can control product pricing, inventory levels, and availability of the store itself. A shopper, through usage of a smartphone app, can access the transactional functions of the store by removing its contents and leaving the premises. It is amazing how simple the experience is to the end user, even though I guarantee the required technology is beautifully complex.

Conclusion

In traditional IoT discussions, we tend to focus on endpoints. I continue to suggest that we as people, both contributors to and consumers of different data, are still the most important nodes on the network. Although we can get much more complex from here, I am positing that the IoT is much more complicated and detailed than just simple measurement devices. I am theorizing that the primary actor in the IoT is still us—people—by way of the access we have to data.

As described in this case study, I now personally collect a significant amount of data about me (or, my devices do). I now know how much I move, how many times my heart beats, how soundly I am sleeping, which roads I have never run, my fastest speed ever on a bike, my logins to the gym, my well-documented coffee addiction, my internet purchases, and my movie-viewing habits. This piece does not even consider the multitude of other data collecting sources now available, including complex social media sites down to simple IoT devices that monitor crockpot cooking time or home surveillance. I believe that if you combined all the data sources I personally populate or interact with, you would be able to describe the person who is me with the probability of creating a real predictive model of future analysis based upon those data sources; maybe even a 3D model. "Past performance is no guarantee of future results," is not a phrase that typically applies to population metrics. People are knowable. I am knowable. I am now fully part of the IoT. And I measure myself so I can get points.

References

Ammar, M., Russello, G., & Crispo, B. (2018). Internet of Things: A Survey on the Security of IoT Frameworks. *Journal of Information Security and Applications, 38*, 8–27.

CMS. (n.d.). Electronic Visit Verification (EVV). Retrieved September 19, 2018, from https://www.medicaid.gov/medicaid/hcbs/guidance/electronic-visit-verification/index.html

Cotton, V., & Patel, M. S. (2018). Gamification Use and Design in Popular Health and Fitness Mobile Applications. *American Journal of Health Promotion*, doi:10.1177/0890117118790394

Dorsey, T. (2018, March 27). 35 State DOTs Are Deploying Drones to Save Lives, Time and Money. Retrieved September 10, 2018, from https://news.transportation.org/Pages/NewsReleaseDetail.aspx?NewsReleaseID=1504

Duffy, J. (2014, June 26). 8 Internet Things That Are Not IoT. Retrieved September 15, 2018, from https://www.networkworld.com/article/2378581/internet-of-things/8-internet-things-that-are-not-iot.html

Huang, W., Chen, H., & Kwon, J. (2018, June). The Impact of Gamification Design on the Success of Health and Fitness Apps. In *The 12th China Summer Workshop on Information Management (CSWIM 2018).*

Indiana Bureau of Motor Vehicles. (2018). BMV Connect. Retrieved September 05, 2018, from https://www.in.gov/bmv/2793.htm

Kaiser Family Foundation. (2018, September 1). Medicaid in Indiana. Retrieved October 3, 2018, from http://files.kff.org/attachment/fact-sheet-medicaid-state-IN

Kranz, M., & IDG Contributor Network. (2018, July 31). The First Step to Starting an Enterprise IoT Project. Retrieved October 2, 2018, from https://www.networkworld.com/article/3293878/internet-of-things/the-first-step-to-starting-an-enterprise-iot-project.html

Petrakis, E. G., Sotiriadis, S., Soultanopoulos, T., Renta, P. T., Buyya, R., & Bessis, N. (2018). Internet of Things as a Service (iTaaS): Challenges and Solutions for Management of Sensor Data on the Cloud and the Fog. *Internet of Things, 3–4,* 156–174.

Sarkar, S., Chatterjee, S., & Misra, S. (2018). Assessment of the Suitability of Fog Computing in the Context of Internet of Things. *IEEE Transactions on Cloud Computing, 6*(1), 46–59.

Statista. (n.d.). Number of Smartphone Users Worldwide 2014–2020. Retrieved September 25, 2018, from www.statista.com/statistics/330695/number-of-smartphone-users-worldwide/

Weber, R. H., & Weber, R. (2010). *Internet of Things* (Vol. 12). Heidelberg, Germany: Springer.

Wortmann, F., & Flüchter, K. (2015). Internet of Things. *Business & Information Systems Engineering, 57*(3), 221–224.

Glossary

Chapter 1: The Yin and Yang of IoT

E-Waste

Once a device stops working or you decide to get another one, where does it go or what do you do with it? Do such devices just pile up? Electronic waste, E-Waste, represents the many ways in which individuals and organizations deal with ICT that is no longer of value to the users.

What Is It?

E-waste describes the buildup of discarded technology or electronic devices.

How Does It Work?

There are many ways that electronic devices can be disposed of, like recycling them or trading them back to the manufacturer for proper disposal. Other tactics are less responsible and harmful to the environment, like throwing them into the trash or just saving them in a box.

One M2M

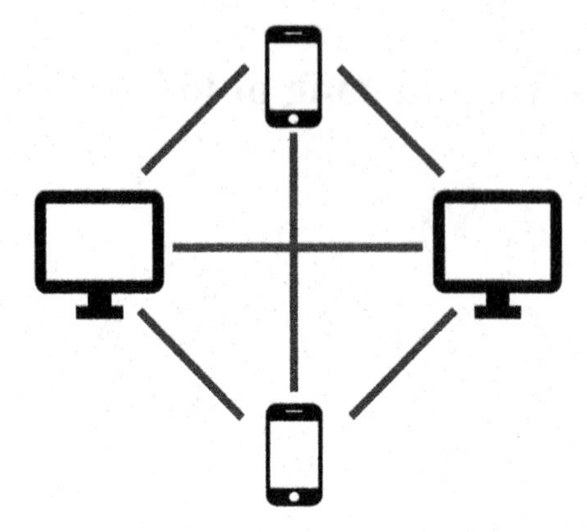

Devices and computing machines connect to each other and work together. What makes such connection possible? *OneM2M* or one Machine-2-Machine.

What Is It?

One M2M is a set of standards for developers that allow multiple devices to seamlessly connect to one another with relative ease. These standards govern all requirements that allow machine-2-machine and Internet of Things (IoT) device interaction.

How Does It Work?

A community of 8 standards development organizations and 200 member groups collaborated to produce the framework needed to support interoperability between IoT and machine technologies (OneM2M, n.d.).

Slow Tech

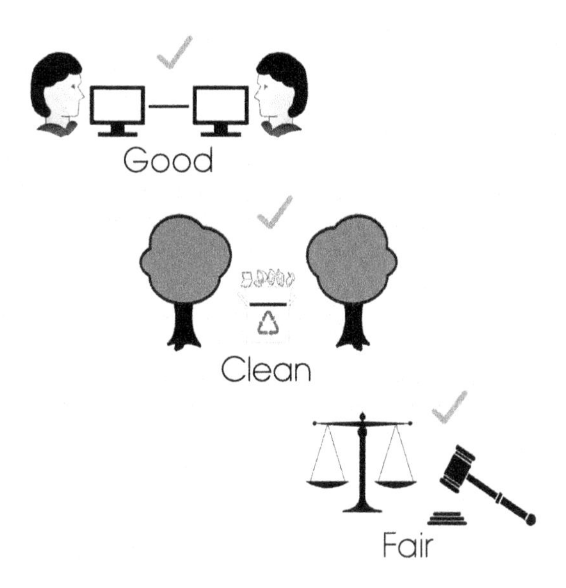

Is there a way for information and communications technology (ICT) to be both progressively innovative and environmentally considerate? The concept of *Slow Tech* works to promote corporate and individual social responsibility with ICT.

What Is It?

Slow Tech is modeled after the Italian Slow Food movement, a concept that emphasizes Good, Clean, and Fair practices. Using the lens of Good, Clean, and Fair technology practices, one can assess the implications of the technology life cycle from one minerals and metals acquisition to manufacturing to usage, to disposal or recycling. The Slow Tech movement promotes more conscious thought on the part of designers, manufacturers, consumers, and other purchasers.

How Does It Work?

The *Good* aspect examines the relation and connection between technology and the people and our planet. *Clean* speaks to the environment and the negative consequences from overproduction and lack of responsible disposal of electronic devices.

Fair analyzes corporate and individual social responsibility and how we should consider manufacturing and mining labor practices, addictive corporate practices (such as with applications), illegal disposal practices, etc.

User Experience

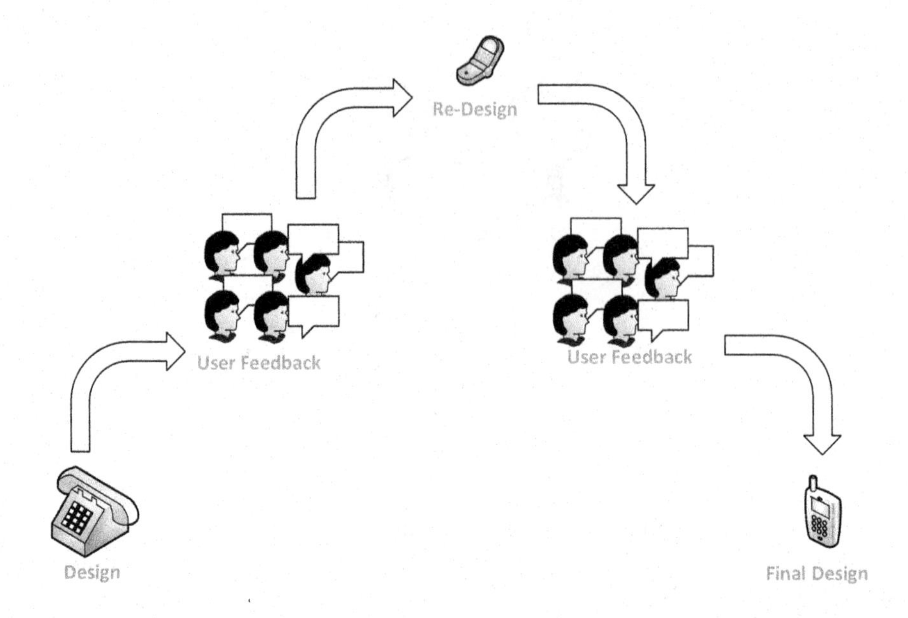

Why is it that certain objects attract more users or market attention than others? Consider, for example, the continuing operating system battle between iOS and Android. Users have preferences based on their *user experiences* with the devices.

What Is It?

User experience (UX) refers to the individual valuation or satisfaction that people have with devices or products from their perspective (not from the designer's perspective). The goal of a good user experience is to exceed user expectations through an analysis of user needs, design requirements, prototypes, and user evaluation and feedback.

How Does It Work?

Product designers conduct a series of surveys or interviews, or use observation, to understand the wants and needs of their users. This requires constant feedback, evaluation and reevaluation, and modeling and remodeling to obtain satisfaction from the user perspective.

Chapter 2: IoT System Integration

Analog Communications

What Is It?

Communications characterized by continuous signals. This includes the auditory and visual signals you hear and see.

How Does It Work?

It is the format biologically required by human perception.

Cloud Computing

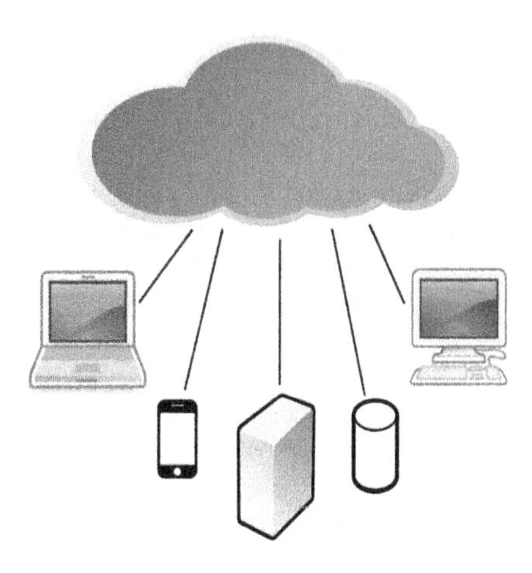

What Is It?

Cloud computing is the practice of locating an organization's computing requirements to server hardware located on the Internet.

How Does It Work?

This can include services such as running operating systems, storage, databases, software, and services like e-mail.

Communication Medium

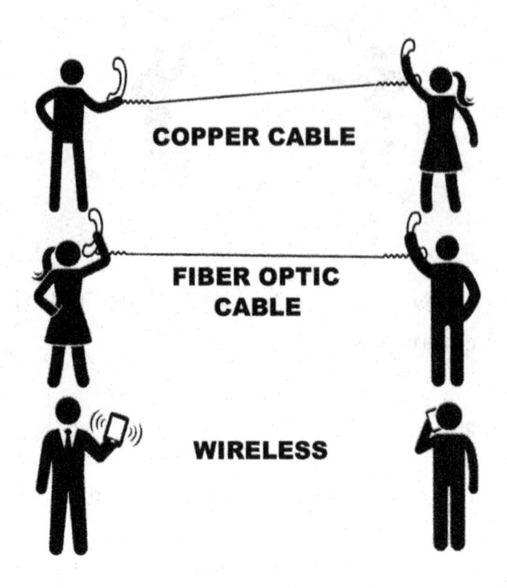

What Is It?

Communication medium refers to the carrier of a signal.

How Does It Work?

The three most common mediums used in data communications are copper cable, fiber optic cable, and through the air (wireless).

Digital Communications

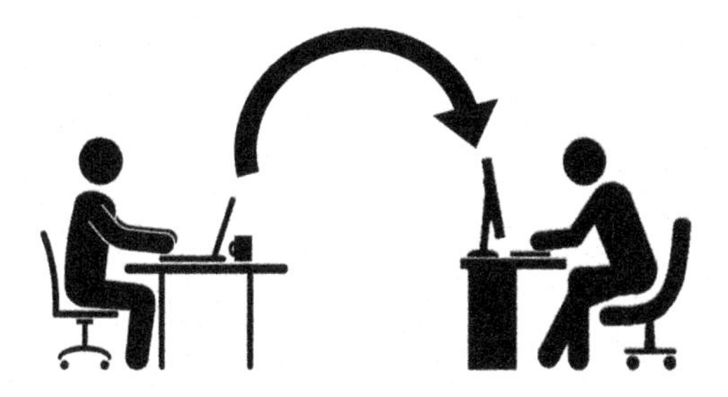

What Is It?

The process of data transmission in a binary format of ones and zeros.

How Does It Work?

Computers and other electronics can only interpret data that is created and transmitted in a binary format.

Embedded System

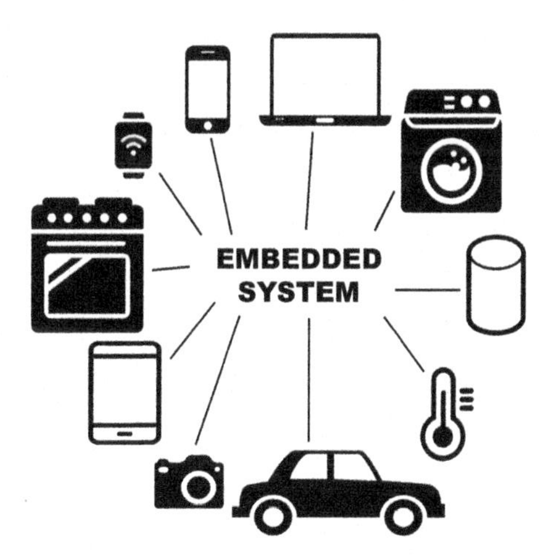

What Is It?

An embedded system is the inclusion of the hardware required to run a computer operating system on a device that is not a traditional computer.

How Does It Work?

It is these small embedded systems that make devices "smart."

IoT-Enabled Solution

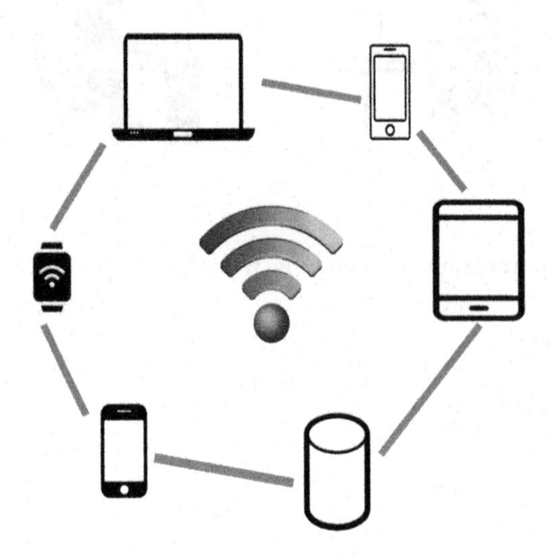

What Is It?

This is a process or service that is made possible through the embedded systems, sensors, communications, and other technologies associated with the IoT.

How Does It Work?

It works through the use of Wireless-Fidelity (Wi-Fi) and communications between devices to send, receive, and process instructions.

Radio Frequency Identification

Radio frequency identification (RFID) is a type of wireless communications.

What Is It?

This technology is frequently used for asset tracking and access control.

How Does It Work?

Items are embedded with an RFID tag that contains identifying information about the item. When the tag comes into close proximity to an RFID reader, that information is transferred wirelessly from the tagged item to the reader.

Chapter 3: IoT Positioning in the Verticals

Business Vertical

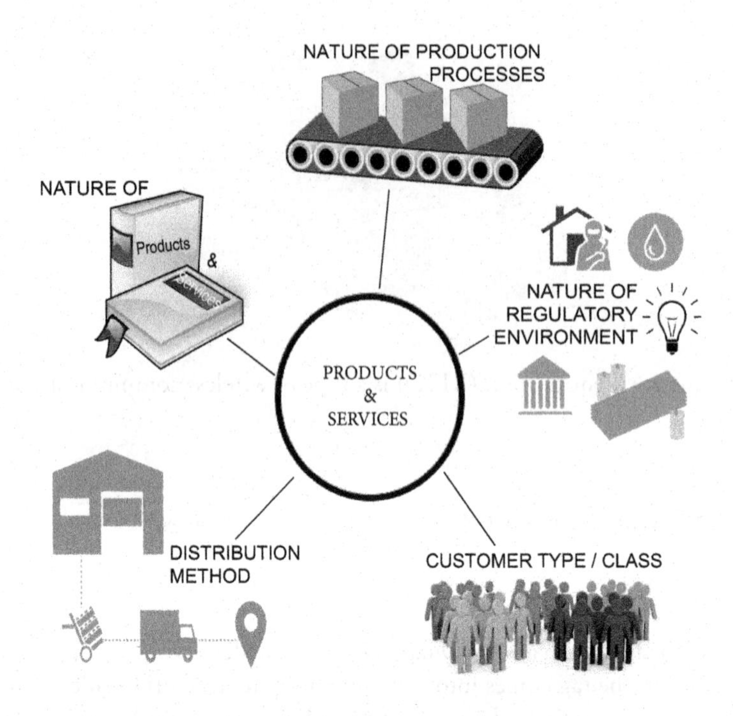

Have you ever wondered how businesses classify their products or services to sell to niche markets? Unique characteristics are used to distinguish and group together related products or services into *Business Verticals*.

What Is It?

A business vertical is a function unique to itself and the enterprise it operates. It assists with the supplying of independent products or services and assumes risks separate from traditional business verticals (GST Council Secretariat, 2016).

How Does It Work?

The business vertical for a product or service may be categorized with the following characteristics as quoted by the CA Club India:

- "The nature of the products or services
- The nature of the production processes
- The type or class of customers for the products or services

- The methods used to distribute the products or provide the services; and
- If applicable, the nature of the regulatory environment, for example, banking, insurance, or public utilities" (Jain, 2017)

The Banking Vertical

The financial industry offers a vast amount of services, and customers have diverse needs. How do banks satisfy the needs of customer groups? Offering services using *Banking Verticals.*

What Is It?

Banking and financial systems create products and services tailored to specific customer groups or niches (The Hindu Business Line, 2014).

How Does It Work?

A bank may offer products and services in a variety of areas that start with banking but also include, insurance, investment services, financial planning, and more (Munchenberg, 2018).

The Information Technology Vertical

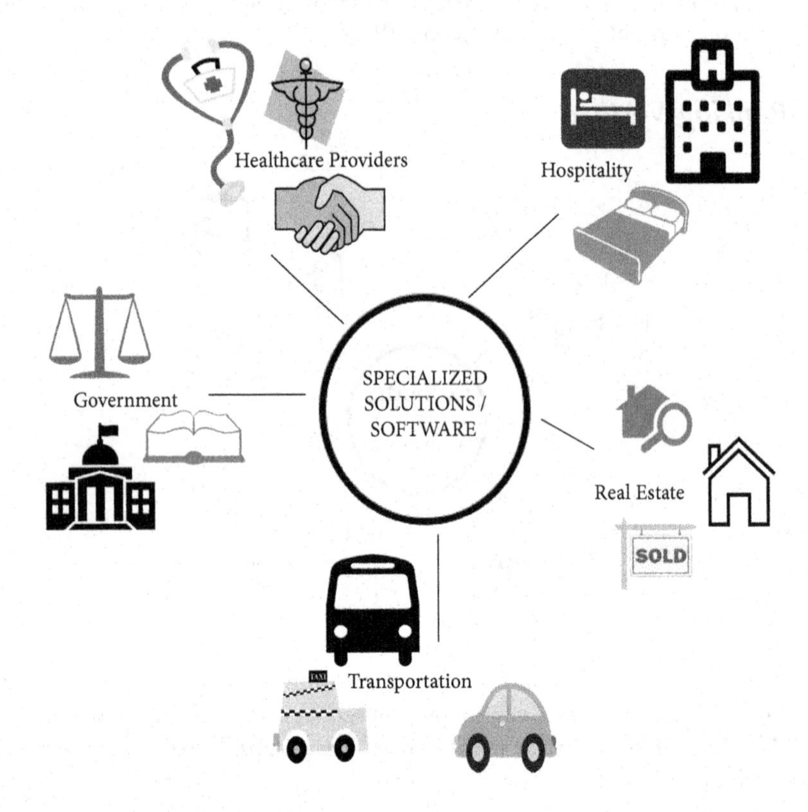

How do companies provide information technology to satisfy the needs of different customer groups? Through the creation of tailored products and services in the information technology (*IT*) *Vertical.*

What Is It?

Creating products and services that fit specific groups of customers in the IT market (Marketing-schools, 2012).

How Does It Work?

An IT provider may create specialized solutions or software to fit the needs of a group of customers or specific industry. Examples of customer groups may include healthcare providers, hospitality, real estate, transportation, or government (Rouse, Vertical Market, 2017).

The Manufacturing Vertical

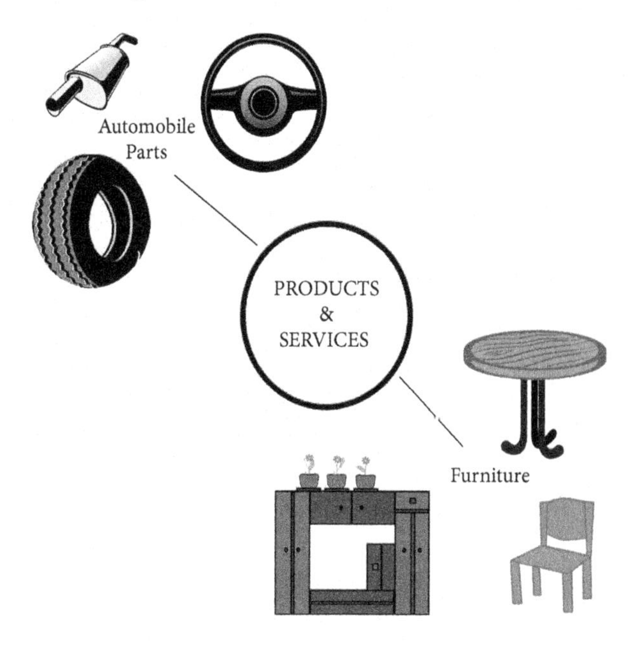

How do manufacturers create solutions to fit the needs of specialized customers? By using *Manufacturing Verticals.*

What Is It?

The manufacturing vertical focuses on creating products and services that fit customers within specific areas of manufacturing or the supply chain (Kumpe & Bolwijn, 1988).

How Does It Work?

By providing manufacturing verticals, products or services may be tailored to different customer groups. Some examples may include manufacturers that can reach customers within niche areas like automobile parts or furniture (Investopedia, 2018).

The Medical Vertical

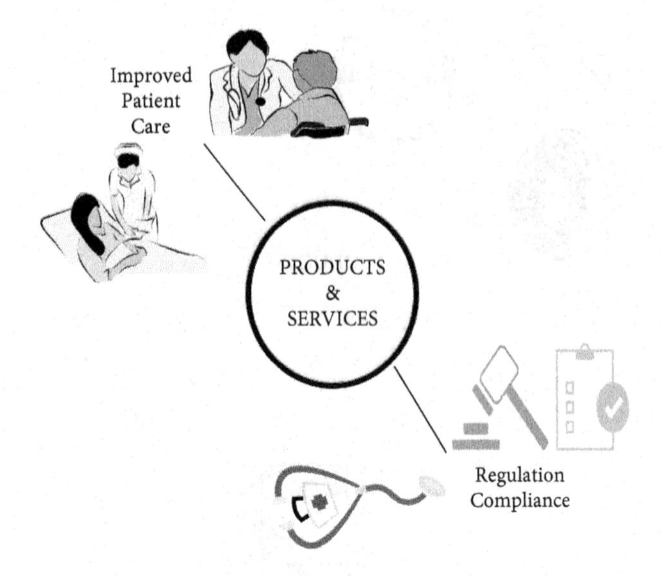

How are products and services customized to fit the needs of patients and health-care providers? The *Medical Vertical* helps channel goods produced to fit customers in this environment.

What Is It?

Organizations create goods and services that suit different providers and user groups within the scope of the medical field (Smith, 2017).

How Does It Work?

The healthcare industry has unique regulations and policies in place. Products and services offered in a medical vertical focus on the users in this space. Some examples of customization may include a focus on regulation compliance or improved patient care (Smith, 2017).

Chapter 4: Emerging Wireless Communication

Dense Wavelength Division Multiplexing and Coarse Wavelength Division Multiplexing

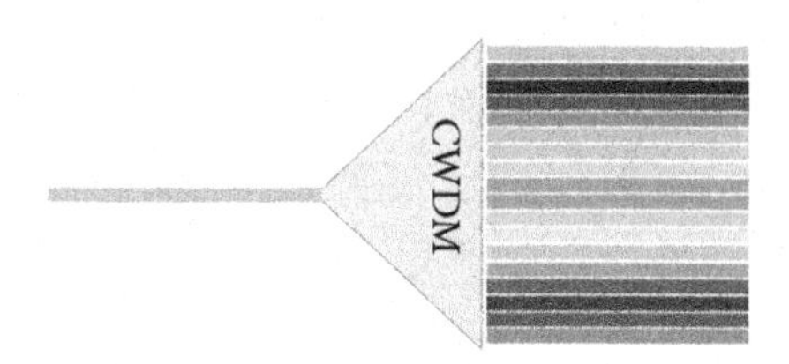

Have you ever wondered how to increase bandwidth over existing fiber networks? What can be done to make fiber networks utilize higher speed protocols? If so, you have been thinking about *Dense Wavelength Division Multiplexing (DWDM)* and *Coarse Wavelength Division Multiplexing (CWDM).*

What Is It?

DWDM and CWDM are equipment that allows separate wavelengths of a single light beam from a laser to be separated into multiple different channels over a dark fiber pair. For CWDM, the technology allows for separation into a maximum of 18 channels. For DWDM, the technology allows for separation into a maximum of 88 channels.

How It Is Used?

This technology is primarily used to dramatically increase bandwidth. It would be the equivalent of running multiple single-modal fiber to increase bandwidth. Rather than having to run multiple cables, DWDM and CWDM allow you to increase bandwidth without the need for extra fiber wires.

IP Address

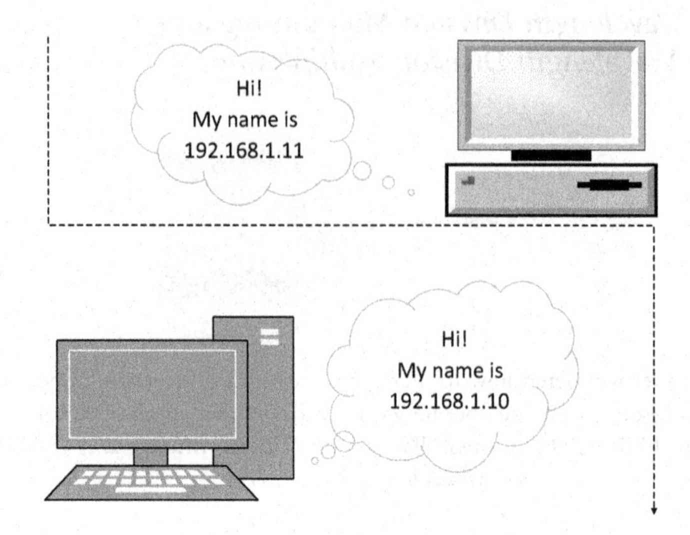

With so many devices connecting to the Internet, it is amazing that every bit of information can be sent to the right place. *IP addresses* are largely responsible for this efficient means of transportation.

What Is It?

The IP address, or Internet Protocol address, of a device is the logical address given so that it can communicate on the Internet. This address can change depending when and where you use the Internet—however, it is unique to your computer, phone, etc. Without IP addresses sending information along, a network would not be safe or reliable.

How Does It Work?

There are two types of IP addresses; IPv4 and IPv6. These two addresses are created using special numbering systems—they can be manipulated in many ways on private networks but are mostly controlled by Internet Service Providers. Both IPv4 and IPv6 work together through a host of different designs. However, IPv6 is newer and more future resistant—this is the system which we will begin to rely on in the future of IoT.

Multiprotocol Label Switching

Have you ever wondered how to speed up the forwarding process of routers? For example, is there a way to make routers forward packets faster, without having to analyze the network-layer header every time? If so, you would be thinking about *Multiprotocol Label Switching (MPLS).*

What Is It?

MPLS uses labels inserted in the addressing headers of packets to transport those packets across the national Wide Area Network facilities. These national networks employing MPLS switches can transport traffic arriving as IP Packets and carried by most layer 2 protocol facilities such as ATM (Asynchronous Transfer Mode), Frame Relay, Ethernet, and simple Point to Point.

How Does It Work?

MPLS is like reading the summary of a book. It takes a lot longer to read an entire book to find the information you are looking for. Instead, you read a shortened version to get the information quicker. This is essentially how routers treat packets with MPLS.

Network Striping

Have you ever wondered how large networks like 5G will get services to IoT devices? For example, how will services get to different IoT device categories? If so, you have been thinking about *Network Striping.*

What Is It?

Network striping provides end-to-end network virtual connection across the access/backhaul, metropolitan, and national backbone network for subscribers. There are separate logical virtual stripes for the different types of IoT categories. This network is structured as multiple logical planes. These planes are then allocated different virtual network resources and made available to the IoT device.

How Does It Work?

Think of network striping as sharing a pizza. Each person should get their own slice, but some should get more than others. This is based off varying factors (size, how much they need to eat to survive, etc.). Instead of pizza, network striping breaks the network into smaller pieces for each device based on requirements for the technology.

Orchestrators

Have you ever wondered what is controlling 5G networks? For example, who is controlling 5G and networking stripes that will support IoT and customized user requirements? If so, you are thinking of *Orchestrators*.

What Is It?

An orchestrator is the general label for the integrated software systems that will control end-to-end network facilities responsible for IoT support and network striping. Orchestrators automate behaviors within a network to meet required software and hardware elements to support services run on that network.

How Is It Used?

An orchestrator in networks functions just like an orchestrator in music. They control the functions of the specific groups. For example, like signaling the brass section to play, a network orchestrator signals which service goes to which IoT device.

Wireless Connectivity

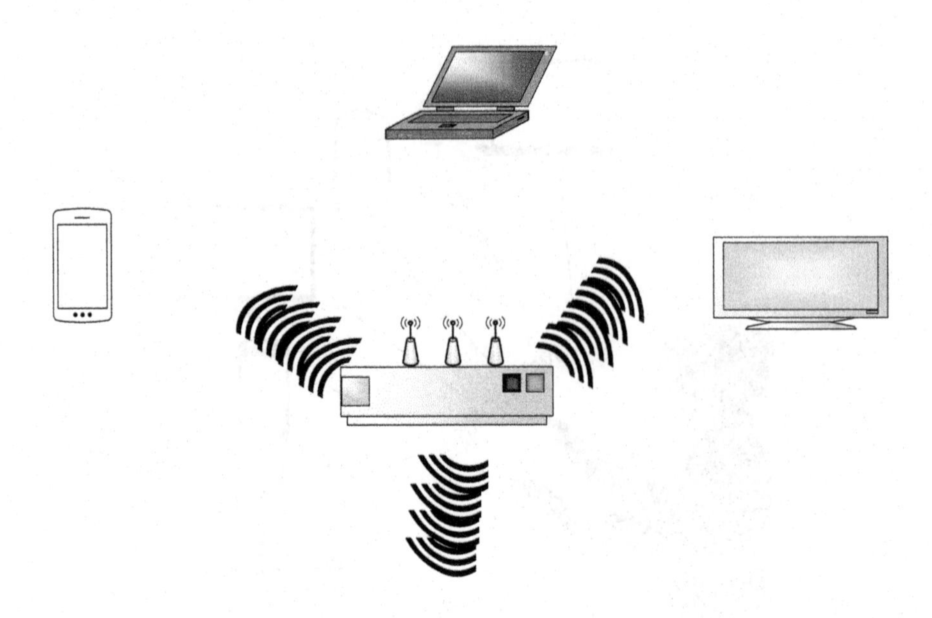

Have you ever wondered what makes your cell phone, computer, or any IoT devices connect to the Internet without cables? If so, you are inquiring about its *wireless connectivity* capabilities.

What Is It?

Essentially, almost every technological device remains connected not only to the Internet but each other, in varied locations, while also sending and receiving data information. Wireless connectivity for IoT devices can occur in the forms of Wi-Fi, low-power wide area networks (LPWAN), light fidelity (Li-Fi), and finally personal area networks (PANs). The wireless part is emphasized by the ability to connect without the use of a cable; it can be free to roam or move as needed around a specific zone. All areas of wireless connectivity provide unique characteristics for IoT devices to keep their connection alive.

How Does It Work?

As an IoT device learns its surroundings, different wireless connectivity techniques are evaluated by the designer of the products or the provider of the network. Certain techniques are better suited for specific devices and locations. Depending on where a device is located, and the extent of wireless connectivity needed, the technique with the strongest compatibility will work to provide network access. For this to function properly, the user must maintain connection to a power source and log into the network as prompted.

Wireless-Fidelity

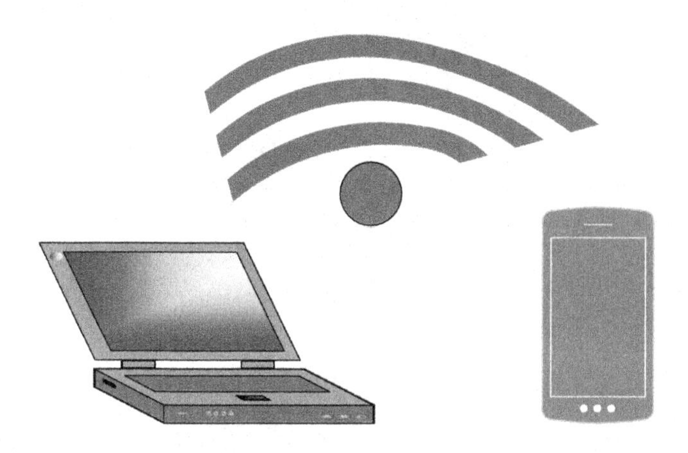

There is a type of connectivity for IoT devices that is the most common. Are you able to guess what it is? If you said *Wireless-Fidelity*, or *Wi-Fi*, you would be correct!

What Is It?

Wi-Fi is the ability for an individual to obtain broadband network access on numerous devices through a network of wireless access points and specific unlicensed radio frequency signals. Wi-Fi in the commercial sense was created in the end of the 20th century by the protocol standard "IEEE 802.11 alpha-numeric designation" (Cablefree).

How Does It Work?

Wi-Fi utilizes a main router hub to provide connections to everyday devices. The user will look for the available connections and sync their devices to the network. Wi-Fi for IoT devices has opened new paths for Internet access convenient to almost anyone.

Wireless Local Area Network

You might think that the Internet functions all together as one piece; but, it is the result of many networks pieced and configured to work together. A wireless local area network (*WLAN*) is just one type of network in the great scheme of the Internet.

What Is It?

In many cases, WLAN is synonymous with the term wi-fi—this is because wi-fi is the most common means of establishing a network that transmits data wirelessly. However, a WLAN can be more broadly understood as any network in a limited geographic location where data is sent and received wirelessly. These networks are commonly set up in homes, college campuses, and nearly any modern establishment.

How Does It Work?

A WLAN can be established using a variety of routers, switches, or other networking tools. These networks are given security settings which allow limited or no access to outsiders. Once your connection is established, networking devices regulate the transmission of any internal or external interactions you have on the network. The simplest WLAN to think of is the wi-fi network you likely have in your home.

Chapter 5: IoT End Devices

Actuator

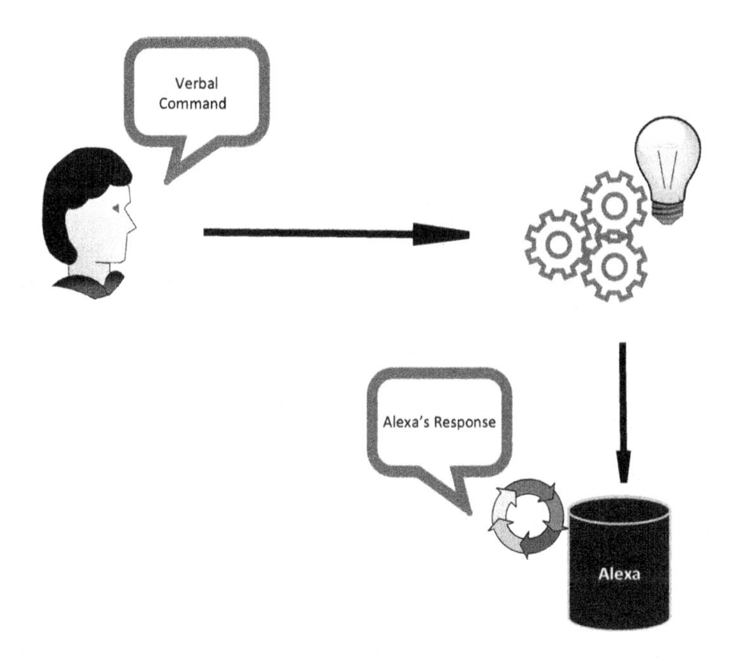

Have you ever wondered what controls the functions of your device? For instance, what allows it to open, close, or move? If so, you have been inquiring about an *Actuator.*

What Is It?

This is a mechanism used to control the functionality of your device. It is a small motor that transfers energy into motion (Actuator, n.d.). The range of motion required for the type of movement or direction specified is produced by an actuator; which draws energy from different sources depending on the system. These energy sources come in many forms: Battery (electric), liquid (hydraulic), compressed air (pneumatic), or heat (thermal) (Rouse, Definition, 2015).

How Does It Work?

You as the user of your "smart device," decide what action you would like your device to perform; open the blinds. When that command is entered, your device sends the command to the actuator. The actuator sends the respective energy source to be transmitted into motion that carries out your request (i.e. opening your blinds).

Analytics

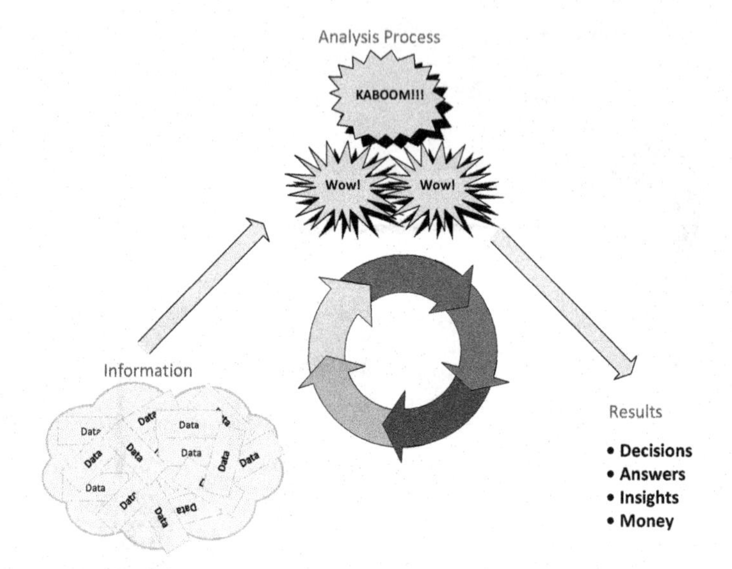

Ever wondered how data is turned into useful information? If so, you are wondering about *Analytics*.

What Is It?

Analytics are a series of methods used to determine patterns within data.

How Is It Used?

Data from various systems or items are collected and processed through a series of steps, algorithms, and tests. Results of these methods define information that is used to explain or discover items otherwise unknown.

Big Data

With the continuous growth and circulation of information flowing throughout the world, there must be a term that defines this exponential growth. What do you call massive amounts of information?... *Big Data*.

What Is It?

Big Data is just that, large amounts of information or data. It also encompasses the volume (magnitude), velocity (rate), and variety (variation of sources) that describe the overall build of the data that is collected (TechAmerica Foundation's Federal Big Data Commission, 2012). So how is this different than "regular data?" The only distinction is the magnitude of data compiled and the increased assistance and use provided to a number of industries to gain insight about a particular subject.

How Does It Work?

Information is collected from various sources (i.e. online, database, servers) and manipulated through a series of analytic techniques to produce an explanation, patter, or trend.

Sensor

Have you ever wondered how your device knows what to do and when? For example, how does your "smart thermostat" know when to decrease or increase the temperature in your home if no one is there? If so, then you are inquiring about a *Sensor.*

What Is It?

Simply put, this is a detection system for IoT devices. The sole purpose of a sensor is to observe its surroundings and respond accordingly to what your device is programmed to do (IoT technology stack part one—from sensors, actuators and gateways to IoT platforms, n.d.). This piece of hardware comes in various detective functions: temperature, pressure, chemical, motion and more (Tracy, 2016). How does it work?

At your home, you have installed a "smart doorbell" with camera capabilities. Therefore, when someone approaches the porch/door, ringing said doorbell, the sensor picks up the motion or presence of a body that sends a signal to the doorbell to populate an image to your smartphone, computer, or tablet indicating there is a "being" at your door.

Smart Sensor

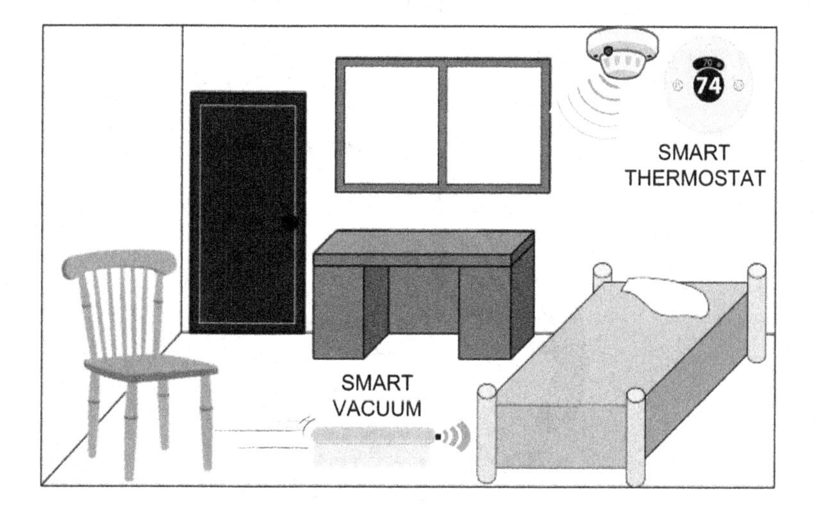

The environment that we live in today contains numerous elements, resources, and energy. What if there was a mechanism that could interpret the components of the environment and produce a specified action? This mechanism is called a *Smart Sensor.*

What Is It?

A smart sensor is a sensor that accounts for the resources, elements, and actions in the surrounding environment and performs accordingly (Rouse & Haughn, IoT Agenda, 2015).

How Is It Used?

The device scans the immediate area reviewing its contents. It compares the contents to the programmed functions to produce an action influenced by the external physical environment and what a person has programmed their device to do if certain conditions are present.

Chapter 6: IoT Architecture and Compatibility with Current Infrastructure

Evolved Packet Core

An important factor in enabling voice and data to live simultaneously on a network is Evolved Packet Core *(EPC)*.

What Is It?

EPC is a method of merging voice and data on a 4G long-term evolution (LTE) network. This convergence of voice and data exists on an IP service architecture and enables features such as VoIP or Voice-Over-Internet Protocol (Rouse, n.d.).

How Does It Work?

EPC combines transfer modes such as synchronous transfer mode and ATM in existing networks. This helps to integrate networks, and create networks, that are both high capacity and high performance that function on an IP-based structure (Techopedia Evolved Packet Core, n.d.).

Fifth-Generation Wireless

If you have questioned what the new development in wireless connection will be—wonder no more—it is fifth-generation (5G) wireless.

What Is It?

Fifth generation is the version of wireless that will expose users to a virtually limitless amount of data faster than ever before. It will perform better than the newest versions of fourth generation (4G) by providing more reliable signals that have less attenuation due to various obstructions (e.g. trees, buildings, air). This form of wireless connection will be the foundation for supporting the efficient functionality of IoT devices (Shankland, 2017).

How Does It Work?

5G wireless surpasses 4G in functionality due to its method for transporting signals. While 4G relies on large high-power cell towers to send signals over long distances, 5G lives on signals sent over shorter distances via small cell stations. The cell stations for 5G will be positioned in closer proximity to each other than cell towers; "located in places such as light poles or building roofs" (Gerwig & Haughn, 2016).

Gateway

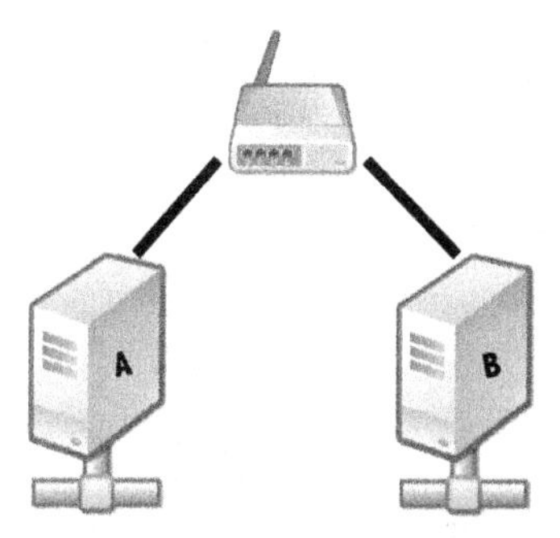

If you have ever wondered how networks are connected, the answer is through a *Gateway*.

What Is It?

A gateway is used for connecting two separate networks or connecting a network to the Internet. Through a gateway, devices on different networks can communicate. The typical device used as a gateway to connect networks is a router. However, there are times where small businesses may use a computer as the designated gateway to access the Internet (Mitchell, 2018).

How Does It Work?

A gateway is a common method for transporting data between networks and to and from the Internet. Various nodes such as servers, modems, or routers can serve as a gateway. While those devices may have different functions, as gateways each device can receive data from a host and transmitting it to a specified destination (WhatIsMyIPAddress.com, n.d.).

Metro Backbone

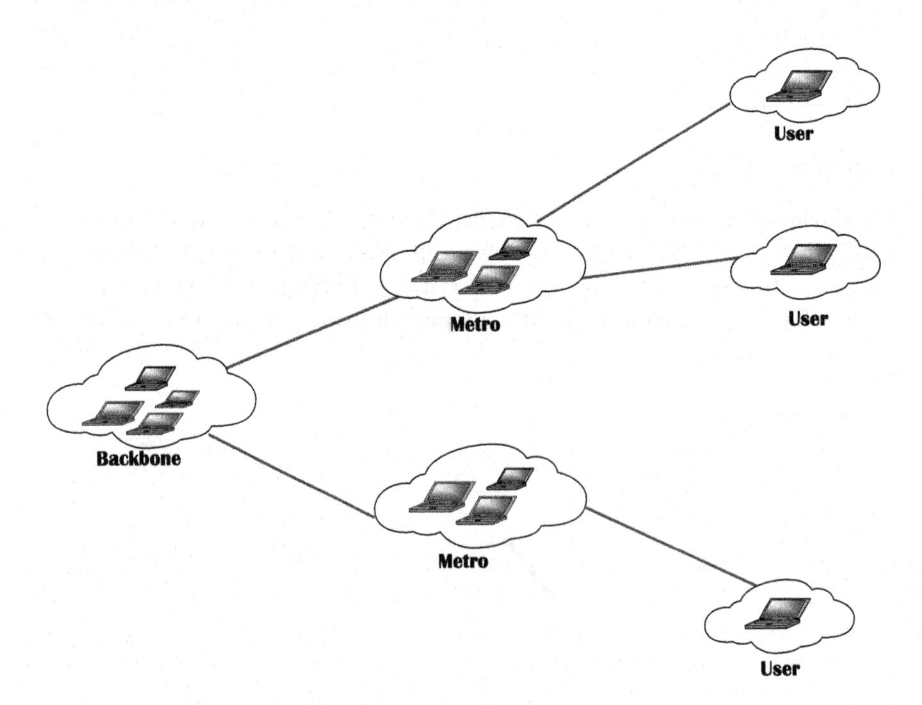

Have you ever wondered how traffic is managed for metropolitan area networks (MAN)? If so, the answer is through its *backbone*.

What Is It?

A metro backbone is a culmination of various nodes and mediums that work together to provide pathways for traffic on a MAN. The networking components of a metro backbone create a bridge that connects networks over long distances (Techopedia, n.d.).

How Does It Work?

For a metro backbone, the goals are to allow multiple local area networks (LAN) to communicate, handle the amount of data transported through those networks, and connect those networks to the Internet or other outside networks. To do that, metro backbones typically use fiber optic cabling to connect networking devices such as hubs, routers, or switches to facilitate the transmission of data (The Network Encyclopedia, n.d.).

Chapter 7: Ethics and Policy of IoT

Ethical Use of Data

Since data often contain information sensitive in nature, there must be standards that drive how data is used. To decipher the morality of data and how it is processed—*Ethical Use of Data* is considered.

What Is It?

A method to manage how information collected by companies and devices is used rightfully to protect users' information. These methods enforce privacy and accountability surrounding data.

How Does It Work?

Organizations and policies are implemented to provide transparency behind the use of an individual's data. This ensures that users truly understand what they are "signing up for" when their information is taken.

Four Laws of Robotics

With the growth of assisted electronics such as robots, governance must be implemented. To control the safety of robotic use—Four Laws of Robotics were created.

What Is It?

The Four Laws of Robotics are a set of regulations that protect the interaction between human beings and robotic machines. These laws make it so that the robotic machines do not commit harmful acts towards humans.

How Does It Work?

When robots and other autonomous systems are created, reference of four laws are used to govern the programming of such devices to ensure safety of their actions. Four laws are listed below:

1. A robot may not injure a human being or, through inaction, allow a human being to come to harm.
2. A robot must obey the orders given to it by human beings, except where such orders would conflict with the First Law.
3. A robot must protect its own existence as long as such protection does not conflict with the First or Second Law.
4. A robot may not harm humanity or, by inaction, allow humanity to come to harm. (Do We Need Asimov's Laws?, 2014)

Opt Out

OPT-IN

OPT-OUT

How can a person remove themselves and their information from company interaction? Through an *Opt-Out* option.

What Is It?

An option that allows the user to decline the use of their information being collected and shared.

How Does It Work?

The user signs a waiver or release form citing their disagreement. The company or vendor of the device logs this information and follows the non-consent of data usage rules.

Vagueness

With the number of devices in circulation, there are many categories of operation. Have you ever read a set of instructions hoping to define what a device is or how to operate it but was left in the same position as before reading them—confused? The state of the instructions is represented by *Vagueness.*

What Is It?

Vagueness is the lack of clarity of a device. There is no outright distinction that allows a device to be put in a specific operational category (i.e. virtual or physical devices).

How Does It Work?

A device is created with a mixed set of characteristics leaving it to be considered a tool of multiple categories.

Chapter 8: Empowering Older Adults with IoT

Enabling Technology

Have you ever wondered what drives change for a user or culture? What is the concept that allows innovation and invention for the next great piece of technology? This process is called *Enabling Technology*.

What Is It?

Enabling technology is characterized as a rapid development and application of unique, generic, or specialized technology within industries and markets. The idea of creating a smartphone feature that is unique and solves a human problem is a form of enabling technology. Combining innovative ideas with technology advances creates applications for enabling technology.

How Does It Work?

The rise of the Internet has allowed small or local business owners to compete with other similar businesses around the world; enabling technology allowed this opportunity to occur. Understanding the supply and demand within a market allows for individuals or companies to create innovative technologies that will help create new business landscapes.

Gerontechnology

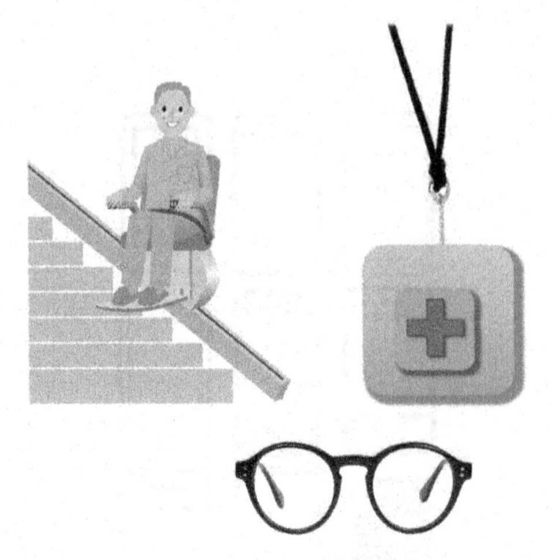

Have you ever wondered if there were designated groups for each societal generation? We have all heard of the following generation categories: Baby boomer, Generation X, and Millennials. The technology-oriented category for the aging adult is called Gerontechnology.

What Is It?

Gerontechnology is research aimed at improving elders' health, social participation, and independent living. Gerontology is the study of aging adults living in our society, while technology is the application of potential new ways to improve daily living and social participation within the aging adult community.

How Does It Work?

Designing and developing beneficial products and services will increase quality of life for aging adults. Gerontechnology research can be applied to disease prevention, endurance, mobility, maintenance of strength, and other physical or cognitive abilities.

Smart Cities and Communities

If you have wondered how many cities and communities are adapting to the rise in urban populations globally, then you are thinking about a *Smart City.*

What Is It?

Smart cities and communities use information and communication technology that will increase operational efficiency, share information with the public, and improve quality of both government services and citizen welfare (Rouse, 2017).

How Does It Work?

One example of a smart city/community initiative is using a cell phone application that connects to parking meters within a city. This application would help drivers find available parking spaces without prolonged circling of crowded city blocks. The smart meter also enables digital payment, so there's no risk of coming up short of coins for the meter (Rouse, 2017).

Smart Device

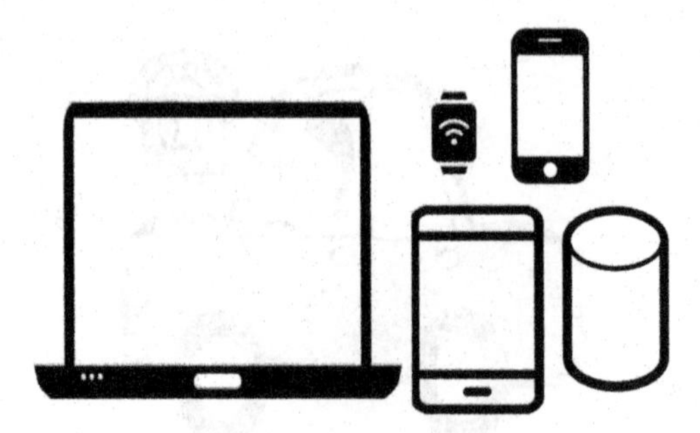

What is a smart device and how do they differ from any other device? With technology advances, many devices used in our daily lives have become interactive and self-directed.

What Is It?

A smart device is an electronic gadget that interacts and connects with its user and other smart devices. Smart devices understand simple commands and help in daily activities (Google search). Predominantly small in size, most smart devices are watches and smartphones (i.e. Samsung Droid, Apple iPhone); however, portable laptops and tablets are considered smart devices as well. The ability to connect devices with the Internet, take notes, and/or listen to music are some commands familiar to a smart device.

How Does It Work?

If you want to search for local restaurants to place a "pick-up order" on your smart device, you will type the request into google or ask "Siri" to find local restaurants. This command allows cell phone towers to use the "GPS" tracker in your phone to find local restaurants. Once you find a restaurant, the application on your smart device will provide current phone number, street address, menu options, etc. After you "call-in" your order, you then can use the same application to receive directions to the restaurant.

Smart Nation

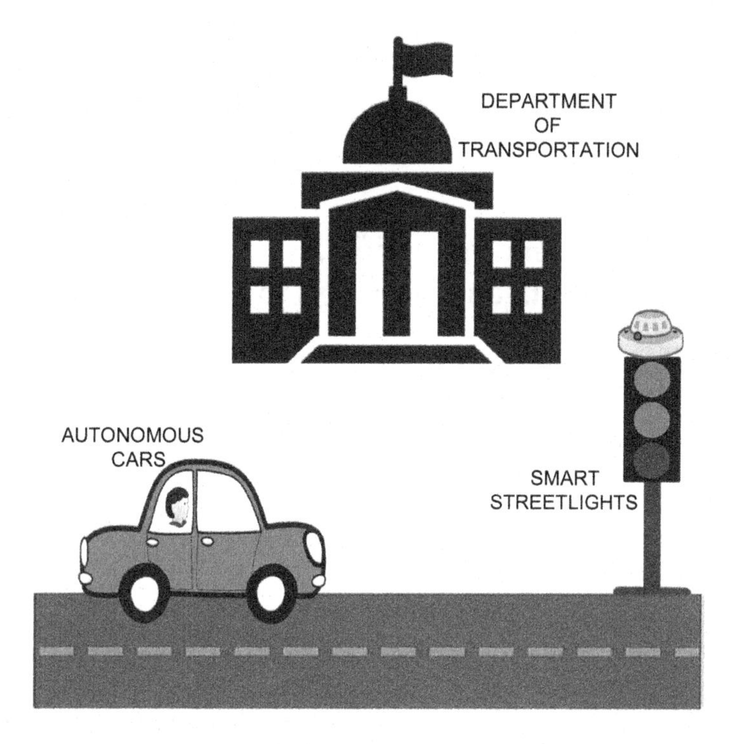

As ICT innovations advance, cities, states, and nations have the opportunity to define and embrace policies and strategies to take advantage of innovations for the benefit of society and the economy. Harnessing digital technologies will improve living conditions and create closer communities; this will allow citizens to achieve a higher quality of life, will create new job opportunities, and will encourage businesses to innovate and grow.

What Is It?

Smart nations will bundle government services from across different agencies to help citizens within the different stages of their lives. Smart nations will include artificial intelligence (AI) and autonomous vehicles to increase reliability of public and private transportation. Smart nations will also incorporate "smart streetlights" and sensors within roadways to be more energy efficient while making citizens more secure. Smart Nations will allow citizens to go cashless with ease within a better online protection platform.

How Does It Work?

Smart nations have the same concept of smart home devices (refrigerators, washers and dryers, and smart phones) at a global level. From window blinds to household water heaters, to crowd/traffic control, Smart nations allow everybody to prepare their daily activities with less disruption or chances of forgetting what needs to be completed.

Chapter 10: Redefining Artificial Intelligence through IoT

Algorithm

$$r_i = \left(c_i - 1 + \sum_{j+k=i} a_j b_k\right) \bmod D$$

$$c_i = \left[(c_i - 1 + \sum_{j+k=i} a_j b_k) / D\right]$$

$$c_0 = 0$$

Have you ever wondered what allows non-intelligent computers, such as robots, to use AI to execute complex tasks? For instance, what allows a robot to walk, run, and converse as if it has a human brain? If so, you have been inquiring about the basic backbone of computing known as an *Algorithm*.

What Is It?

Algorithms contain formulas and act as procedures for computers to conduct a sequence of actions to solve a problem (Rouse, 2017). Algorithms serve as complex logical instructions that are written in formal languages that contain mathematical formulas and various computer commands that allow computers to execute problem-solving tasks. Algorithms act as a set of guidelines or formal rules that enable the computer to logically solve problems and perform the functions it was designed to do.

How Does It Work?

Algorithms have been created to help computers play games, such as chess, to rapidly learn and master the game. In fact, some supercomputers that use AI have only been provided basic rules and have become masters within 24 h by repetitively playing and learning the game in an accelerated fashion (Silver et al., 2017).

Machine Learning

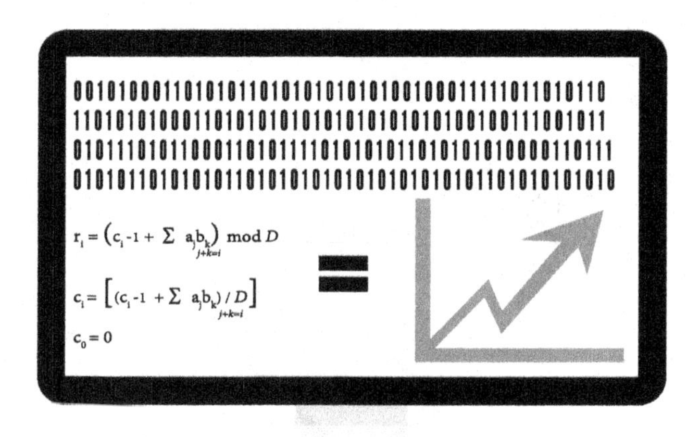

Have you ever wondered what allows a computer to use algorithms to analyze data and make advanced predictions? For instance, what if an organization used AI to analytically study past consumer tendencies to project future spending? If so, you have been inquiring about the scientific practice known as *Machine Learning*.

What Is It?

Machine learning is the complex scientific practice that allows computers to use algorithms on large data sets to extract patterns and create future predictions in an automated fashion (Herbrich, 2017). Machine learning occurs because computers utilize algorithms to parse data, which results in learning to enable the computer to make predictions. As a computer becomes trained, the algorithms allow the computer to learn how to perform tasks and solve problems that it initially was not capable of doing (Copeland, 2016).

How Does It Work?

Organizations use machine learning through the creation and implementation of algorithms on computers by introducing new data sets to identify analytical patterns. Organizations can use machine learning to target new market segments, improve services, conduct deep research, and enhance AI computing capabilities. Machine learning empowers organizations because they can leverage AI to identify patterns and trends in Big Data.

Natural Language Processing

Have you ever wondered what allows non-intelligent computers to listen to a human voice and carry out the request? For instance, if an individual asks a virtual assistant on a smartphone to play a song by a particular artist, why does this happen? If so, you have been inquiring about *Natural Language Processing (NLP).*

What Is It?

NLP is an area of AI that pairs human communication with computers, so they can interpret, understand, and use human language to accomplish a task (SAS, n.d.). Powerful computing resources, complex algorithms, and Big Data allows NLP to occur to assist interactions between humans and computers (SAS, n.d.). NLP allows computers to determine the important details from a communication by interpreting the meaning of a message after hearing or reading an input (SAS, n.d.). The overall goal of NLP is to bridge the communication gap between humans and computers to complete the desired tasks.

How Does It Work?

There are different uses for NLP, which requires different techniques to be used. Algorithms, machine learning, and mathematics act as the foundation for NLP in AI and can be used in industries such as healthcare, business, and finance. Computers using NLP break down human communication into pieces to analyze the segments and create meaning (SAS, n.d.). An example of a human using NLP could be illustrated by someone asking a smart device for the current weather forecast in a specific area. The smart device will analyze the user's verbal request to identify the meaning of the communication. Then, the virtual assistant on the smart device will conduct a search and produce an output that informs the user of the current weather forecast for that specific area and time.

Neural Network

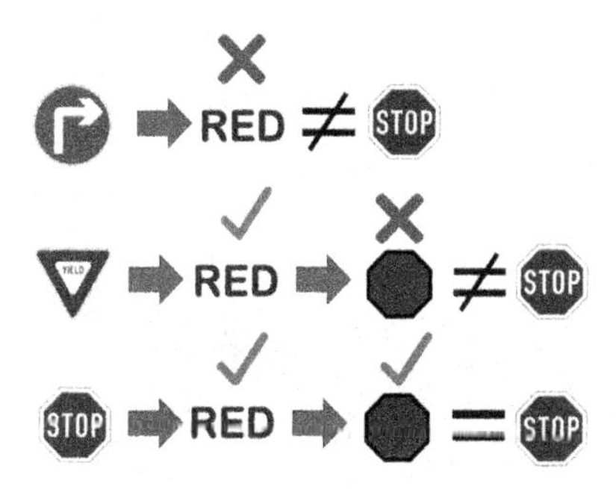

Have you ever wondered what allows AI to continuously learn and improve? For instance, what allows AI to solve complex problems that it initially does not know how to solve? If so, you have been inquiring about a *Neural Network*.

What Is It?

Neural networks can be classified in an area of computer science known as "deep learning." Neural networks are modeled similarly to the human brain where simple processing recognitions are closely interconnected and organized into layers. The algorithms determine if the information is significant in weighting and if it should pass a threshold to continually be evaluated by the neural network's deeper layers (Hardesty, 2017). Since neural networks undergo training where incoming randomized messages are evaluated, inputs are fed into the bottom layer and mathematically evaluated to determine if they should be recognized as significant outputs (Hardesty, 2017). As significant data continues to pass the threshold, the neural network becomes "trained" and can move to the inference stage where it can evaluate and classify information from other untagged data sets (Monroe, 2018). Today, neural networks are a driving force in AI largely due to the advancement of graphical processing units (GPUs) that have allowed neural networks to contain more layers to improve performance (Hardesty, 2017).

How Does It Work?

Neural networks can be trained to complete various tasks, such as identifying stop signs from image files (Copeland, 2016). Initially, an image is chopped and inputted into the first layer where individual neurons evaluate the images to assign weights that will be totaled once the inputted data has been processed and evaluated after being processed by the final layer (Copeland, 2016). In the neural network, each layer will have different tasks that work in an interconnected manner to determine if the pixels of a file represent the shape, contains the letters, and displays the color of a stop sign (Copeland, 2016). As the inputs are evaluated, the final mathematical weighting allows the neural network to use and probabilistically determine if the file contains an image of a stop sign (Copeland, 2016).

Reinforcement Learning

Just like humans, AI machines require invested time to learn and make improvements to its processes. In fact, humans and machines learn in almost the exact same way and are more similar than many people may be comfortable believing. One method of teaching a machine, just like a person, is to use *Reinforcement Learning*.

What Is It?

In reinforcement learning, humans assist in teaching the machines. This involves giving the AI a goal that is not defined with a specific metric. Instead of telling it to simply “improve efficiency” or “find solutions,” AI will run scenarios and report results, which are then evaluated by humans and given feedback. The AI takes the feedback and adjusts the next scenario to achieve better results (Greene, 2017).

How Does It Work?

This method of learning allows machines and software agents to automatically determine the ideal behavior within a specific context, to maximize its performance.

Transfer Learning

Another way machines can learn is through *Transfer Learning.* Once an AI has successfully learned something—like how to determine if an image features a specific breed of dog—it can continue to build on its knowledge on this subject, even if it's currently tasked to work on something else.

What Is It?

This method of machine learning can be better understood if you think about the way humans learn about the world around us. Humans transfer knowledge between tasks every day by applying relevant knowledge from previous learning experiences when encountering a new task (Torrey & Shavlik, 2009).

How Does It Work?

In transfer learning, there are three measures that are used to potentially improve learning in AI. The first is the initial performance that is achievable in the targeted task using only the transferred knowledge compared to the initial performance of another machine. The second measure is the amount of time it takes to fully learn the targeted task, given the transferred knowledge compared to the amount of time to learn it from scratch. The third is the final performance achieved in the targeted task compared to the final level without transfer learning (Torrey & Shavlik, 2009).

Turing Test

As technology advances year after year, many AI experts have concerns about safety as robots become more and more indistinguishable from humans. The *Turing Test*, named after its creator Alan Turing, is a test used to measure intelligence of machines and determine if the test taker is a human or a machine.

What Is It?

While the Turing Test was originally conceived as a way of determining if a human could be fooled by a conversation between a human and an AI, it has since become shorthand for any AI that can fool a person into believing they're seeing or interacting with a real person (Greene, 2017). While some people argue that this test has never been passed by AI up to this point, others say it has several times.

How Does It Work?

Conducting a Turing Test is simple. Put a computer (A) and a human (B) in one location and a human tester (C) in another. If the tester (C) cannot recognize which candidate is human and which is a computer after a series of questions, typically asked in a chatting interface, then the computer successfully passed the Turing test. Turing proposed that if the judge was less than 50% accurate, then the computer must be a passable simulation of a human being and hence, intelligent (The Turing Test, 1999).

Unsupervised Learning

Machines really are capable of learning by themselves, and they can use many layers of data and processing capability to do so. With *Unsupervised Learning*, we don't give the AI an answer, and the goal is to understand the underlying patterns of a data set (Lison, 2012).

What Is It?

Unsupervised learning is where you only have input data and no corresponding output variables. Rather than finding patterns that are predefined, in unsupervised learning, we feed a machine a large amount of data so that it can find whatever patterns it is able to locate (Greene, 2017).

How Does It Work?

There is no "right" answer to unlock in unsupervised learning. Algorithms are left to their own devices to discover, analyze, and present the structure in the data (Brownlee, 2016). This is done when an AI system is presented with unlabeled, uncategorized data, and when the system's algorithms have had no prior training. Unsupervised learning algorithms can perform more complex processing tasks than supervised learning systems but can also be more unpredictable. While an unsupervised learning AI system might, for example, figure out on its own how to sort cats from dogs, it might also add unforeseen and undesired categories to deal with unusual breeds, creating clutter instead of order (Haughn, 2016).

Chapter 11: Launching an IoT Startup

Angel Financing

Have you ever wanted a source for funding an idea you have but didn't know where or how to? If so, *Angel Financing* is one way to get funding!

What Is It?

Angel financing is the act of making private investments such as money, business advice, and contacts into a business to help the business grow.

How Does It Work?

Individuals called "angel investors" provide "angel financing." Angel investors are high-net-worth or wealthy individuals who have an interest in investing in and growing startup businesses and aspiring entrepreneurs. Aside from investments in money, angel investors also provide the entrepreneurs with advice and contacts.

The process of "angel financing" usually begins by entrepreneurs applying for funding—giving their business plan or executive summary to an angel investor or a group of angel investors for screening. After review, the entrepreneurs that show more promise are invited to give a series of presentations to the investors. The businesses that create the greatest interest go into a due diligence review process and screening. In the end, the winner(s) emerges, and the angel investors invest. Angel financing is one of the key drivers for startup and the growth of new businesses. The Angel Capital Association (ACA) lists over 300 U.S. groups in its database.

Articles of Incorporation

Do you want to start and register your business? To do so, you need to file *Articles of Incorporation* with the Office of the Secretary of State.

What Is It?

A set of legal documents needed to register a created corporation and are filed with the Office of the Secretary of State in the state where the business is to operate.

How Does It Work?

To form a business corporation, the business must file Articles of Incorporation within the state in which it is going to operate. These documents are filed so that the business is legally recognized as a corporation. The Articles of Incorporation typically include the name of the corporation, the name and address of the Registered Agent, the number of shares of stock the corporation is authorized to issue, the name and address of all the incorporators, the original signature of the all the incorporators, and a filing fee.

Business Plan

Do you plan to start a business? You should start drawing up your *Business Plan.*

What Is It?

A business plan is primarily writing down the detailed description of a new business, its objectives, and the steps needed to achieve those objectives.

How Does It Work?

When building a business, just like when building a house, a blueprint and its foundation are needed to start and keep it standing while it's being built. A business plan is just like a blueprint or foundation for your business. However, a business plan must be regularly revised because of the changing nature of the business world. Some of the essential sections of a business plan are an executive summary, a mission statement, a background of your company, a background of your industry, a description of your product/service, a competitor analysis, a marketing plan, a financial plan, a management plan, an operating plan, and a timeline.

Capital

Have you thought of how or where to fund your business? Well, what you need is *Capital*.

What Is It?

Capital is how companies invest in their business. Capital is money, assets, or other forms of wealth that can be spent on a business to create even more wealth.

How Does It Work?

Capital can be funded through sources such as personal savings, money from your family and/or friends, through credit, loans from the banks, and from investors like angel investors, and venture capitalists.

Deck (aka Pitch Deck)

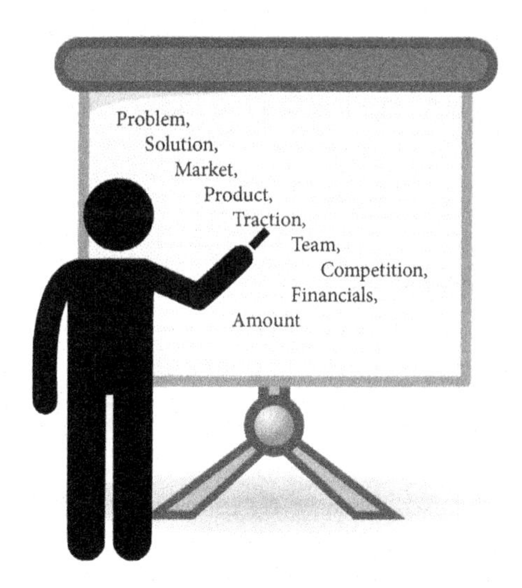

Have you thought about how to pitch your idea or business to stakeholders or investors? A *Pitch Deck* would aid you in pitching your ideas to others.

What Is It?

A pitch deck is an oral presentation of a market opportunity along with presentation slides that persuade potential business partners and/or investors.

How Does It Work?

Pitch decks contain text and visuals that pass across essential information in a clear, simple, and straightforward manner. A pitch deck includes headings such as the problem, solution, market, product, traction, team, competition, financials, and amount being raised.

Due Diligence

If you have ever had to get information about the risks that might be involved in doing business or going through with a transaction, then you have experienced *Due Diligence*!

What Is It?

Due diligence is a thorough investigation of a proposed investment and of different factors involved in that transaction before it is finalized to check out an investment's worthiness.

How Does It Work?

Due diligence can be used when you're in the process of buying a business. Here, you would want to investigate and evaluate the seller's industry and market history, business's reputation, inventory, legal documents, financial statements, and all other values that concern that business.

Family Lifestyle Business

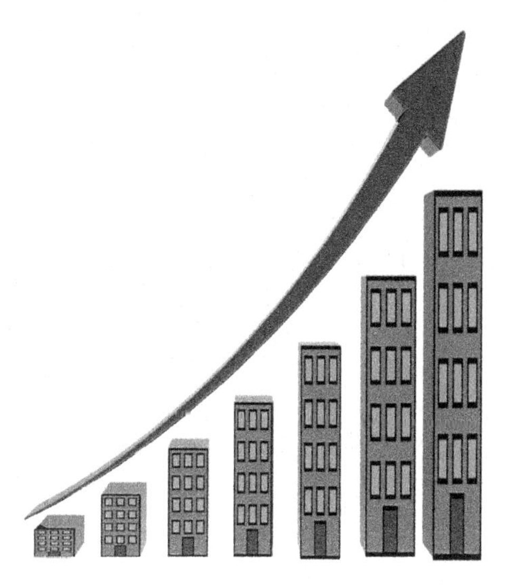

Have you ever been to a local bookstore? If so, then you have been to a *Family Lifestyle Business.*

What Is It?

A family lifestyle business is a small business with a slow and gradual growth plan that is mainly created to provide immediate income to its owners.

How Does It Work?

In a family lifestyle business, the owners usually retain ownership of the business and have control over its operations. Also, a family lifestyle business hopes to become profitable as soon as possible, instead of waiting for years to reach profitability.

Minimum Viable Product

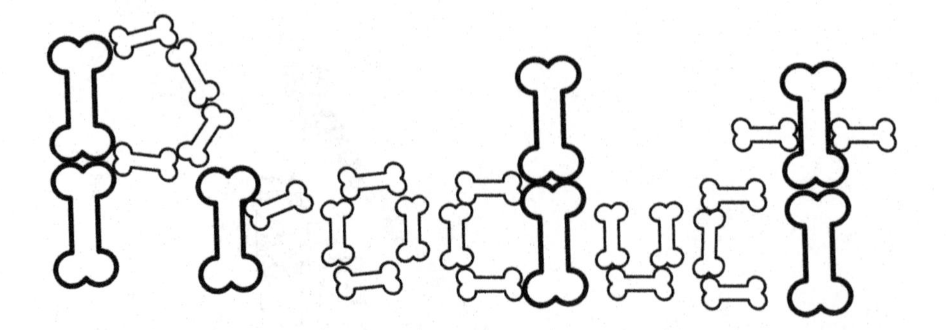

Have you ever wondered how a startup demonstrates its product's value in the early stages of the company's existence? If so, then you are inquiring about a *Minimum Viable Product (MVP).*

What Is It?

An MVP is a "bare bones" version of a company's product that demonstrates its market value and potential for growth. This product should also include development capabilities (Moogk, 2012).

How Does It Work?

An MVP must be developed and tested to add validity to a company product. It must be tested among a group of potential users which can help determine the product's potential for success in the market and determine whether the product can make money at the predicted scale (Divan, 2014).

Nondisclosure Agreement

Have you ever wondered how some companies manage to contain classified information or prevent leakage of company secrets? If so, then you are inquiring about a *Nondisclosure Agreement.*

What Is It?

A Nondisclosure Agreement, often referred to as an NDA, is a legally binding contract between parties that provides confidentiality regarding certain information. NDAs are often used when a company provides a trade secret for development, financial backing, evaluation, and marketing purposes (Nondisclosure Agreement, n.d.).

How Does It Work?

NDAs require that employees or partners protect employers' or partner's trade secrets. Those involved in the NDA are only permitted to share this classified information with those so authorized. They are used in any organization or profession that handles classified information (Wetter, 2017).

Sweat Equity

Have you ever wondered how to appropriately determine the value of a company beyond its financial investments? If so, then you are inquiring about *Sweat Equity.*

What Is It?

Sweat equity is defined as the equity generated in a project due to the hard work and efforts of its owner. A startup company may be more valuable for its time invested from owners than from its capital investment (Downes & Goodman, 2014).

How Does It Work?

Small business owners and entrepreneurs use sweat equity to fund their ventures. In order to accurately calculate the value of sweat equity, value must be placed on the time spent doing an activity or in furthering the development of the business (Sweat Equity, n.d.).

Term Sheet

Have you ever wondered how a company determines and negotiates the logistics of an investment? If so, then you are inquiring about a *Term Sheet.*

What Is It?

A term sheet, also referred to as a letter of intent, memorandum of understanding, or agreement in principle, is a nonbinding contract that determines the conditions in which an investment will be made. This does not finalize the investment but rather outlines the terms of the agreement (Amalyan, 2018).

How Does It Work?

Term sheets are typically negotiated during an investment opportunity. A term sheet can include items such as liquidity preference, redemption, anti-dilution provisions, exclusivity, and more (Zocco, 2007).

Traction

Have you ever wondered how to provide market value to potential investors? If so, then you are inquiring about *Traction*.

What Is It?

Traction is the proof one provides to potential investors that shows people are using and buying a product. It has been defined as "quantitative evidence of market demand" (Ravikant, 2011).

How Does It Work?

Traction is one of the most important factors to demonstrate when negotiating with an investor. Traction helps to summarize the amount of risk that may be involved. The more traction that can be demonstrated, the less of a risk the project is to the potential investor (Jensen, 2014).

Valuation

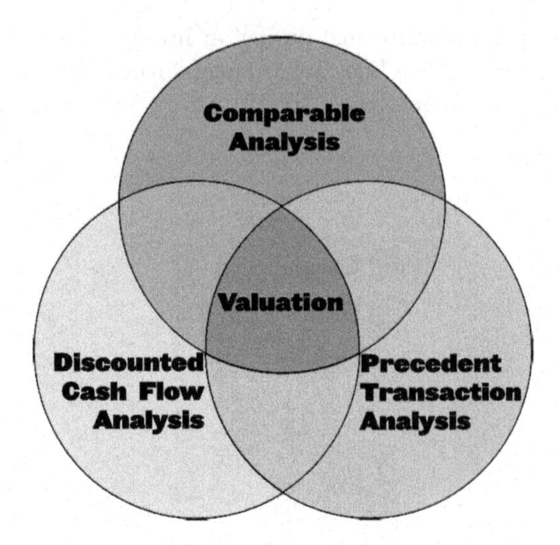

Have you ever wondered how to accurately determine at what a company is valued? If so, then you are inquiring about *Valuation.*

What Is It?

Valuation defines a company's financial worth. Valuation places value or worth on an asset (Downes & Goodman, 2014).

How Does It Work?

Valuation is the process of determining the value or worth of a company. There are three main methods of valuation. Discounted cash flow analysis forecasts the free cash flow into the future. Precedent transaction analysis compares the company with other businesses that have been acquired or sold in the same industry. Comparable analysis evaluates the company against trading multiples of other companies (Valuation Methods, n.d.).

Value Prop

Have you ever wondered what makes a product or service more enticing than others that are offered from different vendors? If so, then you are inquiring about a *Value Proposition (Prop).*

What Is It?

Value prop is the feature or element that makes a product or business uniquely attractive to consumers. These values can be created in more than one element, such as price, quality, and location (Hassan, 2012).

How Does It Work?

Value prop is essential to the success of differentiation of a company. A company must spend a substantial amount of time constructing a value prop so that it may become the "backbone" to develop a value training program for the sales force (Liozu, 2015).

Chapter 13: Security with IoT

Accountability

Everyone that is assigned tasks from managers, colleagues, or others should hold their selves accountable for accomplishing the task. If you do not hold yourself *accountable,* then what will be accomplished at the end of the day?

What Is It?

Accountability is another word for responsibility of assigned tasks. Accountability also can represent one's actions and how an individual is viewed. Accountability is not an example of punishment; it is an example of owning a task until it is completed.

How Does It Work?

An individual or an organization can be evaluated on what they were accountable for (Rouse, 2015). An employee is accountable to complete assigned work responsibilities as well as report to work when they are scheduled to work.

Authentication

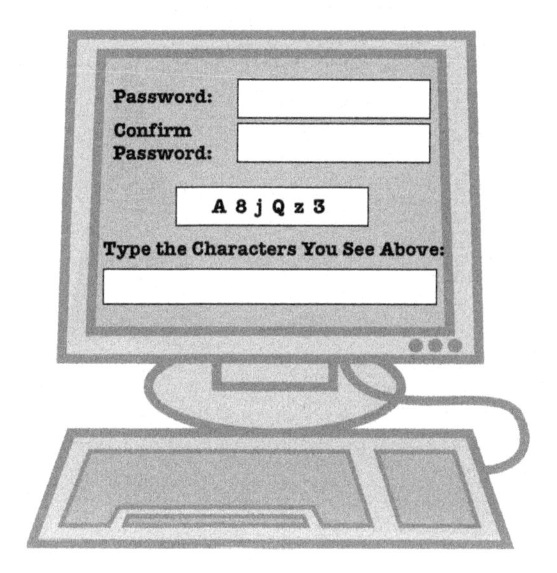

Have you ever wondered how people or machines make sure that information is secure? Then you have been thinking about the *Authentication* process. This process will become more secure and robust as society becomes more technologically advanced.

What Is It?

Authentication is a process that proves whether someone or something is who or what it has declared to be (Rouse, 2015). Authentication can be confirmed by multiple of options: self-identity, someone vouching for the individual, username/password, and/or biometric devices.

How Does It Work?

Authentication works by having usernames, password, biometric scans, etc. in place to make sure the user has permission to access the system that they are trying to get into. Using a debit card that requires a pin number is making sure the individual that is using it is the correct person.

Availability

How do you figure out if someone is available to help on a project or other tasks that you need assistance with? The opposite word of busy is *Available.*

What Is It?

Availability is to be otherwise occupied. To have the bandwidth of working on projects that others cannot due to their lack of "free time."

How Does It Work?

Availability can relate to a computer system booting up properly without any issues to be available right away that allows someone to complete tasks on the system.

Confidentiality

To feel comfortable telling someone a specific topic that you do not want anyone to know is considered having confidence in that individual. If you do not feel comfortable nor confident in people, you should not tell them private information.

What Is It?

Confidentiality is a set of rules, promises, and/or practices that limit the number of individuals with knowledge of the information. Confidentiality is used to keep a conversation or piece of information between a small group of individuals. Individuals have confidence in one another that a specific topic will not be leaked.

How Does It Work?

Confidentiality means keeping something private unless otherwise given permission to release the information. Having a conversation with Human Resources regarding salary is considered confidential information that nobody outside of key individuals need to know this information. In certain cases, breaking confidentiality rules can lead to disciplinary actions within a company or loss of friendships.

Integrity

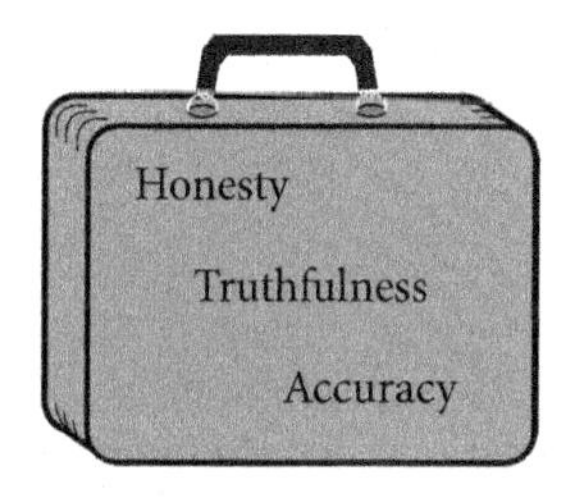

Ever wondered what enforces the value for one to operate in a manner that is held to high esteem? To do the right thing and have trust is to have *Integrity*.

What Is It?

Integrity is a quality of being honest, strong moral principles, and morale uprightness. It is also a process that is applied to data to ensure that it is correct and consistent across its lifecycle.

How Does It Work?

Individuals who are ethically aware and make the right choices to adhere integrity. Business that rely on the accuracy of data within their business systems and applications create procedures to assure the integrity and quality of the data that they are using to run their operations.

Chapter 14: Legal Issues with IoT

Anonymity

When a person cannot be identified then they are viewed as working in *Anonymity.*

What Is It?

The ability to have ones' identity concealed when completing computer tasks.

How Does It Work?

An employee is asked to work on a private project. To keep the project and its workers confidential, the employee assumes an alternate name or credentials that cannot be associated with their own identity.

Data Brokering

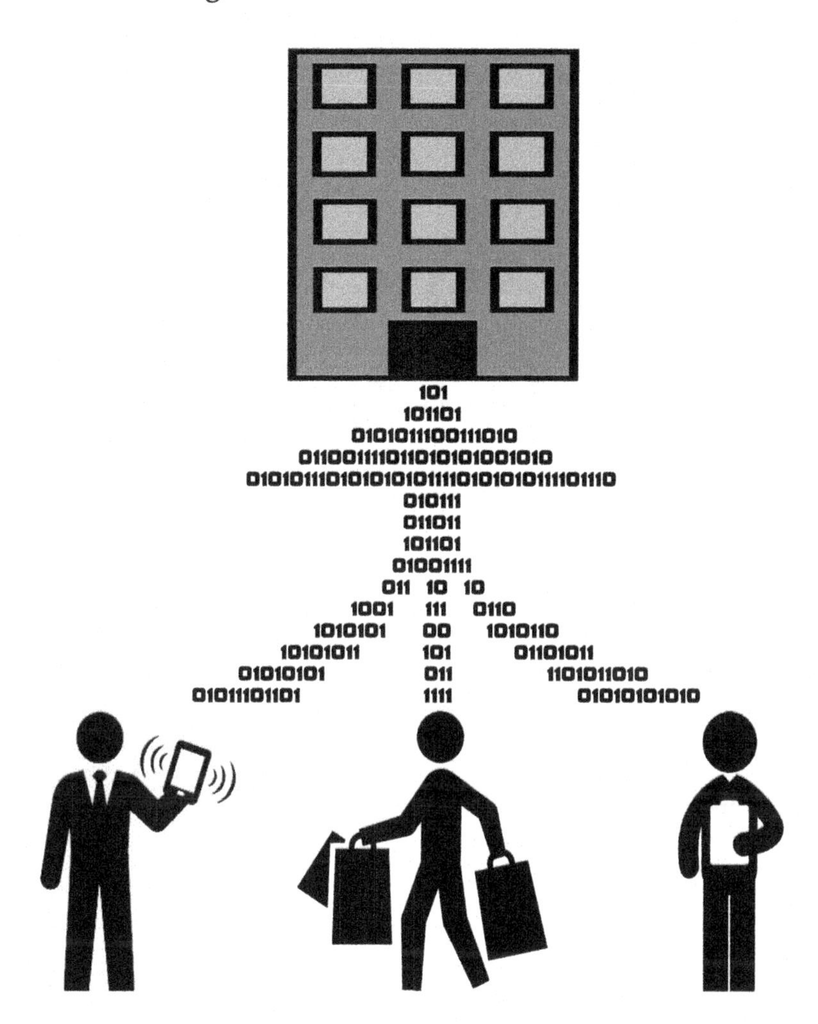

Data is constantly flowing from one entity to another. The process of companies or other entities receiving such information is done through *Data Brokering.*

What Is It?

The collection and selling of consumer data that is used to profile individuals is subject to little control or limits.

How Does It Work?

The information collected through data brokering comes from various outlets (i.e. websites, company systems etc.). Once gathered, individuals or companies use it in the context of their products or services to grow their market share.

Data Masking

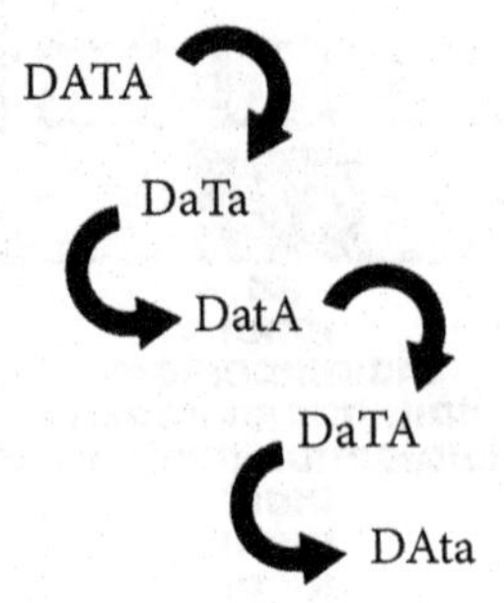

How do companies evaluate their product to gain information to provide feature-enhanced versions (i.e. Windows 8, Windows 9, and Windows 10) without damaging the current product in the process? This is done through *Data Masking*.

What Is It?

This is a method of creating manipulatable copies of original data that is used for testing purposes. Use of the replicated data allows for the original data to remain intact during tests.

How Is It Used?

The values contained in the data are not exact replicas but similar and used to predict outcomes, conduct training, and testing.

Discrimination

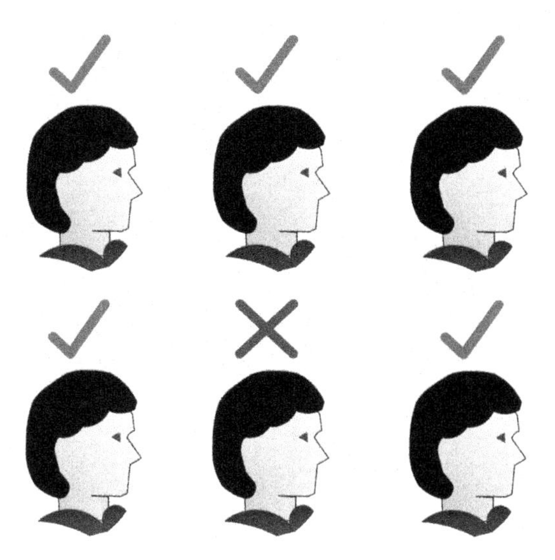

Have you ever been denied access to something or the ability to participate in an opportunity because of a quality that you did or did not have? If so, then this is a form of *Discrimination*.

What Is It?

The use of associations or algorithms in predictive analytics to make decisions that may have a direct and negative impact on an individual or groups of individuals.

How Is It Used?

These determinations can be made in both the public and private sectors and could include screening tools that impact a person's ability to board an airplane, obtain a job, or secure a credit card or loan, for example.

Chapter 15: IoT: A Business Perspective

Accounting

How do businesses keep track of whether or not they are clearing a profit? They apply accounting practices.

What Is It?

The business function that tracks financial information in a business.

How Is It Used?

Businesses usually either have an in-house accounting staff or outsource their accounting activities for a firm that specializes in providing such services. Regular reporting tools enable companies to monitor the financial aspects of their businesses in real time.

Automation

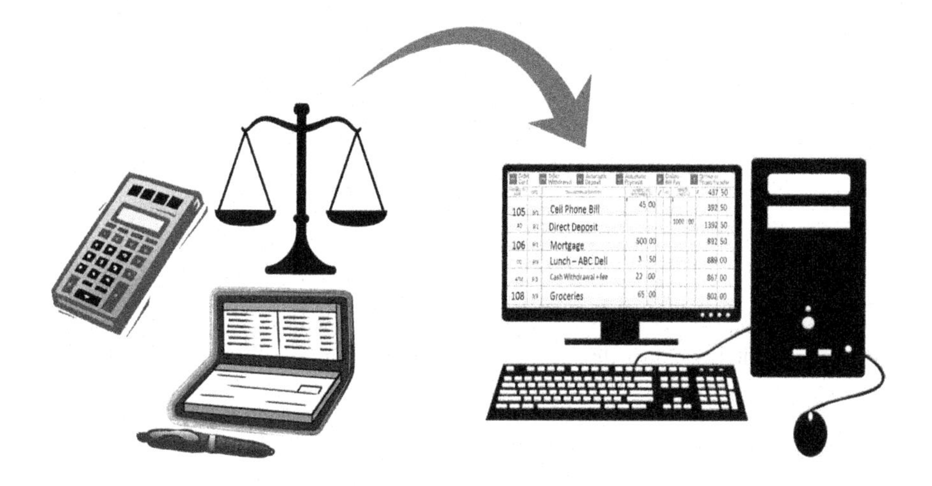

Have you ever wondered how we are able to do things more rapidly and with greater accuracy? Automation is the means.

What Is It?

Automation is the application of computerization to manual processes to make them more efficient.

How Is It Used?

Balancing a checkbook once took a bit of effort along with the use of a hand-held calculator. Now one can balance a checkbook in minutes through automation provided by banking applications.

Chipped

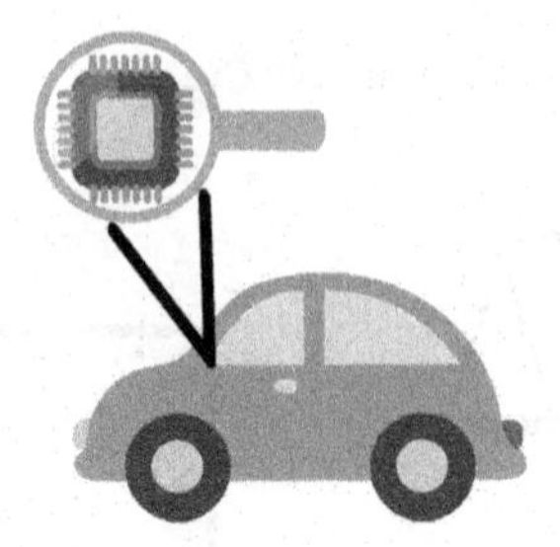

Product or solution designers create programmable computer chip solutions that can be embedded within a device to support some form of automation.

What Is It?

A term used to describe something that has an embedded computer chip.

How Does It Work?

The design of a device such as an automobile involves complex planning for the automation of services that were once mechanically based. For example, most contemporary automobiles have embedded computer chips that support functions such as cruise control, temperature controls, braking, and parking.

Drone

Have you ever wondered what to call an aircraft that has no pilot? It is a drone.

What Is It?

An unmanned flying device often equipped with a camera.

How Does It Work?

Drones rely on communication and connectivity between a controller and the device as well as a ground-based controller to issue instructions to the device to maintain flight.

Geographic Information Systems

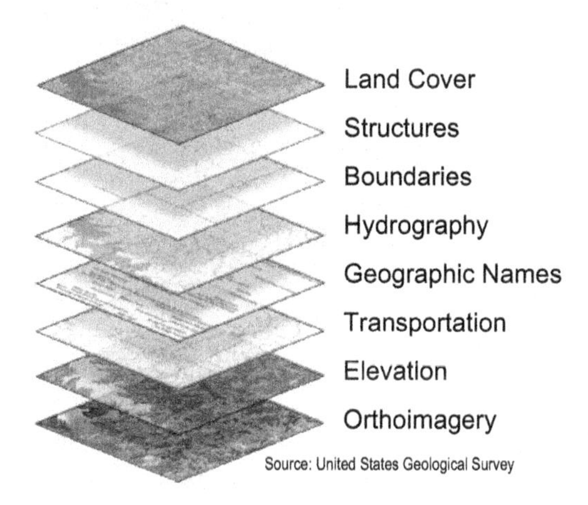

This acronym stands for Geographic Information Systems (GIS).

What Is It?

A framework for providing geographic information.

How Does It Work?

When you use the navigation system in a vehicle to submit a destination address and follow the automated directions to that location, you are using a GIS system.

Logistics

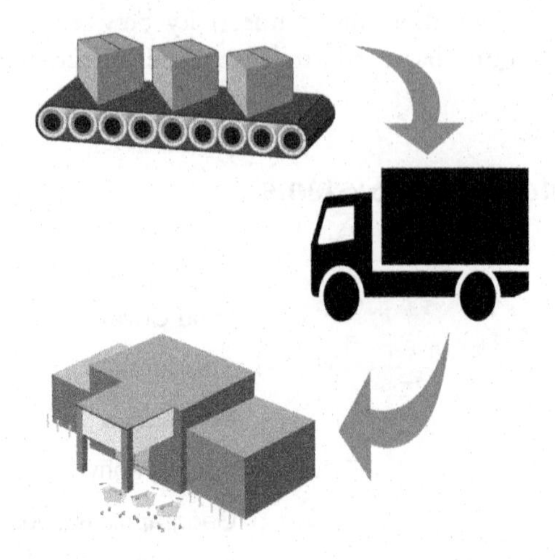

How do businesses like Walmart manage to keep stock on their shelves?

What Is It?

The business moving goods from one location to another.

How Does It Work?

Strategic and tactical planning goes into creating workflows to determine how goods can be moved from one location to another, such as Walmart's distribution hubs and trucking services to shift products to storefronts in a timely manner.

Production

Companies and organizations design units that focus on the skills, technologies, and processes needed to manufacture a product.

What Is It?

The business function that manufactures a good or service, sometimes called "Operations."

How Does It Work?

Production is managed as a functional business unit in most companies and is staffed by individuals with the skills and experience to innovate and manage the operational processes.

RFID

This acronym stands for radio frequency identification.

What Is It?

A device that contains a chip and antenna used for identification purposes.

How Does It Work?

Radio waves are used to read information stored on a tag associated with an object. That information is then available for processing for a specific business objective.

Supply Chain

How does a new smartphone make it to your pocket? The answer is through a complex supply chain that handles the devices from the point of design through manufacturing and to delivery.

What Is It?

An interorganizational linkage that facilitates logistics efficiency.

How Does It Work?

Companies build relationships with vendors and suppliers that align with the product and business needs. They manage these relationships to assure the right components or services are available when needed to support their business operations.

References

Actuator. (n.d.). Retrieved January 21, 2018, from Technopedia: www.techopedia.com/definition/17043/actuator

Amalyan, Nataly D. 2018. Start-up First Term Sheet: Stumbling Blocks to Avoid. *Biznes Inform*. Volume 1. Issue 480.

Brownlee, Jason. "Supervised and Unsupervised Machine Learning Algorithms." Machine Learning Mastery. September 22, 2016. Accessed April 20, 2018. https://machinelearningmastery.com/supervised-and-unsupervised-machine-learning-algorithms/

Copeland, Michael. "What's the Difference between Artificial Intelligence, Machine Learning, and Deep Learning?" July 29, 2016. Accessed May 19, 2018. https://blogs.nvidia.com/blog/2016/07/29/whats-difference-artificial-intelligence-machine-learning-deep-learning-ai/

Divan, Deepak. 2014. Validating a Business Model for a Minimum Viable Product [Entrepreneur Viewpoint]. *IEEE Power Electronics Magazine.* Volume 1. Issue 4.

Do we need Asimov's laws? (2014, May 16). Retrieved May 5, 2018 from www.technologyreview.com/s/527336/do-we-need-asimovs-laws/

Downes, John and Jordan Elliot Goodman. 2014. Terms. Dictionary of Finance and Investment Terms.

Gerwig, K., & Haughn, M. (2016). 5G. Retrieved May 5, 2018 from https://searchnetworking.techtarget.com/definition/5G

Greene. "A Glossary of Basic Artificial Intelligence Terms and Concepts." The Next Web. September 08, 2017. Accessed April 20, 2018. https://thenextweb.com/artificial-intelligence/2017/09/10/glossary-basic-artificial-intelligence-terms-concepts/

GST Council Secretariat. (2016, November). Retrieved May 19, 2018, from Central Board of Indirect Taxes & Customs: www.cbec.gov.in/resources/htdocs-cbec/gst/draft-model-gst-law-25-11-2016.pdf

Hardesty, Larry. "Explained: Neural Networks." MIT News. April 14, 2017. Accessed May 19, 2018. http://news.mit.edu/2017/explained-neural-networks-deep-learning-0414

Hassan, Almoatazbillah. 2012. The Value Proposition Concept in Marketing: How Customers Perceive the Value Delivered by Firms—A Study of Customer Perspectives on Supermarkets in Southampton in the United Kingdom. *International Journal of Marketing Studies.* Volume 4. Issue 3. Accessed May 15, 2018. www.ccsenet.org/journal/index.php/ijms/article/view/17696

Haughn, Matthew. "What Is Unsupervised Learning?" WhatIs.com. December 2016. Accessed May 20, 2018. https://whatis.techtarget.com/definition/unsupervised-learning

Herbrich, Ralf. "Machine Learning at Amazon." WSDM '17: *Proceedings of the Tenth ACM International Conference on Web Search and Data Mining,* February 2, 2017, 535. Accessed May 19, 2018. doi:10.1145/3018661.3022764.

Investopedia. (2018). Vertical market. Retrieved May 21, 2018, from www.investopedia.com/terms/v/verticalmarket.asp

IoT technology stack part one—from sensors, actuators and gateways to IoT platforms. (n.d.). Retrieved January 21, 2018, from i-Scoop: www.i-scoop.eu/internet-of-things-guide/iot-technology-stack-devices-gateways-platforms/#IoT_devices_sensors

Jain, P. (2017, January 27). Definition of business vertical. Retrieved May 20, 2018, from www.caclubindia.com/articles/definition-of-business-vertical-28976.asp

Jensen, Tyler. How to Determine Market Traction for Your Startup. 2014. Accessed May 15, 2018. https://thestartupgarage.com/determine-market-traction-startup/

Kumpe, T., & Bolwijn, P. T. (1988, March). Manufacturing: the new case for vertical integration. Retrieved May 22, 2018, from Harvard Business Review: https://hbr.org/1988/03/manufacturing-the-new-case-for-vertical-integration

Liozu, Stephan. 2015. Building a Distinctive and Compelling Value Proposition. *The Journal of Professional Pricing.* Accessed May 15, 2018. http://stephanliozu.com/wp-content/uploads/2015/04/Liozu-Building-a-Distinctive-and-Compelling-Value-Proposition-Q1-2015.pdf

Lison, Pierre. "An Introduction to Machine Learning." University of Oslo. October 3, 2012. Accessed April 20, 2018. http://folk.uio.no/plison/pdfs/talks/machinelearning.pdf

Marketing-schools. (2012). Vertical marketing: explore the strategy of vertical marketing. Retrieved May 21, 2018, from Marketing-schools.org: www.marketing-schools.org/types-of-marketing/vertical-marketing.html

Mitchell, B. (2018). Learning the meaning of a network gateway. Retrieved from www.lifewire.com/definition-of-gateway-817891

Monroe, Don. 2018. Chips for Artificial Intelligence. *Communications of the ACM*. Volume 61. Issue 4: 15–17. Accessed May 19, 2018. doi:10.1145/3185523.

Moogk, Dobrila Rancic. 2012. Minimum Viable Product and the Importance of Experimentation in Technology Startups. *Technology Innovation Management Review*. Volume 10: 17.

Munchenberg, S. (2018). Consumer benefits of well-governed vertically integrated finacial services businesses. Retrieved May 20, 2018, from Australian Banking Association: www.ausbanking.org.au/media/aba-blog/consumer-benefits-of-well-governed-vertically-integrated-financial-services

OneM2M. (n.d.). Retrieved from IoT Security Summit: https://tmt.knect365.com/iot-security/sponsors/one-m2m

Ravikant, Naval. The Anatomy of a Fundable Startup. Speech, San Fransisco, June 15, 2011. Accessed May 15, 2018. https://vimeo.com/25392719?PID=6155063&SID=jh876gje0a00orcn01sry&cjevent=1109c52e588811e8832f01680a240614

Rouse, Margaret (2015, September 30). Definition. Retrieved January 21, 2018, from TechTarget: IoT Agenda: http://internetofthingsagenda.techtarget.com/definition/actuator

Rouse, Margaret (n.d.). Evolved packet core (EPC). Retrieved from https://searchtelecom.techtarget.com/definition/Evolved-Packet-Core-EPC

Rouse, Margaret (2017, August). Vertical market. Retrieved May 21, 2018, from Tech Target: https://searchitchannel.techtarget.com/definition/vertical-market

Rouse, Margaret, Fouad, Tawfiq, and Ali. "Algorithm." WhatIS. Accessed May 19, 2018. https://whatis.techtarget.com/definition/algorithm

Rouse, Margaret, and Haughn, Matthew (2015, October). IoT Agenda. Retrieved April 15, 2018, from https://internetofthingsagenda.techtarget.com/definition/smart-sensor

Smith, S. (2017, April 25). Tips for getting started in healthcare vertical markets. Retrieved May 22, 2018, from Search IT Channel: https://searchitchannel.techtarget.com/news/450417499/Tips-for-getting-started-in-healthcare-vertical-markets

SAS. "Natural Language Processing: What It Is and Why It Matters." SAS. Accessed May 19, 2018. www.sas.com/en_us/insights/analytics/what-is-natural-language-processing-nlp.html#nlphowitworks

Shankland, S. (2017). Get ready for 'Unlimited Data' of 5G networks in 2019. Retrieved from www.cnet.com/news/5g-phone-networks-could-ease-data-limit-worries/

Silver, David, Thomas Hubert, Julian Schrittwieser, Ioannis Antonoglou, Matthew Lai, Arthur Guez, Marc Lanctot, Laurent Sifre, Dharshan Kumaran, Thore Graepel, Timothy Lillicrap, Karen Simonyan, Demis Hassabis. "Mastering Chess and Shogi by Self-Play with a General Reinforcement Learning Algorithm." December 5, 2017. Accessed May 19, 2018. https://arxiv.org/abs/1712.01815

Techopedia. (n.d.). Backbone. Retrieved from www.techopedia.com/definition/3158/backbone

Techopedia. (n.d.). Evolved packet core (EPC). Retrieved from www.techopedia.com/definition/31764/evolved-packet-core-epc

TechAmerica Foundation's Federal Big Data Commission. (2012). *Demystifying Big Data: A Practical Guide to Transforming the Business of Government.* Washington, DC: TechAmerica Foundation. Retrieved from https://bigdatawg.nist.gov/_uploadfiles/M0068_v1_3903747095.pdf

The Hindu Business Line. (2014, January 7). Bank design—horizontal & vertical. Retrieved May 22, 2018, from The Hindu Business Line: www.thehindubusinessline.com/money-and-banking/Bank-design-%E2%80%94-horizontal-amp-vertical/article20709277.ece

The Network Encylopedia. (n.d.). What is backbone (in computer networking)? Retrieved from www.thenetworkencyclopedia.com/entry/backbone/

"The Turing Test." Artificial Intelligence. 1999. Accessed April 20, 2018. www.psych.utoronto.ca/users/reingold/courses/ai/turing.html

Torrey, Lisa, and Jude Shavlik. "Transfer Learning." University of Wisconsin. June 2009. Accessed May 20, 2018. http://ftp.cs.wisc.edu/machine-learning/shavlik-group/torrey.handbook09.pdf

Tracy, P. (2016, December 16). Sensor types and their IoT use cases. Retrieved January 21, 2018, from RCR Wireless News: Intelligence on All Things Wireless: www.rcrwireless.com/20161206/internet-of-things/sensor-iot-tag31-tag99

Wetter, Olive E. 2017. Improving the effectiveness of nondisclosure agreements by strengthening concept learning. *R & D Management.* Volume 47. Issue 2.

WhatIsMyIPAddress.com. (n.d.). What is a gateway and what does it do? Retrieved from https://whatismyipaddress.com/gateway

Zocco, Dennis. 2007. Fantasynet Venture Capital Term Sheet Negotiation. *Journal of the International Academy for Case Studies.* Volume 15. Issue 5.

n.d. Nondisclosure Agreement: Definition from Nolo's Plain English Law Dictionary. Accessed May 15, 2018. www.law.cornell.edu/wex/nondisclosure_agreement

n.d. Valuation Methods. Corporate Finance Institute. Accessed May 15, 2018. https://corporatefinanceinstitute.com/resources/knowledge/valuation/valuation-methods/

n.d. Sweat Equity. Corporate Finance Institute. Accessed May 15, 2018. https://corporatefinanceinstitute.com/resources/knowledge/valuation/sweat-equity/

Index

N

O

U

V

W

Z